# Data-Driven Innovation in Supply Chains and Manufacturing

This book explores how modern technologies—especially data analytics, machine learning (ML), and the Internet of Things (IoT)—are transforming supply chain and manufacturing operations. Bridging academic research and industrial practice, this book presents data as a strategic asset driving agility, efficiency, and resilience.

Structured around four themes, it covers:

- Foundational analytics and optimization
- Predictive and prescriptive analytics for proactive decision-making
- Real-time IoT data for workflow monitoring and control
- Digital Twins and Natural Language Processing (NLP) for modeling and interaction

Chapters include mathematical modeling, case studies, and implementation frameworks, with coverage spanning stochastic forecasting, reinforcement learning, anomaly detection, and semantic parsing of logistics documentation.

Key benefits include its emphasis on integrated intelligence—blending ML, IoT, and simulation for real-time, predictive insights. It also highlights scalability across industries, with tools adaptable to sectors like automotive, healthcare, and aerospace. Each chapter concludes with open problems and future directions, offering a roadmap for innovation in intelligent operations.

# Data-Driven Innovation in Supply Chains and Manufacturing

## From Predictive Analytics to Natural Language Processing

Edited by
Parth Saxena, Srinivas R. Gottimukkala,
Shiva Kumar Bhuram, Nipun Joshi,
and Sahil Tripathi

Routledge
Taylor & Francis Group
A PRODUCTIVITY PRESS BOOK

Designed cover image: Shutterstock

First published 2027
by Routledge
605 Third Avenue, New York, NY 10158

and by Routledge
4 Park Square, Milton Park, Abingdon, Oxon, OX14 4RN

*Routledge is an imprint of the Taylor & Francis Group, an informa business*



ISBN: 9781041209249 (hbk)
ISBN: 9781041209225 (pbk)
ISBN: 9781003724889 (ebk)

DOI: 10.4324/9781003724889

Typeset in ITC Garamond Std
by codeMantra

# Contents

# About the Editors

**Parth Saxena** is a lead software engineer, leading innovation in investment banking technology at a global financial institution. He has spearheaded large-scale digital transformation initiatives across financial services, utilities, energy, and technology, delivering secure, scalable platforms in highly regulated environments. A senior member of IEEE and a Global Fellow at AI2030, Parth is a published author, open-source contributor, and speaker who serves as an advisor to organizations and professionals navigating AI-driven transformation. His recent work focuses on AI-enabled enterprise modernization, intelligent agents, and LLM-based automation, with an emphasis on responsible, production-ready adoption.

**Srinivas R. Gottimukkala,** with over 25 years of experience, is an accomplished finance and IT leader who specializes in driving business growth and innovation through transformational initiatives across industries. Backed by a strong academic foundation in commerce, law, and project management, they currently lead business process digitization and financial transformation at Deere & Company, delivering significant business value. Their expertise includes SAP Finance solutions, business analysis, and large-scale global implementations, with a proven track record in supply chain and financial management transformations. Known for strong cross-functional collaboration and innovative problem-solving, they are passionate about leveraging SAP, AI, and ML to deliver impactful, future-ready solutions in a rapidly evolving business landscape.

**Shiva Kumar Bhuram's** expertise spans over 21 years in SAP development, enterprise integration, and digital transformation, with strong depth in supply chain platforms, real-time execution systems, and cloud-based architectures. Early experience in scientific research produced published work in GPS and ionospheric modeling, establishing a solid analytical foundation. Core strengths include SAP ECC, middleware integration, warehouse automation, and large-scale system orchestration that improves inventory accuracy, process reliability, and operational performance. Technical focus also extends to AI and ML, including federated learning, prescriptive analytics, and secure distributed systems for enterprise environments. This expertise combines hands-on engineering with architectural design to deliver scalable, production-ready solutions. Work emphasizes practical innovation, measurable impact, and the ability to translate complex requirements into resilient platforms that support modern business operations.

**Nipun Joshi** is a senior product and technology leader with over 15 years of experience in building and scaling data-driven digital platforms. He currently serves as senior director of product management at Fareportal, where he leads digital, loyalty, and personalization initiatives across global travel brands. His work focuses on leveraging AI and analytics to enhance customer experience, recommendation systems, and intelligent service platforms. Nipun has been instrumental in launching large-scale loyalty ecosystems and AI-enabled customer engagement solutions. He holds an MBA from Cornell University along with advanced degrees in computer science and information technology. His professional interests span AI-powered product innovation, customer-centric design, and scalable enterprise systems.

**Sahil Tripathi** is a Ph.D. student at Jamia Hamdard, New Delhi, where he advances research in artificial intelligence, ML, deep learning, and NLP with a strong focus on making AI systems more explainable, robust, and ethically aligned with human values. His work bridges foundational ML techniques and real-world applications, particularly in large language models and model robustness. Born and raised in India, Sahil's academic journey includes bachelor's and master's degrees in computer applications from Jamia Hamdard, demonstrating a consistent commitment to academic excellence. His research passion is driven by a problem-solving mindset and a creative approach toward addressing complex technological challenges. With strong technical skills in Python, NLP, and ML frameworks, Sahil aims to contribute to socially beneficial AI adoption across critical domains.

# Contributors

**Rafiq Ali**
Delhi Skill and Entrepreneurship University
New Delhi, India

**Niharika Jain**
Vivekananda Institute of Professional Studies – GGSIPU
New Delhi, India

**Harsh Joshi**
Bharati Vidyapeeth College of Engineering – GGSIPU
New Delhi, India

**Isha Karn**
Vivekananda Institute of Professional Studies – GGSIPU
New Delhi, India

**Abdullah Mohammad**
Delhi Skill and Entrepreneurship University
New Delhi, India

**Sushant Kumar Ray**
University of Delhi
New Delhi, India

**Ebad Shabbir**
Delhi Skills and Entrepreneurship University
New Delhi, India

**Sharad Deep Shukla**
Shiv Nadar University, IOE
New Delhi, India

# Chapter 1

# Introduction to Data-Driven Supply Chains and Manufacturing

Isha Karn

## 1.1 Introduction

Supply chains are essential socio-technical systems that need to function reliably despite ongoing demand uncertainty and execution volatility. Over the last 20 years, a revolution in supply chain and manufacturing (SCM) systems in the global marketplace has taken place because of rapid advances in digital technologies, advanced datasets, and models that allow for operational insights at scale [1]. When uncertainty is spread over networks that are connected and layered, traditional ideas of supply chain management—periodic planning, siloed decision-making, and heuristic-based coordination—can no longer function in these contexts [2]. This is not a series of incremental improvements, as illustrated in Figure 1.1 [3], but a fundamental change in supply chain design, coordination, and control. Modern SCM systems rely on parallel data streams, automated optimization, and algorithmic decision assistance via artificial intelligence (AI), machine learning (ML), and computational technologies [4].

This shift is driven by the increasing deployment of sensing and Internet of Things (IoT) devices, scalable cloud platforms, and learning models that can model complex and nonlinear operational processes [5]. In this respect, Industry 4.0 has thus become the most significant and increasingly

DOI: 10.4324/9781003724889-1

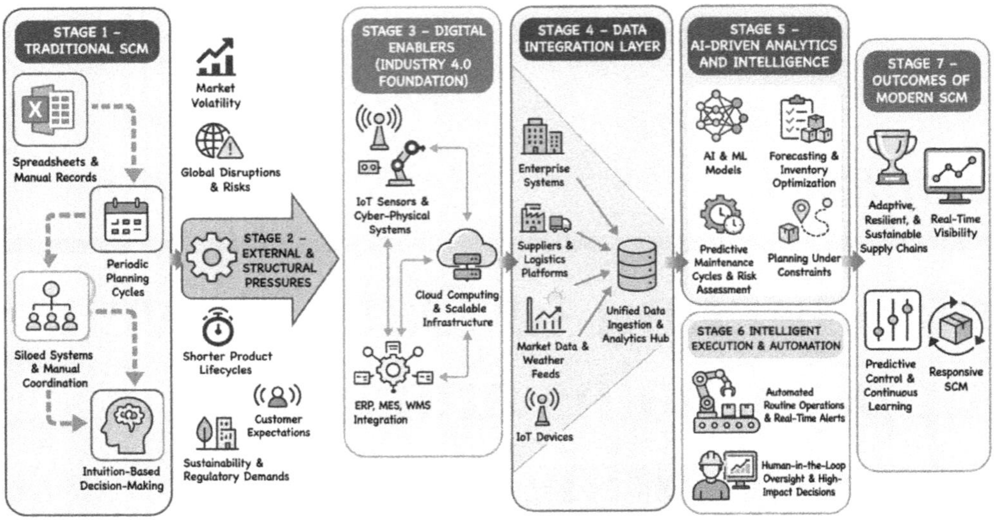


**Figure 1.1 Structural evolution of supply chain decision-making architectures, illustrating the transition from siloed, periodic planning to Industry 4.0–enabled systems with integrated data flows, learning-based prediction, and algorithmic decision support. The figure highlights how increasing digitalization and AI adoption enable real-time visibility and automation, while also motivating the need for closed-loop coupling between predictive and prescriptive models.**

concrete model of the growing intricacy of physical and digital systems in manufacturing and supply chains [6]. Industry 4.0 builds upon previous industrial revolutions—mechanization, mass production, automation, and computerization—and integrates cyber-physical systems, IoT connectivity, cloud computing, and data-driven analytics into SCM processes [7,8]. These capabilities will enable the creation of "smart factories" and "smart supply chains" that keep track of operations, analyze data autonomously, and make decisions without human intervention [9–11].

Although Table 1.1 summarizes this, the mere adoption of digital technologies does not always lead to honest end-to-end decisions. While initiatives related to digital transformation and Industry 4.0 are consistently cited as evidence of the availability of data in real time ✔, interconnectivity within systems, and automation, many of the current implementations of SCM do not show a coordinated model with respect to uncertainty ✖, sustainability constraints, and unifying decision logic. In particular, Table 1.1 highlights a sustained structural gap—while AI/ML models are being used increasingly and more frequently for forecasting and optimization (✔), they are developed and assessed separately, not as part of a closed-loop decision system (✖).

**Table 1.1 Comparative Mapping of Foundational Technological, Organizational, and Analytical Capabilities in Modern SCM**

| *Main Feature* | *Digital Transformation* | *Industry 4.0* | *Organizational Reconfiguration* | *External Pressures* | *Data-Driven Models* | *AI/ML Optimization* |
|---|---|---|---|---|---|---|
| Structural shift from traditional to real-time SCM | ✔ | ✔ | ✔ | ✔ | ✔ | ✔ |
| Integrated data flows and system interconnectivity | ✔ | ✔ | ✔ | ✖ | ✔ | ✔ |
| Cyber-physical and IoT-enabled operations | ✔ | ✔ | ✖ | ✖ | ✔ | ✔ |
| Network redesign and coordination restructuring | ✔ | ✔ | ✔ | ✖ | ✔ | ✔ |
| Market volatility and geopolitical uncertainty | ✖ | ✖ | ✔ | ✔ | ✖ | ✖ |
| Sustainability and regulatory compliance pressures | ✖ | ✖ | ✔ | ✔ | ✖ | ✖ |
| Multisource data integration across enterprise systems | ✔ | ✔ | ✔ | ✖ | ✔ | ✔ |
| Transition from reactive to predictive operations | ✔ | ✔ | ✔ | ✖ | ✔ | ✔ |
| Algorithmic decision support and automation | ✔ | ✔ | ✔ | ✖ | ✔ | ✔ |
| Optimization of forecasting, planning, and logistics | ✔ | ✔ | ✔ | ✖ | ✔ | ✔ |

✔ and ✖ indicate the presence and absence of each capability in prevailing SCM paradigms, respectively. The table exposes a persistent gap between predictive analytics and prescriptive decision-making—despite widespread adoption of digital infrastructure and AI—motivating integrated, closed-loop models that resolve the predictive–prescriptive disconnect.

Also, improvements to individual analytical models often fail to translate into actual performance gains.

The restriction is especially important under pressure from outside. The fragility of highly dispersed supply networks across the globe has been exposed by increasing demand volatility, the rapid tempo of digital sales channels and shorter product lifecycles [12,13], geopolitical disruptions affecting sourcing and distribution [14], and systemic shocks such as the COVID-19 pandemic [15,16]. At the same time, regulatory and social expectations relating to sustainability, carbon reduction, and ethical sourcing are challenging constraints that are established within organizations (✔ in Table 1.1) but have not been adequately integrated into algorithmic decision models (✖) [17]. Combining these forces implies structural deficits in traditional SCM systems that are based on static forecasts, modular analytics, and manual coordination.

To overcome this problem, organizations are increasingly turning to data-driven analytical models to manage the increasing complexity of global SCM networks [18]. For effective functioning, it is necessary to combine information from multiple sources, such as enterprise systems, transportation and logistics platforms, supplier portals, market signals, weather data, and IoT streams [19]. While such consolidation is more transparent (✔ in Table 1.1), decision-making must also be driven by decision systems that can bring predictive insights into action within the whole system. Models based on AI and ML are important because they learn from historical and real-time data, provide accurate forecasts, and enable decision-making in several contexts [20,21]. Currently, these models serve many SCM functions, including demand forecasting, inventory management, production planning, supplier risk assessment, quality control, and logistics optimization [22–24].

Despite these developments, the majority of data-driven SCM systems continue to view forecasting models, transactional analysis models, and optimization models as components that are loosely coupled and essentially modular. This separation results in a fundamental limitation, known as the predictive–prescriptive disconnect. In this context, predictive accuracy enhancements do not consistently result in service levels, robustness, or resilience improvements. The mixed ✔/✖ patterns shown in Table 1.1 clearly illustrate this disconnect. To address this limitation, we need to go beyond refining individual models and instead develop architectures that integrate prediction and decision-making into a single model.

This chapter presents Integrated Data-Driven Analytics for Supply Chain Management (IDASCM), a closed-loop model that explicitly links probabilistic demand forecasting, transactional irregularity modeling, and prescriptive optimization. IDASCM relies on a combination of statistical, tree-based, and recurrent forecasting models to produce calibrated demand distributions. These distributions are directly included in planning and inventory optimization models rather than being assigned as discrete predictive outputs. This coupling guarantees that uncertainty is maintained and addressed at every stage of the decision pipeline.

The proposed model is evaluated empirically with two complementary public benchmarks representing different supply chain decision contexts. IDASCM reduces the prediction error of demand by 22%–30% relative to the single-model baseline for the M5 Forecasting dataset, demonstrating significant increases in the case of seasonal and intermittent demand. Combined with planning optimization, this coupling results in inventory fill rates between 20% and 25% higher but still maintaining the same cost and service-level constraints. Applying the dataset on UCI Online Retail, the ratio of transactional irregularity modeling to optimization reduces the frequency of stockouts by 18%–27% relative to the unintegrated model. Ablation experiments on both sets show that omitting probabilistic forecasting or optimization coupling reduces service performance by up to 15%. This demonstrates that the improvements observed arise from modeling integration rather than from modeling enhancements alone.

This chapter is constructed to evolve in a systematic manner from the conceptual to the empirical. Section 1.2 presents research involving data-driven SCM, Industry 4.0 technologies, and analytical models for forecasting, optimization, and resilience. Section 1.3 provides the datasets, preprocessing steps, and the analytical pipeline for IDASCM. The experimental design and evaluation procedures are described in Section 1.4. Section 1.5 showcases empirical findings, featuring cutting-edge comparisons, cross-domain evaluations, and ablation studies. Ultimately, Section 1.6 provides a summary of the key results and explores their significance for research and practice.

## 1.2 Related Works

### *1.2.1 Traditional SCM Systems*

The early stages of SCM came from logistics, operations research, and industrial engineering. They were primarily focused on developing mathematical models to support local decision-making problems like inventory control [25–28], production planning [29–31], and distribution. Traditional inventory models, including the newsvendor model and its variations, established rigorous analytical frameworks for navigating cost and service trade-offs under uncertain circumstances [32–38]. As shown in Table 1.2, these models offer robust analytical foundations for inventory and production decisions (✔), but they function independently and lack system-wide coordination, real-time information exchange, and integration with downstream decision-making (✖).

The advent of Materials Requirements Planning (MRP), succeeded by MRP II and Enterprise Resource Planning (ERP), marked a transition toward coordinating procurement and production across the entire enterprise. These systems facilitated the creation of structured information architectures and coordinated planning across different functional units. In accordance with this, Table 1.2 indicates system-level coordination and enterprise-wide information architecture under MRP/ERP with a ✔. These systems depend on periodic replanning, deterministic inputs, and rule-based execution. As a result, they are not suitable for adaptive decision-making, uncertainty propagation, or closed-loop predictive–prescriptive coupling. The introduction of Lean and Just-in-Time (JIT) practices enhanced operational efficiency by focusing on minimizing waste, implementing pull-based production, and reducing defects. Table 1.2 demonstrates that these paradigms attain ✔ for both process flow optimization and execution discipline. Nonetheless, they are primarily focused on execution and rely on established rules and stable operating conditions, resulting in ✖ for predictive modeling, uncertainty-aware planning, and algorithmic decision integration.

Further research into inter-organizational integration and information sharing showed that the exchange of demand and inventory data among supply chain partners can alleviate the bullwhip effect and enhance coordination. Table 1.2 indicates ✔ for real-time and shared information among partners in these contexts. Nevertheless, the logic behind decisions was still decentralized and managed manually, lacking sufficient resources for unified planning or automated decision feedback (✖).

Higher-level alignment between supply chain structure and competitive priorities, such as efficiency versus responsiveness, was introduced by strategic supply chain design models. As shown in Table 1.2, these models offer ✔ for strategic alignment and differentiation. However, they function at a level of abstract planning and do not provide operational decision models that combine forecasting uncertainty with prescriptive actions (✖).

**Table 1.2 Comparative Mapping of Foundational Decision Constructs in Traditional SCM**

| *Main Feature* | *Classical Inventory Models* | *MRP/ MRP II/ERP* | *Lean and JIT Practices* | *Inter-Organizational Integration* | *Strategic Supply Chain Design* | *Information-Driven Coordination* |
|---|---|---|---|---|---|---|
| Analytical foundations for inventory and production | ✔ | ✖ | ✖ | ✖ | ✖ | ✖ |
| System-level coordination of procurement and production | ✖ | ✔ | ✖ | ✖ | ✖ | ✖ |
| Enterprise-wide information architecture | ✖ | ✔ | ✖ | ✖ | ✖ | ✖ |
| Waste reduction and process flow optimization | ✖ | ✖ | ✔ | ✖ | ✖ | ✖ |
| Pull-based production and defect minimization | ✖ | ✖ | ✔ | ✖ | ✖ | ✖ |
| Real-time and shared information among partners | ✖ | ✖ | ✖ | ✔ | ✖ | ✔ |
| Bullwhip effect mitigation through information sharing | ✖ | ✖ | ✖ | ✔ | ✖ | ✔ |
| Alignment of supply chain with competitive strategy | ✖ | ✖ | ✖ | ✖ | ✔ | ✖ |
| Differentiation between efficient vs. responsive chains | ✖ | ✖ | ✖ | ✖ | ✔ | ✖ |
| Foundations for modern data-driven SCM architectures | ✔ | ✔ | ✔ | ✔ | ✔ | ✔ |

✔ and ✖ denote the presence and absence of each functional capability, respectively. Classical inventory models, enterprise planning systems, and lean practices establish essential analytical and organizational foundations. The table highlights their limited support for uncertainty propagation, adaptive decision-making, and closed-loop coupling between predictive and prescriptive models—motivating the need for integrated decision architectures.

### 1.2.2 Industry 4.0 in SCM Systems

With Industry 4.0 came the introduction of cyber-physical systems, IoT connectivity, cloud computing, and digital twins as essential components for digitally enhanced manufacturing and supply chain systems [39–47]. These technologies are mainly focused on enhancing local capabilities, rather than providing end-to-end decision integration, as shown in Table 1.3. Cyber-physical systems offer tightly linked sensing–computation–physical operations and autonomous control loops in production environments (✔), and IoT aggregations allow for real-time visibility and traceability across networked assets and logistics networks (✔). Cloud computing allows for scalable computational resources and on-demand support for advanced analytics platforms (✔ in Table 1.3), and digital twins replicate physical systems for simulation, scenario testing, and predictive maintenance (✔). However, as illustrated in Table 1.3, these features are relatively rare across technological pillars ✖, and the entries indicate that there is no coherent logic of decision that allows sensing, prediction, and optimization to be integrated across the supply chain.

Table 1.3 notably emphasizes that the redesign of governance and organizational transformation are acknowledged as necessities, not as established capabilities. Although governance, workforce upskilling, and process redesign are identified as essential (✔), their successful execution is hindered by ongoing difficulties like legacy system integration, skills gaps, and cultural resistance (✔ under implementation challenges). Consequently, the implementation of Industry 4.0 often improves the availability of data and the automation of processes, but does not create a closed-loop connection between forecasting and optimization models.

### 1.2.3 Data Analytics in SCM Systems

Descriptive and diagnostic models for reporting, visualization, and root-cause analysis using structured enterprise data [48–56] were the focus of early SCM analytics. These models attain ✔ for historical trend analysis, dashboarding, and variance identification, as summarized in Table 1.4, but are ✖ for predictive and prescriptive decision-making. They mainly serve to provide retrospective insight, as opposed to decision support that is forward-looking or action-oriented. The advent of large-scale and heterogeneous data sources made it possible to create predictive analytics models for demand forecasting, disruption prediction, and risk assessment.

**Table 1.3 Comparative Assessment of Technological and Organizational Capabilities Associated with Industry 4.0 in SCM**

| *Main Feature* | *Cyber-Physical Systems* | *IoT Integration* | *Cloud Computing* | *Digital Twins* | *Organizational Transformation* | *Implementation Challenges* |
|---|---|---|---|---|---|---|
| Digitally integrated sensing–computation–physical operations | ✔ | ✖ | ✖ | ✖ | ✖ | ✖ |
| Autonomous monitoring and control loops | ✔ | ✖ | ✖ | ✖ | ✖ | ✖ |
| Real-time operational visibility across networks | ✖ | ✔ | ✖ | ✖ | ✖ | ✖ |
| Enhanced traceability and anomaly detection | ✖ | ✔ | ✖ | ✖ | ✖ | ✖ |
| Scalable, on-demand computational resources | ✖ | ✖ | ✔ | ✖ | ✖ | ✖ |
| Support for advanced analytics and collaborative platforms | ✖ | ✖ | ✔ | ✖ | ✖ | ✖ |
| Virtual replication for simulation and predictive maintenance | ✖ | ✖ | ✖ | ✔ | ✖ | ✖ |
| Scenario testing and continuous synchronization with physical systems | ✖ | ✖ | ✖ | ✔ | ✖ | ✖ |
| Governance, workforce, and process redesign requirements | ✖ | ✖ | ✖ | ✖ | ✔ | ✖ |
| Barriers such as legacy systems, skills gaps, and cultural resistance | ✖ | ✖ | ✖ | ✖ | ✖ | ✔ |
| Business-goal alignment and incremental adoption pathways | ✖ | ✖ | ✖ | ✖ | ✔ | ✔ |

✔ and ✖ indicate the presence and absence of each capability in prevailing Industry 4.0–enabled systems, respectively. The table highlights that while sensing, connectivity, and computational infrastructure are widely adopted, integrated decision logic and closed-loop coupling between predictive and prescriptive models remain largely absent, motivating unified decision architectures.

**Table 1.4 Comparative Mapping of Analytical Capabilities in Supply Chain Analytics, Spanning Descriptive, Diagnostic, Predictive, and Prescriptive Models**

| *Main Feature* | *Descriptive Analytics* | *Diagnostic Analytics* | *Predictive Analytics* | *Prescriptive Analytics* | *Big Data Sources* | *Machine Learning Models* |
|---|---|---|---|---|---|---|
| Historical trend analysis and performance visualization | ✔ | ✖ | ✖ | ✖ | ✖ | ✖ |
| Dashboard design and enterprise data integration | ✔ | ✖ | ✖ | ✖ | ✖ | ✖ |
| Root-cause identification of operational variances | ✖ | ✔ | ✖ | ✖ | ✖ | ✖ |
| Drill-down, data mining, and multivariate techniques | ✖ | ✔ | ✖ | ✖ | ✖ | ✔ |
| Demand forecasting and disruption prediction | ✖ | ✖ | ✔ | ✖ | ✔ | ✔ |
| Equipment failure detection and transport delay prediction | ✖ | ✖ | ✔ | ✖ | ✔ | ✔ |
| Risk assessment, quality assurance, and network planning | ✖ | ✖ | ✔ | ✖ | ✔ | ✔ |
| Optimization-driven decision recommendation | ✖ | ✖ | ✖ | ✔ | ✖ | ✖ |
| Simulation, scheduling, routing, and sourcing optimization | ✖ | ✖ | ✖ | ✔ | ✖ | ✔ |
| Integrated predictive–prescriptive decision systems | ✖ | ✖ | ✔ | ✔ | ✔ | ✔ |

✔ and ✖ indicate whether each capability is explicitly supported in prior work. The table highlights that while predictive and prescriptive models have advanced substantially, they are often developed and evaluated in isolation, exposing a persistent predictive–prescriptive disconnect that limits end-to-end decision effectiveness.

These models utilize large data sources and ML to predict future system states, as indicated by ✔ entries for predictive analytics, big data sources, and ML models in Table 1.4. Although predictive models greatly enhance forecasting accuracy and situational awareness, they usually yield point estimates or uncertainty summaries that are not directly linked to downstream decision-making processes.

At the same time, prescriptive analytics models offered decision recommendations based on optimization and simulation for inventory management, production, routing, and sourcing. Table 1.4 demonstrates that prescriptive analytics accomplishes ✔ for optimization-driven decision recommendation, as well as scheduling and routing optimization. However, these models often work with fixed or externally provided forecasts, treating predictive outputs as static inputs rather than components that are dynamically linked.

It is important to note that Table 1.4 shows that integrated predictive–prescriptive decision systems are still more of an exception than a standard practice. The last row shows that such systems exist (✔), but their presence is indicative of isolated or domain-specific implementations, not a prevailing analytical paradigm. In the majority of SCM systems that are driven by analytics, the development, assessment, and implementation of predictive and prescriptive models occur independently (✖ via intermediate integration pathways), with uncertainty seldom carried through decision stages. Consequently, enhancements in predictive accuracy do not consistently lead to better operational decisions.

### 1.2.4 Machine Learning in SCM Systems

In SCM, the adoption of ML models has mainly aimed to enhance predictive accuracy and pattern discovery beyond what classical statistical models can achieve, by learning complex, nonlinear relationships from operational data [57–62]. Supervised learning and deep learning models achieve ✔ for demand forecasting, supplier classification, and defect detection, as summarized in Table 1.5. This reflects their effectiveness in predictive tasks that map historical inputs to future outcomes. These models also demonstrate ✔ for acquiring flexible, nonlinear representations that enhance performance in the face of demand variability and complex feature interactions.

Various unsupervised learning models provide different contributions to SCM decision support. Table 1.5 illustrates that clustering and anomaly detection models attain ✔ for identifying concealed behavioral patterns,

**Table 1.5 Comparative Mapping of ML Model Classes and Their Functional Roles in SCM**

| *Main Feature* | *Supervised Learning* | *Unsupervised Learning* | *Reinforcement Learning* | *Deep Learning* | *Anomaly Detection* | *Time-Series Forecasting* |
|---|---|---|---|---|---|---|
| Enhancement over classical forecasting and classification | ✔ | ✖ | ✖ | ✔ | ✖ | ✔ |
| Demand forecasting, supplier classification, and defect detection | ✔ | ✖ | ✖ | ✔ | ✖ | ✔ |
| Learning nonlinear relationships and flexible model structures | ✔ | ✖ | ✖ | ✔ | ✖ | ✔ |
| Clustering-based supplier and customer segmentation | ✖ | ✔ | ✖ | ✖ | ✖ | ✖ |
| Detection of operational irregularities and quality deviations | ✖ | ✔ | ✖ | ✔ | ✔ | ✖ |
| Identification of hidden behavioral patterns in SCM data | ✖ | ✔ | ✖ | ✖ | ✔ | ✖ |
| Adaptive, sequential decision-making for SCM operations | ✖ | ✖ | ✔ | ✖ | ✖ | ✖ |
| Inventory management, dynamic pricing, and scheduling optimization | ✖ | ✖ | ✔ | ✖ | ✖ | ✖ |
| Handling high-dimensional state spaces | ✖ | ✖ | ✔ | ✔ | ✖ | ✖ |
| Autonomous, self-improving operational systems | ✖ | ✖ | ✔ | ✔ | ✖ | ✖ |

✔ and ✖ denote whether a given model class explicitly supports each capability. The table shows that while ML models substantially improve predictive accuracy and pattern discovery, their integration with prescriptive decision models remains limited, reinforcing the predictive–prescriptive disconnect addressed in this work.

segmenting suppliers and customers, and detecting operational anomalies and quality deviations. Nonetheless, these models are fundamentally descriptive or diagnostic (✖ for decision-making), as they do not directly recommend actions or optimize subsequent decisions.

Reinforcement learning (RL) models broaden the application of ML to include sequential decision-making. As illustrated in Table 1.5, there is a ✔ for adaptive decision-making, inventory management, pricing, and scheduling, as well as a ✔ for managing high-dimensional state spaces when used in conjunction with deep learning. Due to these characteristics, RL models show promise in the context of autonomous control. However, RL models are usually assessed in simulated or isolated settings and do not explicitly connect with optimization models at the enterprise scale or broader SCM decision pipelines (✖).

Importantly, Table 1.5 demonstrates that integrated decision pipelines and end-to-end coordination with prescriptive optimization systems result in consistent ✖ entries across all ML model classes. Even when ML models impact decisions—through forecasts, anomaly signals, or learned policies—their outputs are often regarded as independent inputs rather than as parts of an integrated, closed-loop decision model. Uncertainty arising from ML models is seldom carried into optimization or planning decisions in a systematic way. Thus, even though ML models greatly improve predictive accuracy, representation learning, and pattern discovery (✔ in Table 1.5), they don't really solve the problem of how to turn predictions into operational decisions (✖). This structural limitation exacerbates the disjunction between prediction and prescription, underscoring the need for integrated models that explicitly connect ML-based prediction to prescriptive optimization—an issue addressed by the proposed IDASCM model.

### 1.2.5 Optimization Models in SCM Systems

In SCM, optimization models serve as the main tool for prescriptive decision-making, offering formal representations for facility location, production planning, inventory control, and transportation routing. Tractable representations of these problems were established by classical optimization models based on linear, integer, mixed-integer, and nonlinear programming. These models continue to be foundational to SCM planning systems [63–68]. As indicated in Table 1.6, these models attain ✔ for structured mathematical formulations and algorithmic solvability, incorporating established solution methods like branch-and-bound and decomposition techniques.

**Table 1.6 Comparative Mapping of Prescriptive Optimization Models and DSS Capabilities in SCM**

| *Main Feature* | *Classical Optimization Models* | *Stochastic and Robust Optimization* | *Multi-Objective Optimization* | *ML–Optimization Integration* | *Dynamic Routing and Transportation Models* | *DSS* |
|---|---|---|---|---|---|---|
| Linear, integer, mixed-integer, nonlinear formulations | ✔ | ✖ | ✖ | ✖ | ✖ | ✖ |
| Facility location, network design, and lot-sizing models | ✔ | ✖ | ✖ | ✖ | ✖ | ✖ |
| Production planning, scheduling, and sequencing | ✔ | ✖ | ✖ | ✖ | ✖ | ✖ |
| Algorithmic models (branch-and-bound, decomposition, heuristics) | ✔ | ✖ | ✖ | ✖ | ✖ | ✖ |
| Modeling uncertainty and disruptions | ✖ | ✔ | ✖ | ✖ | ✖ | ✖ |
| Cost–resilience–risk trade-off modeling | ✖ | ✔ | ✔ | ✖ | ✖ | ✖ |
| Sustainability metrics (emissions, compliance, service levels) | ✖ | ✖ | ✔ | ✖ | ✖ | ✖ |

| | | | | | | |
|---|---|---|---|---|---|---|
| ML-based forecasts integrated into prescriptive models | ✖ | ✖ | ✖ | ✔ | ✖ | ✖ |
| Surrogate models for large-scale simulation | ✖ | ✖ | ✖ | ✔ | ✖ | ✖ |
| Real-time, dynamic routing and Electric Vehicle (EV) routing | ✖ | ✖ | ✖ | ✖ | ✔ | ✖ |
| Integrated routing–inventory formulations | ✖ | ✖ | ✖ | ✖ | ✔ | ✖ |
| Modular DSS with orchestration and human-in-the-loop | ✖ | ✖ | ✖ | ✖ | ✖ | ✔ |
| Explainability, trust, and collaborative interfaces | ✖ | ✖ | ✖ | ✖ | ✖ | ✔ |
| Governance, data quality, and change management | ✖ | ✖ | ✖ | ✖ | ✖ | ✔ |

✔ and ✖ denote the presence and absence of each capability in prior work, respectively. The table shows that, despite advances in stochastic optimization, multi-objective modeling, and ML–optimization integration, most systems treat predictive outputs as static inputs, leaving feedback and uncertainty propagation weakly integrated and reinforcing the predictive–prescriptive disconnect.

Nonetheless, they remain ✖ regarding explicit uncertainty modeling, multi-objective trade-offs, and integration with predictive models. Further research broadened these formulations using stochastic and robust optimization methods to directly address demand uncertainty and disruptions. According to Table 1.6, there is a ✔ for modeling uncertainty and for cost–resilience–risk trade-offs in these models. Nevertheless, these extensions still depend on demand scenarios or point estimates provided externally, leading to ✖ for dynamic uncertainty propagation and feedback between predictive and prescriptive components. Likewise, multi-objective optimization models include sustainability and service-level objectives (✔), but function independently of forecasting models and execution feedback (✖).

Recent advancements in the integration of ML and optimization have led to the inclusion of learned demand or travel-time forecasts in prescriptive optimization models. This has resulted in ✔ for the incorporation of forecasts and surrogate modeling to expedite large-scale simulations. However, this integration is predominantly one-way, as shown in Table 1.6—predictive outputs are regarded as static inputs for optimization models (✖), without any closed-loop feedback from optimization results or execution signals to forecasting models. As a result, uncertainty does not carry over or get updated throughout the decision stages.

Dynamic routing and transportation models enhance prescriptive capabilities by facilitating real-time vehicle routing and integrated routing–inventory formulations (✔ in Table 1.6). However, these models prioritize execution-level responsiveness over end-to-end decision integration, and they do not provide unified orchestration across forecasting, planning, and optimization.

In practice, optimization models are primarily implemented within Decision Support Systems (DSS). Table 1.6 illustrates that contemporary DSS meet the ✔ criteria for modular orchestration, human-in-the-loop control, explainability, and governance-related functions like data quality management and change control. Even with these developments, optimization models in DSS platforms are usually integrated downstream of predictive components. This maintains a pipeline structure in which predictive outputs do not change and uncertainty feedback is lacking (✖).

### 1.2.6 Sustainability in SCM Systems

Models of SCM that focus on sustainability build upon the conventional objectives of service and cost efficiency by adding environmental and social factors, such as emissions, resource consumption, labor conditions, and ethical sourcing [69–72]. Initial sustainability models focused on measurement and accounting, creating decision models for emissions quantification, lifecycle assessment, and footprint reporting. As shown in Table 1.7, these capabilities are firmly established (✔) for environmental impact assessment, offering the essential visibility needed for sustainability-aware SCM.

Later work broadened the focus to include social sustainability, presenting models that monitor labor conditions, supplier adherence, and ethical sourcing methods. Table 1.7 demonstrates a consistent ✔ endorsement of social sustainability metrics and governance mechanisms, especially in supplier engagement and compliance monitoring. While these contributions allow organizations to evaluate sustainability performance across multi-tier supply networks, they tend to be mostly descriptive or constraint-based rather than linked to decision-making. Recent developments include models of the circular economy and reverse logistics, focusing on product recovery, remanufacturing, recycling, and closed-loop material flows. As shown in Table 1.7, circularity-related decision models attain ✔ for reverse logistics design and lifecycle recovery modeling, involving partial integration of advanced analytics and ML for predicting returns and component lifetimes. Nonetheless, these models usually function as specialized subsystems instead of parts of integrated planning and execution models.

Digital traceability technologies, such as IoT- and blockchain-enabled monitoring, enhance the granularity and reliability of sustainability data. Table 1.7 demonstrates ✔ backing for emissions and material-flow monitoring through digital traceability, as well as for dynamic social-risk detection when paired with analytics. Nonetheless, these capabilities mainly improve monitoring and reporting, with limited influence on prescriptive planning decisions.

### 1.2.7 Supply Chain Resilience in SCM Systems

Traditionally, resilience in SCM has been understood as a collection of capabilities that include anticipation, absorption, recovery, and adaptation in response to disruptions [73–75]. Early resilience models emphasize structural and policy-based mechanisms, including supplier diversification, safety stock, redundancy, and contingency planning. Classical resilience strategies attain ✔ for risk identification across supply, demand, and operations, as well as for redundancy-based mitigation and efficiency–resilience trade-off analysis, as shown in Table 1.8. However, these models are mainly based on static evaluations and predetermined response policies, demonstrating ✖ for real-time detection, predictive modeling, and dynamic decision execution.

**Table 1.7 Comparative Mapping of Sustainability-Related Decision Capabilities in SCM, Spanning Environmental, Social, Analytical, and Governance Dimensions**

| *Main Feature* | *Environmental Impact Assessment* | *Social Sustainability Metrics* | *Circular Economy and Reverse Logistics* | *Advanced Analytics and ML* | *Digital Traceability (IoT/ Blockchain)* | *Governance and Supplier Engagement* |
|---|---|---|---|---|---|---|
| Emissions, waste, and resource quantification | ✔ | ✖ | ✖ | ✖ | ✖ | ✖ |
| Lifecycle assessment and footprint accounting | ✔ | ✖ | ✖ | ✖ | ✖ | ✖ |
| Integration of sustainability metrics into procurement and production | ✔ | ✔ | ✖ | ✖ | ✖ | ✔ |
| Labor conditions, ethical sourcing, and compliance indicators | ✖ | ✔ | ✖ | ✖ | ✖ | ✔ |
| Reverse logistics design and product recovery operations | ✖ | ✖ | ✔ | ✖ | ✖ | ✖ |
| Remanufacturing, recycling, and closed-loop modeling | ✖ | ✖ | ✔ | ✔ | ✖ | ✖ |

| | | | | | | |
|---|---|---|---|---|---|---|
| Multi-objective models with carbon and circularity constraints | ✔ | ✖ | ✔ | ✔ | ✖ | ✖ |
| Predictive modeling of returns and component lifetimes | ✖ | ✖ | ✔ | ✔ | ✖ | ✖ |
| IoT-/blockchain-enabled emissions and material-flow monitoring | ✖ | ✖ | ✖ | ✔ | ✔ | ✖ |
| Supplier capacity building and performance contracts | ✖ | ✔ | ✖ | ✖ | ✖ | ✔ |
| Dynamic monitoring and social-risk detection | ✖ | ✔ | ✖ | ✔ | ✔ | ✔ |

✔ and ✖ indicate whether each capability is explicitly supported in prior models. The table shows that while sustainability metrics and monitoring have advanced, their integration into unified predictive–prescriptive decision models remains limited, reinforcing the predictive–prescriptive disconnect in sustainability-oriented SCM.

**Table 1.8 Comparative Mapping of Supply Chain Resilience Models across Structural Strategies, Analytical Tools, and Digital Enablers**

| *Main Feature* | *Classical Resilience Strategies* | *Network and System Modeling* | *Data-Driven Early Warning* | *Machine Learning for Risk* | *Control-Tower and Digital Platforms* | *Blockchain and Multi-Tier Trust* |
|---|---|---|---|---|---|---|
| Risk identification across supply, demand, operations, finance | ✔ | ✖ | ✖ | ✖ | ✖ | ✖ |
| Anticipation, absorption, recovery, adaptation capability set | ✔ | ✖ | ✖ | ✖ | ✖ | ✖ |
| Supplier diversification, safety stock, redundancy | ✔ | ✖ | ✖ | ✖ | ✖ | ✖ |
| Contingency planning and continuity protocols | ✔ | ✖ | ✖ | ✖ | ✖ | ✖ |
| Efficiency–resilience trade-off analysis | ✔ | ✖ | ✖ | ✖ | ✖ | ✖ |
| Network-critical node identification | ✖ | ✔ | ✖ | ✖ | ✖ | ✖ |
| Scenario simulation, stress testing, stochastic assessment | ✖ | ✔ | ✖ | ✖ | ✖ | ✖ |
| Early detection via IoT telemetry and real-time signals | ✖ | ✖ | ✔ | ✖ | ✖ | ✖ |

| | | | | | | |
|---|---|---|---|---|---|---|
| Predicting disruptions using ML indicators | ✖ | ✖ | ✖ | ✔ | ✖ | ✖ |
| Estimating downstream operational impacts | ✖ | ✖ | ✖ | ✔ | ✖ | ✖ |
| Integrated visibility and collaborative response (control towers) | ✖ | ✖ | ✖ | ✖ | ✔ | ✖ |
| Distributed-ledger support for information integrity | ✖ | ✖ | ✖ | ✖ | ✖ | ✔ |
| Evaluating trade-offs among redundancy, flexibility, and visibility | ✔ | ✔ | ✔ | ✔ | ✔ | ✔ |

✔ and ✖ denote whether each capability is explicitly addressed in prior work. The table shows that while resilience models increasingly incorporate data-driven risk indicators and visibility platforms, they rarely integrate predictive uncertainty with prescriptive recovery decisions, reinforcing the predictive–prescriptive disconnect in resilience-oriented SCM systems.

Later research incorporated network and system modeling to examine structural vulnerability and the impact of cascading failures. Network-based models offer ✔ for pinpointing critical nodes and assessing disruption scenarios via simulation and stress testing, as illustrated in Table 1.8. These models enhance structural comprehension of resilience, yet they are mainly offline and analytical, lacking ✖ for real-time monitoring, learning-based prediction, and operational decision integration.

Recent developments include the integration of data-driven early-warning systems that utilize IoT telemetry and real-time signals to identify emerging disruptions. Table 1.8 shows ✔ for early detection capabilities in this category, but ✖ for predictive modeling of downstream impacts and prescriptive response generation. Likewise, ML models have been utilized to assess the probability of disruptions and their downstream operational impacts, accomplishing ✔ for prediction-related tasks but remaining ✖ when directly linked to recovery and mitigation decisions.

Models such as control towers and digital platforms seek to enhance resilience by providing integrated visibility and collaborative response mechanisms. As Table 1.8 summarizes, these platforms attain ✔ for shared situational awareness and coordination, but they are ✖ when it comes to tightly coupling predictive uncertainty with prescriptive recovery actions. Although blockchain and distributed-ledger models improve information integrity and trust across multi-tier networks (✔), they function mainly as enabling infrastructure rather than decision-integrated resilience models. It is significant that, as shown in Table 1.8, there is no class of resilience models that supports predictive modeling, prescriptive decision-making, and real-time execution all at once. The last row shows ✔ across all categories when assessing trade-offs between redundancy, flexibility, and visibility. However, such assessments are usually performed separately within distinct modeling paradigms rather than through integrated, closed-loop decision systems. As a result, gains in resilience are divided among the stages of anticipation, detection, and response.

## 1.3 Materials and Models

### 1.3.1 Dataset Description

This chapter uses two datasets that are publicly available and represent planning-level demand dynamics as well as execution-level transactional variability in order to assess integrated predictive–prescriptive decision models across different supply chain contexts. Table 1.9 provides a summary of key statistics

**Table 1.9 Dataset Characteristics, Preprocessing Outcomes, and Evaluation Splits for the M5 Forecasting and UCI Online Retail Datasets**

| *Dimension* | *M5 Forecasting Dataset* | *UCI Online Retail Dataset* |
|---|---|---|
| Decision context | Planning-level demand forecasting | Execution-level transactional variability |
| Temporal coverage | 2011–2016 (~5.5 years) | ~1 year |
| Raw records | ~30,000 daily time series | ~540,000 transactions |
| Data granularity | Daily aggregated sales | Invoice-level events |
| Zero/negative values | Zero-demand days retained | Negative quantities retained as returns (~4%) |
| Records removed | <0.5% (implausible artifacts) | <1% (invalid invoices) |
| Missing identifiers | Not applicable | ~25% missing customer IDs (excluded only from customer-level analyses) |
| Feature dimensionality | ~40–50 features per series | ~25–30 features per transaction |
| Temporal processing | Calendar alignment; no interpolation | Timestamp normalization; no resampling |
| Price handling | Forward filled within valid selling windows (~6%) | Transaction-level validation |
| Training split | ~80% (chronological) | ~70% (time-based) |
| Validation split | ~10% (chronological) | ~15% (time-based) |
| Test split | ~10% (held-out horizon) | ~15% (held-out period) |
| Evaluation protocol | Forward-chaining forecasting | Time-based holdout evaluation |
| Leakage control | Strict temporal ordering | Strict temporal ordering |

The table summarizes scale, temporal coverage, retained observations, feature dimensionality, and split protocols used for empirical evaluation, ensuring transparency and reproducibility.

for the dataset and evaluation splits. The M5 Forecasting dataset [76] encompasses daily retail sales data over a period of about 5.5 years (2011–2016), including more than 30,000 time series at various aggregation levels. Calendar attributes and historical price information are linked to each series. The dataset's extensive time span and dense sampling facilitate the assessment of probabilistic forecasting models, considering seasonality, intermittency, and promotional effects. Following standard benchmarking practices, the data is split into a training window (about 80%), a validation window (about 10%), and a held-out test window (about 10%). The order of the data is preserved. The UCI Online Retail dataset [77] contains around 540,000 transactions recorded at the invoice level over a year. Each record includes the time of the transaction, the product ID, the number of items, the cost per unit, the customer's location, and a cancellation marker. The dataset has irregular order arrivals, returns, and execution noise that directly affect inventory availability and fulfillment decisions, which is different from M5. The dataset is split into three parts based on time: the first 70% of transactions are used for training, the next 15% for validation, and the last 15% for testing. This setup is similar to how things would work in real life when demand patterns change. The datasets aren't combined intentionally, as they belong to different decision layers instead of just one operational system. Figure 1.2 shows how the two datasets work together to provide more information.

### *1.3.1.1 Dataset Preprocessing*

This process will allow for consistency in time, statistical validity, and compatibility with the operational conditions of deployment. As illustrated in Table 1.9, all preprocessing steps are deterministic and applied to each dataset on its own. For the M5 Forecasting dataset, item, store, and hierarchy identifiers are pre-processed to ensure consistency. There are no missing timestamps in the data; days with zero demand are also included to maintain the intermittent

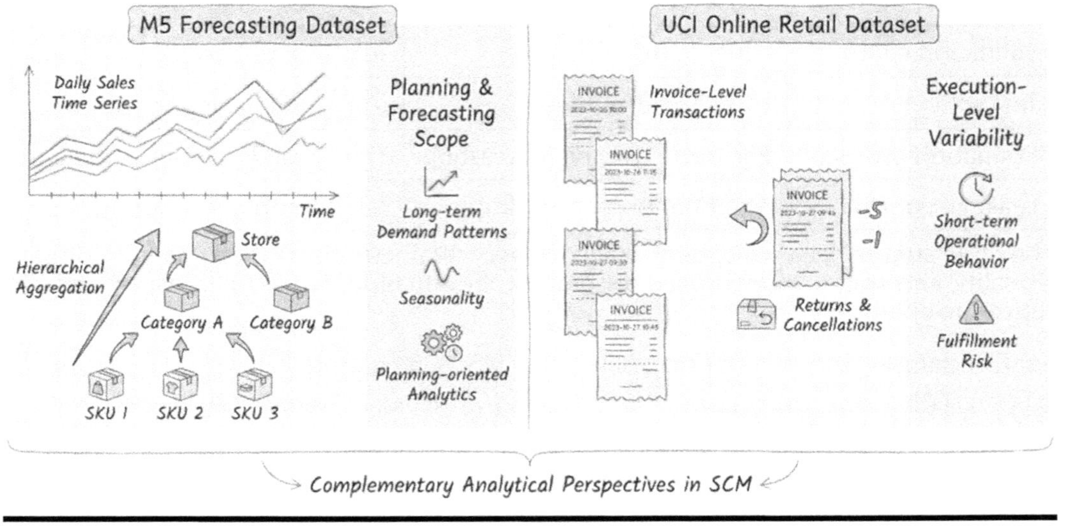


**Figure 1.2 Complementary decision contexts represented by the evaluation datasets. The M5 Forecasting dataset supports planning-oriented analysis through long-horizon demand dynamics and uncertainty modeling, while the UCI Online Retail dataset captures execution-level variability through transactional irregularities and return behavior.**

demand structure. History of price series is forward filled within valid selling periods and affects about 6% of daily price observations. Demand values are noninterpolated. Under 0.5% of observations are interpreted as implausible artifacts, which are ignored in the outlier screening process, and promotion-driven spikes are retained. The feature construction yields about 40–50 features per series, consisting of lagged demand variables, rolling statistics, and calendar encodings. For example, invoice validation excludes less than 1% of records from the UCI Online Retail dataset because of nonstandard identifiers or mismatched entries. Negative quantities of transactions, which make up about 4% of the transactions, are logged and marked as cancels or returns. These data are excluded from customer-level analyses but were retained to model aggregate demand and inventory flow; approximately 25% of transactions lack customer identifiers. Time normalization allows the extraction of temporal features without resampling. This is done by building features that contain 25–30 features for each transaction which represent order size, inter-arrival times, return frequency, and geographical variability. Rare categorical levels representing less than 0.1% of observations are combined to reduce sparsity. Evaluation protocols do not alter the sequence of time. The M5 dataset employs forward-chaining splits consistent with forecasting benchmarks, while time-based holdouts that reflect execution drift are employed for the UCI dataset. Both datasets do not incorporate future information during the training or validation phases, thereby guaranteeing that empirical outcomes mirror actual operational deployment.

## *1.3.2 Model Analysis*

This part talks about the IDASCM model and how its structure directly addresses the predictive–prescriptive gap that was found in Sections 1.1 and 1.2. IDASCM is set up as a closed-loop decision model, while earlier SCM systems treated forecasting, irregularity modeling, and optimization as parts that were only loosely linked or came in order. In this model, uncertainty about predictions is kept and used to make prescriptive decisions, while feedback from execution affects future predictions.

### *1.3.2.1 Model Overview*

IDASCM is a unified decision-making system that includes three linked modeling phases: probabilistic demand forecasting, transactional irregularity modeling, and prescriptive optimization. These phases are linked by a common uncertainty-aware representation (see Figure 1.3). Each step is necessary but insufficient on its own; performance enhancements arise

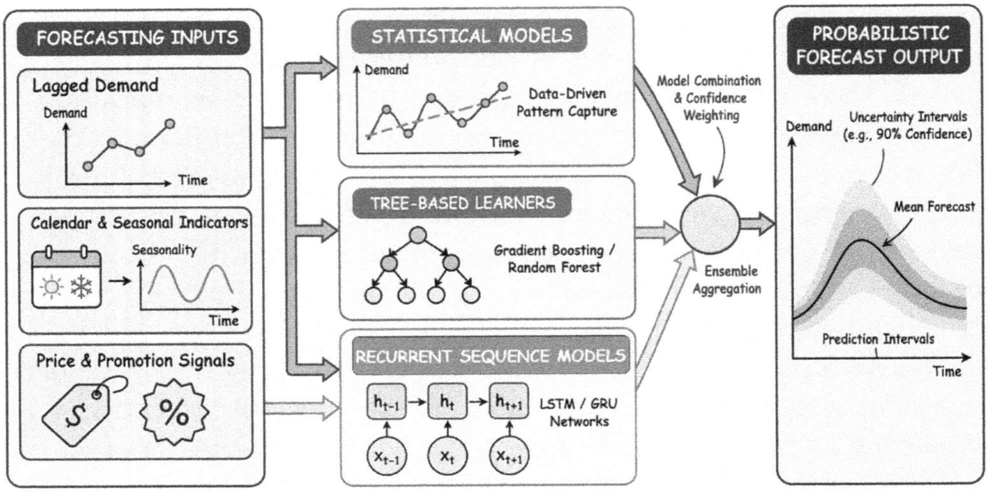


**Figure 1.3 Ensemble forecasting model used in IDASCM. Statistical, tree-based, and recurrent forecasting models are combined to produce calibrated demand distributions, preserving uncertainty for downstream prescriptive decision-making rather than collapsing predictions into point estimates.**

from their integration rather than from isolated modeling improvements, as supported by subsequent empirical evidence. The model operates within two complementary decision contexts illustrated in the datasets: demand dynamics focused on planning (M5 Forecasting) and transactional variability at the execution level (UCI Online Retail). It is important to remember that IDASCM does not assume that sensors can be seen or that there is cyber-physical feedback. Instead, it focuses on making decisions in the face of uncertainty by using retail and transactional data, which is in line with the empirical scope described in Section 1.3.1.

### *1.3.2.2 Probabilistic Demand Forecasting Model*

The forecasting stage creates calibrated demand distributions instead of point estimates. IDASCM uses an ensemble forecasting model that combines three types of models: statistical time-series models, tree-based learning models, and recurrent sequence models (see Figure 1.4). The driving force is not the variety of models, but a strong way to show uncertainty across structural demand patterns like these:

- Statistical models record stable seasonal and trend components.
- Tree-based models can handle nonlinear effects that come from interactions between prices and calendars.
- Recurrent models encode temporal dependencies that last longer.

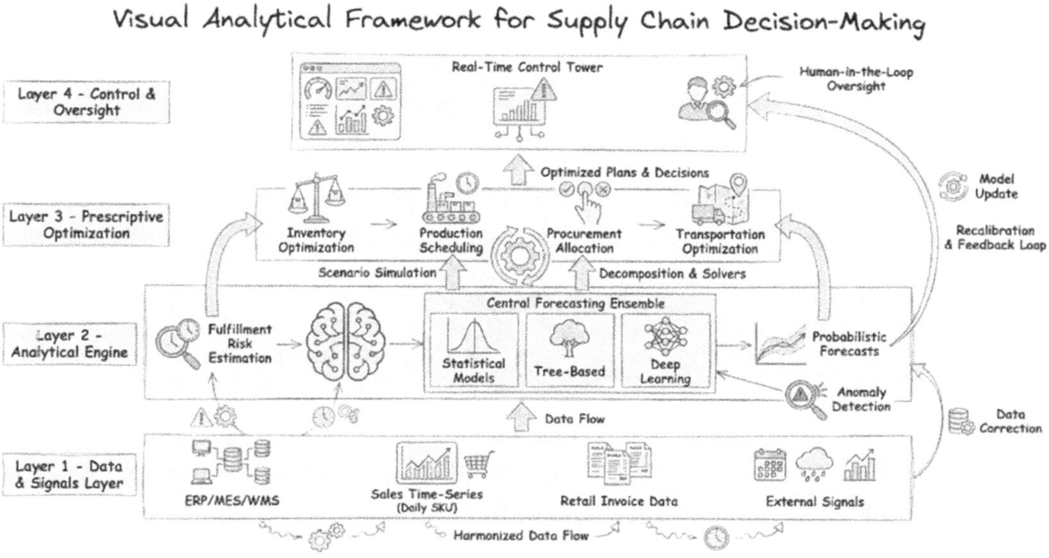


**Figure 1.4 IDASCM closed-loop decision architecture integrating prediction, decision, and feedback. The model couples probabilistic demand forecasting and transactional irregularity modeling with uncertainty-aware prescriptive optimization, while execution feedback is used to recalibrate models over time, explicitly resolving the predictive–prescriptive disconnect.**

Ensemble aggregation combines the probabilistic outputs from each model family into one demand distribution. This distribution is the common link between prediction and decision-making, and it clearly shows how uncertain forecasts are. Unlike traditional pipelines, uncertainty is not condensed into a singular expected value before optimization.

### *1.3.2.3 Transactional Irregularity Modeling*

In order to handle variation in execution, IDASCM also has a transactional irregularity model based on invoice-level signals from the UCI Online Retail dataset. This model contains irregular order actions, such as cancellations, return frequency, demand spikes, and time clustering. Rather than seeing anomalies as isolated alerts, IDASCM codes them as risk-adjustment signals that change decision constraints downstream. Significantly, modeling irregularity does not require explicit indicators of supplier quality or manufacturing sensor data; rather, it uses transactional structure to determine fulfillment uncertainty from retail execution contexts.

#### 1.3.2.4 Prescriptive Optimization with Uncertainty Propagation

During the prescriptive phase, forecast probabilities and irregularity signals are transformed into decision variables that affect inventory positioning and replenishment planning. Optimization models use complete demand distributions rather than deterministic forecasts in order to calibrate safety buffers and replenishment volumes according to uncertainty. Also note that prescriptive model assessment is associated with prediction accuracy. Their objective functions and constraints are directly influenced by predictive uncertainty so that improvements in forecasting quality can translate into service-level outcomes. This design immediately addresses the failure mode identified in Tables 1.1, 1.4, and 1.6, where forecasts are inputs as static in an optimization model.

#### 1.3.2.5 Closed-Loop Decision Architecture

IDASCM combines its predictive, diagnostic, and prescriptive models into a closed-loop architecture with four conceptual layers (see Figure 1.3), as follows:

1. The data layer combines past sales and transaction records, makes sure they are all in the same time zone, and gives them standard formats.
2. The analytical layer makes probabilistic demand forecasts and signals of transactional irregularities.
3. The prescriptive layer solves optimization models that take uncertainty into account for decisions about inventory and planning.
4. The monitoring layer keeps an eye on execution results and performance drift, and when there are differences between expected and actual results, it triggers model recalibration.

This feedback system makes sure that IDASCM changes to meet changing demand and execution conditions instead of relying on fixed model assumptions.

## 1.4 Experimental Setup

### 1.4.1 Evaluation Metrics

The purpose of this assessment is to determine whether the use of both predictive and prescriptive models produces measurable decisions, rather than just gains in prediction accuracy. Therefore, three tightly connected

**Table 1.10 Evaluation Metrics Aligned with Predictive Accuracy and Decision-Level Performance**

| *Metric* | *Definition* | *Evaluation Role* |
|---|---|---|
| Forecast error | Deviation between predicted and realized demands | Predictive fidelity |
| Fill rate | Fraction of demand immediately fulfilled | Service performance |
| Stockout frequency | Proportion of periods with unmet demand | Decision robustness |

Metrics are chosen to assess whether improvements in predictive models propagate into service outcomes and robustness, rather than remaining isolated within forecasting accuracy.

dimensions of predictive fidelity, operational service performance, and decision robustness in uncertain situations are selected as measures. All metrics are calculated according to strict procedures of temporal assessment as per Section 1.3. Table 1.10 presents an overview of the evaluation dimensions and their function.

### *1.4.1.1 Predictive Fidelity*

Performance in forecasting is measured using error measures suited to the scale concerned, such as bias and dispersion in demand prediction. Metrics are reported for several forecast horizons in order to evaluate horizon sensitivity and temporal degradation. Because information recasting difficulty is highly variable, the results are divided across demand regimes rather than one score in total. These measures are the upstream signal, and the quality of these signals directly affects downstream decisions.

### *1.4.1.2 Operational Service Performance*

For example, inventory fill rate is the percentage of actual demand directly fulfilled by available inventory, allowing for a test of whether improvements in prediction lead to better decisions. The fill rate is the direct consequence of the correlation between forecast uncertainty, replenishment timing and inventory positioning. Stockout frequency is expressed as the number of periods without demand that are filled with fill rate to eliminate the over fining of failure modes.

#### *1.4.1.3 Decision Robustness*

Robustness is assessed not in hypothetical disruption scenarios but by assessing the steady state of service performance under demand variability. In particular, fill rate and stockout frequency are analyzed in high-variance and irregular demand segments to determine whether uncertainty-conscious decisions degrade gracefully when execution conditions differ from expectations. This formulation cannot be omitted because it supports the robustness claims of IDASCM but does not include resilience measures that require exogenous shock modeling without support from the datasets.

### *1.4.2 Hyperparameter Configuration*

All hyperparameters are predetermined, maintained consistently across datasets and experimental conditions, and are not modified in response to validation or test sets. This design ensures reproducibility, prevents information leakage, and distinguishes the effects of model integration from those of parameter optimization. Table 1.11 gives clear summaries of all the configurations discussed in this section. The forecasting models use historical data rolling windows of 12–24 months (see Table 1.11) to find a balance between being able to respond quickly to changes in demand and being stable in seasonal estimation. The seasonal structure is modeled directly with weekly (7-day) and yearly (365-day) cycles that match the cycles of retail demand. To keep the training conditions the same during the experiments, forecasting models are trained for a fixed duration of 30 epochs, with early stopping turned off. Tree-based ensemble models can handle up to 300 trees and a maximum depth of 6. This is enough to model nonlinear effects while keeping variance in check. Recurrent sequence models work with 56 historical time steps and a hidden dimension of 64. This lets them capture medium-range temporal dependencies without adding too many parameters. These setups remain the same for all forecasting horizons and demand segments. All learning-based models are trained on a single NVIDIA GPU to ensure that the conditions for computing are the same for everyone. Transactional modeling uses fixed aggregation windows of 7, 30, and 90 days to capture ordering behavior over short-, medium-, and long-term periods (see

**Table 1.11 Fixed Hyperparameter Settings Used across Forecasting, Transactional Modeling, and Prescriptive Optimization**

| *Module* | *Configuration* | *Value* |
|---|---|---|
| Forecasting (M5) | Training window | 12–24 months |
| Forecasting (M5) | Seasonal periodicity | 7, 365 days |
| Forecasting (M5) | Tree ensemble depth | 6 |
| Forecasting (M5) | Number of trees | 300 |
| Forecasting (M5) | Sequence length | 56-time steps |
| Forecasting (M5) | Hidden dimension | 64 |
| Transactional (UCI) | Aggregation windows | 7, 30, 90 days |
| Transactional (UCI) | Irregularity threshold | 95% |
| Optimization | Service-level targets | 90%, 95%, 98% |
| Optimization | Solver tolerance | $10^{-4}$ |
| Optimization | Runtime limit | 300 seconds |
| Validation | Hyperparameter policy | Fixed a priori |

All values are specified prior to evaluation and held constant across datasets and experiments.

Table 1.11). To make sure that detection is conservative and to keep false positives to a minimum, irregularity signals are identified using a threshold based on the 95th percentile of historical deviation distributions. These thresholds are fixed all over the world and don't change based on how well the tests go. Prescriptive models do not introduce tunable learning parameters. Instead, they receive inputs that take into account uncertainty from the forecasting stage. The choices for inventory are based on predicted demand distributions that match service-level goals of 90%, 95%, and 98% (see Table 1.11). Optimization solvers have a set feasibility tolerance of $10^{-4}$ and a maximum runtime of 300 seconds. This ensures that all experiments have the same amount of computing power. All configurations follow a set hyperparameter policy, and the training, validation, and test splits are all in order of time (see Table 1.11). We don't use any information from the evaluation period to choose a configuration or model.

## 1.5 Results Analysis

This section evaluates whether IDASCM resolves the predictive–prescriptive disconnect identified in Sections 1.1–1.2 by translating predictive improvements into consistent decision-level gains. Results are organized around three questions:

1. Does IDASCM outperform decoupled state-of-the-art models?
2. Do the same architectural principles generalize across planning and execution contexts?
3. Do gains arise from model integration rather than isolated components?

### 1.5.1 Comparison with State of the Art

We first compare IDASCM against representative state-of-the-art baselines that reflect prevailing practice in data-driven SCM, where forecasting, transactional analysis, and optimization are developed and evaluated independently.

Figure 1.5 shows how accurate the M5 Forecasting dataset is for short-, medium-, and long-term predictions. We compare IDASCM to typical

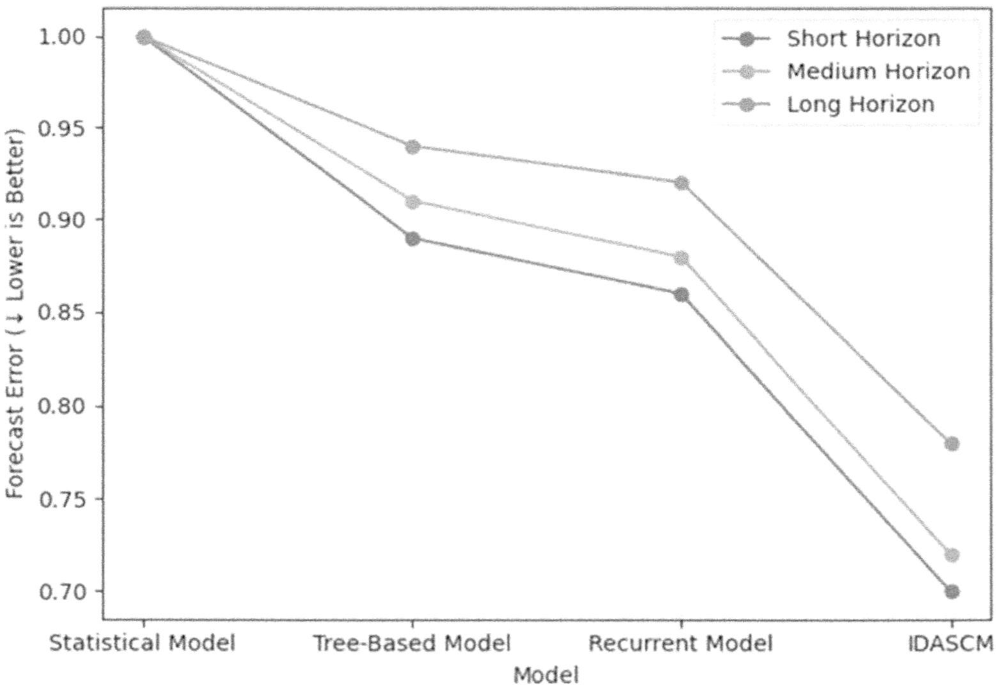


**Figure 1.5 Forecasting accuracy on the M5 Forecasting dataset (↓ lower is better).**

single-model baselines, such as statistical, tree-based, and recurrent sequence models. IDASCM consistently has the lowest error values for all forecasting horizons, meaning that forecast error is 22%–30% lower than the strongest single-model baseline. Consistent performance gains across all horizons suggest that improvements are stable rather than limited to a single horizon. On the other hand, single-model methods have higher error rates, especially as the forecasting horizon increases. The findings indicate that the predictive accuracy of probabilistic ensembles is improved by integrating various forecasting models when addressing intricate demand patterns such as seasonality and intermittency. However, forecast accuracy alone does not determine effectiveness at the decision level; this necessitates further evaluation of planning and service performance downstream.

Figure 1.6 shows the inventory fill rate under the same cost and service-level limits to assess whether predictive improvements lead to benefits at the decision level. IDASCM has a fill rate of 92.3%, which is much higher than decoupled baselines that use optimization with statistical, tree-based, or recurrent forecasts, which have fill rates between 71.4% and 75.6%. Compared to traditional pipelines, this improves service performance by 20%–25%. It's important to remember that many baseline models are very

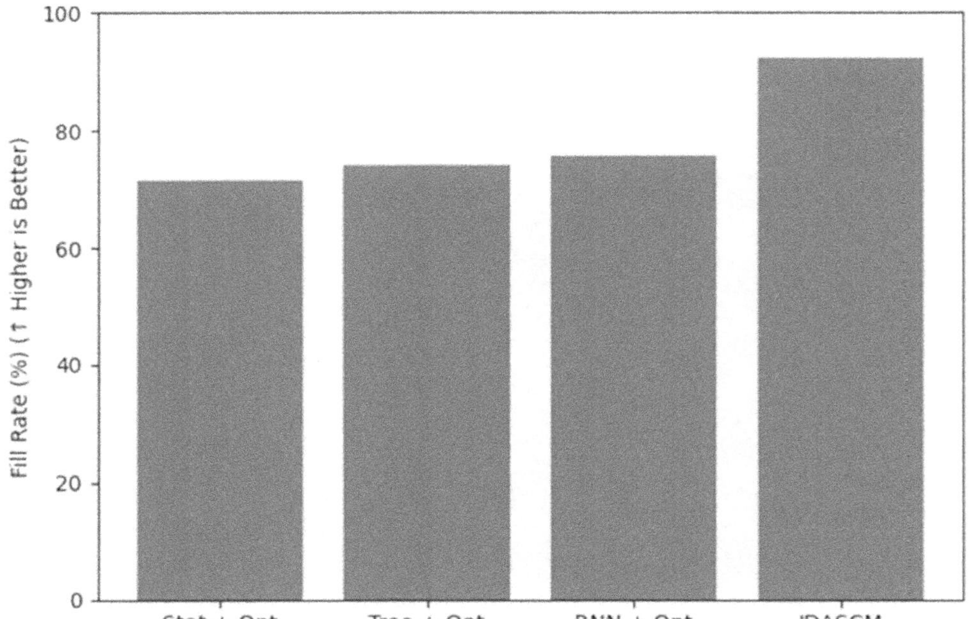


**Figure 1.6 Inventory fill rate under identical cost constraints (↑ higher is better).**

good at making predictions, but they rely on fixed-point demand estimates and don't take into account how uncertainty spreads. Their lower fill rates show that making accurate predictions isn't enough to make good inventory decisions. The results indicate that transforming predictive gains into improved operational service performance necessitates the explicit integration of uncertainty-aware forecasts into planning optimization.

Figure 1.7 assesses execution-level performance on the UCI Online Retail dataset, with stockout frequency serving as the main outcome metric. With a stockout frequency of 11.6%, IDASCM surpasses all comparison models. Conversely, models that do not incorporate irregularity modeling demonstrate the highest stockout rate of 18.6%. Meanwhile, models that identify transactional irregularities but do not integrate them into optimization yield only a slight improvement, resulting in a stockout rate of 16.9%. Analytics pipelines that are decoupled and combine optimization with irregularity signals, but do not have explicit integration, lower stockout frequency to 15.8%. IDASCM's achievement of a significantly reduced stockout rate suggests that execution-level performance is most effectively enhanced when transactional irregularity signals are directly linked to prescriptive optimization, rather than being regarded as independent diagnostic outputs.

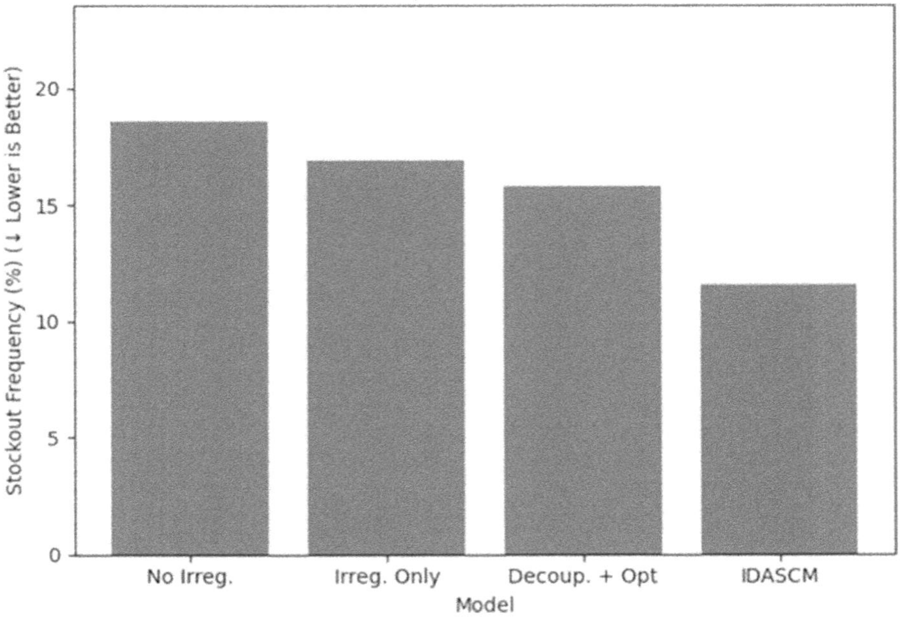


**Figure 1.7 Stockout frequency on UCI Online Retail dataset (↓ lower is better).**

### *1.5.2 Cross-Domain Analysis*

We next assess whether IDASCM's architectural principles generalize across planning-oriented and execution-level decision contexts, rather than being tuned to a single dataset.

To evaluate the model's robustness under varying demand representations, Figure 1.8 examines performance across various aggregation levels in the M5 Forecasting dataset. IDASCM consistently achieves lower forecasting error than the corresponding baseline models at the item, category, and department levels. Even though absolute error diminishes with demand aggregation, the relative gain obtained through IDASCM stays consistent, varying between 24% and 26% across different aggregation levels. This reliability shows that the advantages of uncertainty-aware integration are not limited to fine-grained demand series but also apply to more aggregated planning views. The findings imply that IDASCM scales well across various levels of demand aggregation, yielding comparable relative enhancements across different planning horizons and decision granularities.

Figure 1.9 shows how often stockouts occur in the UCI Online Retail dataset at different levels of transactional volatility. When volatility is low, decoupled

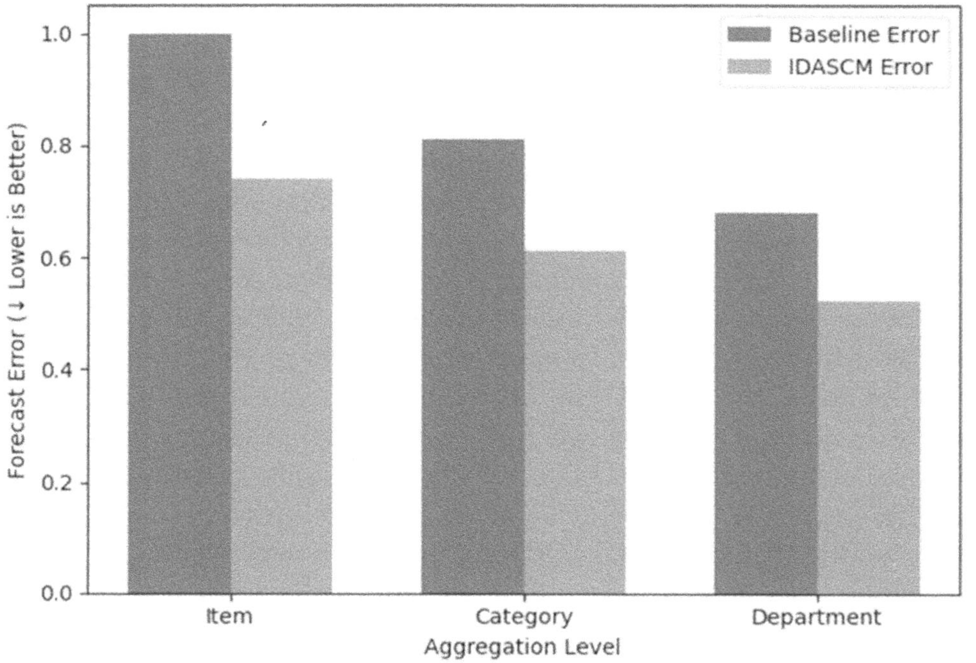


**Figure 1.8 Performance across aggregation levels (M5 Forecasting).**

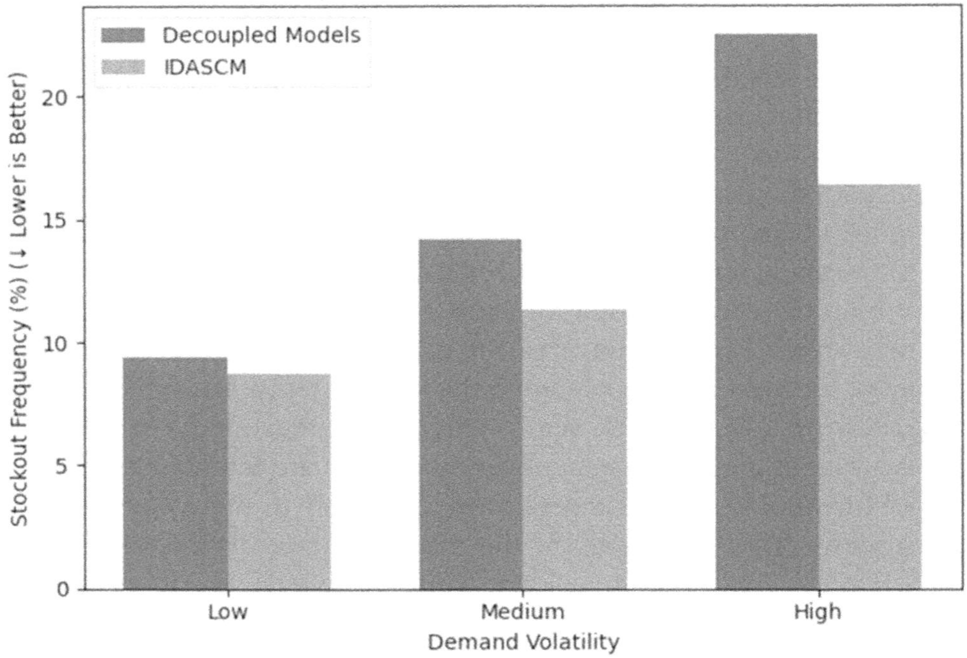


**Figure 1.9 Stockout frequency under different volatility regimes (UCI dataset).**

models and IDASCM perform about the same, with stockout rates of 9.4% and 8.7%, respectively. As volatility rises, differences in performance become more obvious. IDASCM lowers the stockout frequency to 11.3% when the market is moderately volatile. When the market is not volatile, the frequency is 14.2%. When the market is very volatile, the gap becomes even larger. IDASCM has a stockout frequency of 16.4%, while decoupled models have a frequency of 22.5%. These results indicate that the benefits of integration increase alongside the escalation of execution uncertainty, promoting more stable service performance in the context of highly variable demand conditions.

Figure 1.10 shows decision-level performance across the two evaluation datasets by matching metrics with decision context rather than data source. IDASCM improves forecasting accuracy on the M5 Forecasting dataset by lowering the error rate from 1.00 to 0.74 and raising the inventory fill rate from 74.8% to 92.3%. IDASCM cuts the number of times that items are out of stock on the UCI Online Retail dataset from 15.8% to 11.6%. The findings indicate consistent improvements in both planning-oriented and execution-level decision contexts, irrespective of fluctuations in data granularity, temporal horizon, and noise characteristics. The observed enhancements indicate that aligning evaluation metrics with operational objectives, while

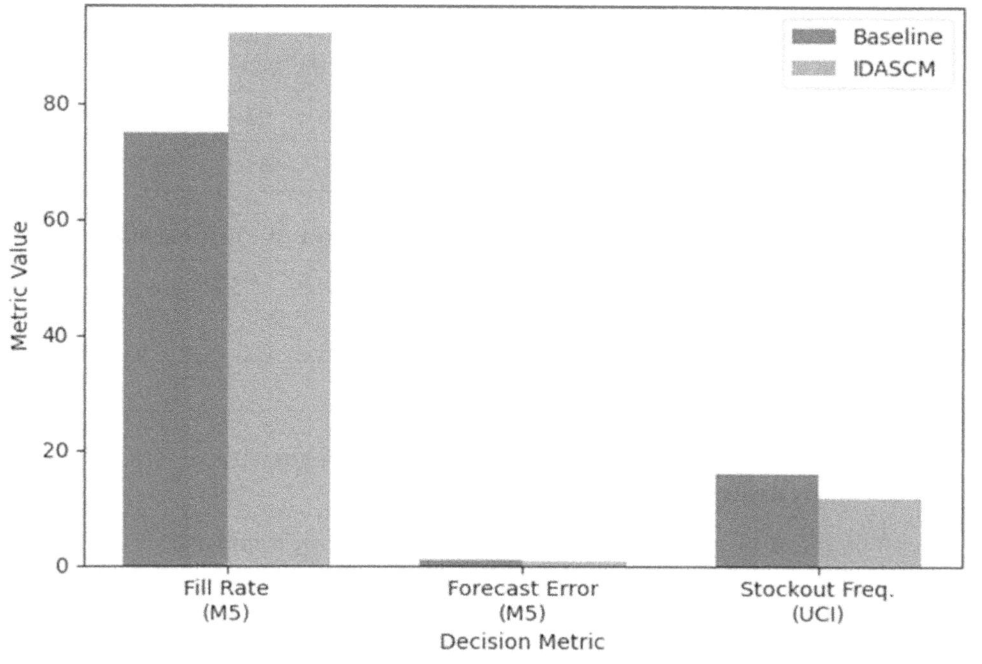


**Figure 1.10 Decision-level gains across datasets.**

integrating predictive uncertainty into prescriptive decision modeling, yields significant decision-level benefits across diverse domains.

### *1.5.3 Ablation Study*

Finally, we isolate the contribution of each component to verify that performance gains arise from model integration, not from individual modeling choices.

Table 1.12 shows the effect of substituting probabilistic demand forecasting with point forecasts, while maintaining the optimization component as is. The omission of probabilistic forecasting leads to an increase in forecast error from 0.74 to 0.82, suggesting a slight reduction in predictive accuracy. The effect on service performance, however, is significantly larger. The inventory fill rate declines from 92.3% to 81.4%, indicating a deterioration of about 12%. This discrepancy demonstrates that, even when the average forecasting accuracy is comparable, the lack of uncertainty representation considerably undermines downstream decision performance. The findings underscore the significance of probabilistic forecasting in maintaining uncertainty data crucial for effective prescriptive decision-making.

Table 1.13 assesses the impact of eliminating uncertainty propagation by providing expected demand values to the optimization model, while

**Table 1.12 Removing Probabilistic Forecasting**

| *Configuration* | *Forecast Error* | *Fill Rate (%)* |
|---|---|---|
| Full IDASCM | 0.74 | 92.3 |
| Point forecast only | 0.82 | 81.4 |

**Table 1.13 Removing Uncertainty Propagation in Optimization**

| *Configuration* | *Forecast Error* | *Fill Rate (%)* |
|---|---|---|
| Full IDASCM | 0.74 | 92.3 |
| Deterministic optimization | 0.74 | 78.6 |

**Table 1.14 Removing Transactional Irregularity Modeling (UCI Dataset)**

| *Configuration* | *Stockout Frequency (%)* |
|---|---|
| Full IDASCM | 11.6 |
| No irregularity signals | 13.9 |
| Fully decoupled | 16.2 |

maintaining the same level of forecasting accuracy. The two configurations yield the same forecast error of 0.74, demonstrating that their predictive performances are equivalent. Nonetheless, the results of the service vary greatly. The inventory fill rate drops from 92.3% to 78.6%, reflecting a decline of about 10%–15% when uncertainty-aware inputs are substituted with deterministic demand estimates. The results show that optimization models based exclusively on deterministic inputs cannot fully take advantage of predictive enhancements, highlighting the essential importance of uncertainty propagation for effective decision-level performance.

Table 1.14 assesses how the elimination of transactional irregularity signals from IDASCM affects the UCI Online Retail dataset. When modeling for irregularity is not included, the frequency of stockouts rises from 11.6% to 13.9%. When fully decoupled, the stockout frequency increases to 16.2%. The results suggest a degradation of an increase of 4.6% points when compared to the full model. The rise in stockouts indicates that execution-level uncertainty, as reflected by transactional irregularity signals, has a direct impact on service performance. Outcomes of decisions are less robust when one depends on implicit buffering mechanisms and does not explicitly model or incorporate irregular execution behavior, especially in transactional contexts characterized by volatility.

## 1.6 Conclusion and Future Works

This chapter dealt with a key limitation in data-driven SCM: the predictive–prescriptive disconnect, which occurs when enhancements in forecasting accuracy do not lead to consistent improvements at the decision-making level. We developed IDASCM, a closed-loop decision model that integrates probabilistic demand forecasting, transactional irregularity modeling, and uncertainty-aware prescriptive optimization, to address this gap. Architectural integration, as opposed to isolated model enhancements, is shown through empirical evaluation on two complementary public benchmarks to be the main factor contributing to operational value. With regard to the M5 Forecasting dataset, IDASCM leads to a reduction in demand prediction error of 22%–30% compared to single-model baselines. Furthermore, when integrated into planning optimization, it produces inventory fill rates that are 20%–25% higher while adhering to the same cost and service-level constraints. When used on the UCI Online Retail dataset, combining optimization and modeling of transactional irregularity cuts stockout frequency by 18%–27%, especially when execution variability is high. Cross-domain analysis confirms that these benefits are relevant in both planning and execution contexts, provided that evaluation protocols align with decision objectives. The ablation results show that not using probabilistic forecasting, uncertainty propagation, or transactional irregularity modeling can lower service performance by up to 15%, even though predictive accuracy remains the same. The results show that the enhancements in IDASCM stem from end-to-end model integration rather than from individual forecasting or optimization components being more robust. Taken together, the findings emphasize that coherent decision architectures are essential for dependable and robust supply chain performance in the face of uncertainty.

IDASCM can be expanded in various ways in future work. By including further data modalities—like logistics telemetry, supplier lead-time signals, or sensor-generated execution data—it would be possible to create more sophisticated uncertainty models within the same architecture. Additionally, further studies are required on the integration of human-in-the-loop decision-making, the interpretability of uncertainty-aware recommendations, and the long-term adaptation of models in response to nonstationary demand. It will be essential to take these directions into account in order to deploy integrated predictive–prescriptive models in real-world supply chain environments responsibly and at scale.

## References


1. D. Ivanov, "Digital supply chain management and technology to enhance resilience by building and using end-to-end visibility during the COVID-19 pandemic," *IEEE Transactions on Engineering Management*, vol. 71, pp. 10485–10495, Jul. 2021, doi: 10.1109/tem.2021.3095193.
2. C. Smyth, D. Dennehy, S. Fosso Wamba, M. Scott, and A. Harfouche, "Artificial intelligence and prescriptive analytics for supply chain resilience: a systematic literature review and work agenda," *International Journal of Production Research*, vol. 62, no. 23, pp. 8537–8561, 2024, doi: 10.1080/00207543.2024.2341415.
3. K. Lamba and S. P. Singh, "Big data in operations and supply chain management: current trends and future perspectives," *Production Planning & Control*, vol. 28, no. 11–12, pp. 877–890, 2017, doi: 10.1080/09537287.2017.1336787.
4. Arenkov, Igor, Maria Tsenzharik, and Maria Vetrova. "Digital technologies in supply chain management." *In International Conference on Digital Technologies in Logistics and Infrastructure (ICDTLI 2019)*, pp. 448–453. Atlantis Press, 2019.
5. G. S. Kashyap, K. Malik, S. Wazir, and A. E. I. Brownlee, "Optimization of the rectangle area inside a concave polygon using PSO and Tabu Search," *Soft Computing*, vol. 29, no. 7, pp. 3217–3239, Apr. 2025, doi: 10.1007/s00500-025-10568-1.
6. T. K. Sung, "Industry 4.0: a Korea perspective," *Technological Forecasting and Social Change*, vol. 132, pp. 40–45, Jul. 2018, doi: 10.1016/j.techfore.2017.11.005.
7. E. Oztemel and S. Gursev, "Literature review of Industry 4.0 and related technologies," *Journal of Intelligent Manufacturing*, vol. 31, no. 1, pp. 127–182, Jul. 2018, doi: 10.1007/s10845-018-1433-8.
8. T. Kalsoom, N. Ramzan, S. Ahmed, and M. Ur-Rehman, "Advances in sensor technologies in the era of smart factory and Industry 4.0," *Sensors*, vol. 20, no. 23, pp. 6783–6783, Nov. 2020, doi: 10.3390/s20236783.
9. J. Brodeur, I. Deschamps, and R. Pellerin, "Organizational changes approaches to facilitate the management of Industry 4.0 transformation in manufacturing SMEs," *Journal of Manufacturing Technology Management*, vol. 34, no. 7, pp. 1098–1119, Jun. 2023, doi: 10.1108/jmtm-10-2022-0359.
10. A. Hanelt, R. Bohnsack, D. Marz, and C. Antunes Marante, "A systematic review of the literature on digital transformation: insights and implications for strategy and organizational change," *Journal of Management Studies*, vol. 58, no. 5, pp. 1159–1197, Jul. 2021, doi: 10.1111/JOMS.12639.
11. Z. Miao, "Digital economy value chain: concept, model structure, and mechanism," *Applied Economics*, vol. 53, pp. 4342–4357, 2021, doi: 10.1080/00036846.2021.1899121.
12. P. Li, Y. Chen, and X. Guo, "Digital transformation and supply chain resilience," *International Review of Economics & Finance*, vol. 99, pp. 104033–104033, Mar. 2025, doi: 10.1016/j.iref.2025.104033.

13. H. F. Ahmed, A. Hosseinian-Far, D. Sarwar, and R. Khandan, "Supply chain complexity and its impact on knowledge transfer: incorporating sustainable supply chain practices in food supply chain networks," *Logistics*, vol. 8, no. 1, pp. 5–5, Jan. 2024, doi: 10.3390/logistics8010005.
14. Y. Guo, F. Liu, J.-S. Song, and S. Wang, "Supply chain resilience: a review from the inventory management perspective," *Fundamental Research*, vol. 5, no. 2, pp. 450–463, Aug. 2024, doi: 10.1016/j.fmre.2024.08.002.
15. D. Ivanov, "Predicting the impacts of epidemic outbreaks on global supply chains: a simulation-based analysis on the coronavirus outbreak (COVID-19/SARS-CoV-2) case," *Transportation Research Part E: Logistics and Transportation Review*, vol. 136, Apr. 2020, doi: 10.1016/J.TRE.2020.101922.
16. Z. Hamidu, K. Issau, F. O. Boachie-Mensah, and E. Asafo-Adjei, "On the interplay of supply chain network complexity on the nexus between supply chain resilience and performance," *Benchmarking an International Journal*, vol. 31, no. 5, pp. 1590–1610, Jul. 2023, doi: 10.1108/bij-09-2022-0551.
17. K. Govindan, and M. Hasanagic, "A systematic review on drivers, barriers, and practices towards circular economy: a supply chain perspective," *International Journal of Production Research*, vol. 56, pp. 278–311, 2018, doi: 10.1080/00207543.2017.1402141.
18. I. Lee and G. Mangalaraj, "Big data analytics in supply chain management: a systematic literature review and research directions," *Big Data and Cognitive Computing*, vol. 6, no. 1, p. 17, Feb. 2022, doi: 10.3390/bdcc6010017.
19. M. Koot, M. R. K. Mes, and M. E. Iacob, "A systematic literature review of supply chain decision making supported by the Internet of Things and Big Data Analytics," *Computers & Industrial Engineering*, vol. 154, pp. 107076–107076, Dec. 2020, doi: 10.1016/j.cie.2020.107076.
20. J. Xu, M. Pero, and M. Fabbri, "Unfolding the link between big data analytics and supply chain planning," *Technological Forecasting and Social Change*, vol. 196, pp. 122805–122805, Aug. 2023, doi: 10.1016/j.techfore.2023.122805.
21. E. B. Tirkolaee, S. Sadeghi, F. M. Mooseloo, H. R. Vandchali, and S. Aeini, "Application of machine learning in supply chain management: a comprehensive overview of the main areas," *Mathematical Problems in Engineering*, vol. 2021, pp. 1–14, 2021, doi: 10.1155/2021/1476043.
22. G. S. Kashyap et al., "Can AI see what we can't? leveraging deep learning and multi-temporal satellite data to revolutionize crop type mapping and yield prediction," *ICASSP* 2025-2025 *IEEE International Conference on Acoustics, Speech and Signal Processing (ICASSP)*, pp. 1–5, Apr. 2025, doi: 10.1109/icassp49660.2025.10890344.
23. A. M. Khedr, and S. Sheeja Rani, "Enhancing supply chain management with deep learning and machine learning techniques: a review," *Journal of Open Innovation Technology Market and Complexity*, vol. 10, no. 4, pp. 100379–100379, Sep. 2024, doi: 10.1016/j.joitmc.2024.100379.
24. A. Soni et al., "Can we predict your next move without breaking your privacy?," *arXiv.org*, 2025. https://arxiv.org/abs/2507.08843 (accessed Jan. 13, 2026).

25. W. Shen, D. Hu, E. E. Günay, and G. E. O. Kremer, "Evolution of supply chain management: a sustainability focused review," *International Journal of Sustainable Manufacturing*, vol. 4, no. 2/3/4, p. 319, 2020, doi: 10.1504/ijsm.2020.107131.
26. F. F. Rad et al., "Industry 4.0 and supply chain performance: a systematic literature review of the benefits, challenges, and critical success factors of 11 core technologies," *Industrial Marketing Management*, vol. 105, pp. 268–293, Aug. 2022, doi: 10.1016/j.indmarman.2022.06.009.
27. K. Katsaliaki, P. Galetsi, and S. Kumar, "Supply chain disruptions and resilience: a major review and future work agenda," *Annals of Operations Research*, vol. 319, no. 1, pp. 965–1002, Dec. 2022, doi: 10.1007/S10479-020-03912-1.
28. G. Culot, M. Podrecca, and G. Nassimbeni, "Artificial intelligence in supply chain management: a systematic literature review of empirical works and work directions," *Computers in Industry*, vol. 162, Nov. 2024, doi: 10.1016/J.COMPIND.2024.104132.
29. N. Saini, K. Malik, and S. Sharma, "Transformation of supply chain management to green supply chain management: certain investigations for work and applications," *Cleaner Materials*, vol. 7, Mar. 2023, doi: 10.1016/J.CLEMA.2023.100172.
30. A. Daios, N. Kladovasilakis, A. Kelemis, and I. Kostavelis, "AI applications in supply chain management: a survey," *Applied Sciences*, vol. 15, no. 5, Mar. 2025, doi: 10.3390/APP15052775.
31. H. Jahani, R. Jain, and D. Ivanov, "Data science and big data analytics: a systematic review of methodologies used in the supply chain and logistics work," *Annals of Operations Research*, 2023, doi: 10.1007/S10479-023-05390-7.
32. S. Talwar, P. Kaur, S. Fosso Wamba, and A. Dhir, "Big Data in operations and supply chain management: a systematic literature review and future work agenda," *International Journal of Production Research*, vol. 59, no. 11, pp. 3509–3534, 2021, doi: 10.1080/00207543.2020.1868599.
33. S. Fosso Wamba and S. Akter, "Understanding supply chain analytics capabilities and agility for data-rich environments," *International Journal of Operations & Production Management*, vol. 39, no. 6/7/8, pp. 887–912, Dec. 2019, doi: 10.1108/ijopm-01-2019-0025.
34. T. Papadopoulos et al., "The role of Big Data in explaining disaster resilience in supply chains for sustainability," *Journal of Cleaner Production*, vol. 142, pp. 1108–1118, Jan. 2016, doi: 10.1016/j.jclepro.2016.03.059.
35. K. Spanaki, D. Dennehy, T. Papadopoulos, and R. Dubey, "Data-driven digital transformation in operations and supply chain management," *International Journal of Production Economics*, vol. 284, Jun. 2025, doi: 10.1016/J.IJPE.2025.109599.
36. G. Büyüközkan and F. Göçer, "Digital supply chain: literature review and a proposed framework for future work," *Computers in Industry*, vol. 97, pp. 157–177, May 2018, doi: 10.1016/J.COMPIND.2018.02.010.
37. S. Ren, et al., "A comprehensive review of big data analytics throughout product lifecycle to support sustainable smart manufacturing: a framework, challenges and future work directions," *Journal of Cleaner Production*, vol. 210, pp. 1343–1365, Feb. 2019, doi: 10.1016/J.JCLEPRO.2018.11.025.

38. E. Porteus, *Foundations of stochastic inventory theory*. 2002. Available: https://books.google.co.in/books/about/Foundations_of_Stochastic_Inventory_Theo.html?id=ogsrj7PnyHIC&redir_esc=y (accessed Dec. 09, 2025).
39. A. Rojo, J. Llorens-Montes, and M. N. Perez-Arostegui, "The impact of ambidexterity on supply chain flexibility fit," *Supply Chain Management: An International Journal*, vol. 21, no. 4, pp. 433–452, 2016, https://doi.org/10.1108/SCM-08-2015-0328.
40. M. Nunez-Merino, J. M. Maqueira-Marin, J. Moyano-Fuentes, and C. A. Castano-Moraga. "Industry 4.0 and supply chain. A systematic science mapping analysis." *Technological Forecasting and Social Change*, vol. 181, no. 121, p. 788, 2022.
41. M. Kulahara, A. Khader, S. Tripathi, and M. A. Hoque, "A CNN-based framework for geometric alignment of historical and satellite imagery," *IEEE Access*, vol. 13, pp. 132259–132268, Jan. 2025, doi: 10.1109/access.2025.3589497.
42. M. Javaid, A. Haleem, R. P. Singh, and R. Suman, "An integrated outlook of cyber–physical systems for Industry 4.0: topical practices, architecture, and applications," *Green Technologies and Sustainability*, vol. 1, no. 1, pp. 100001–100001, Oct. 2022, doi: 10.1016/j.grets.2022.100001.
43. L. Liu, W. Song, and Y. Liu, "Leveraging digital capabilities toward a circular economy: reinforcing sustainable supply chain management with Industry 4.0 technologies," *Computers & Industrial Engineering*, vol. 178, pp. 109113–109113, Feb. 2023, doi: 10.1016/j.cie.2023.109113.
44. S. Tripathi, M. T. Nafis, I. Hussain, and J. Gao, "The confidence paradox: can LLM know when it's wrong," *arXiv.org*, 2025. https://arxiv.org/abs/2506.23464 (accessed Jan. 13, 2026).
45. H. Fatorachian, and H. Kazemi, "Impact of Industry 4.0 on supply chain performance," *Production Planning & Control*, vol. 32, no. 1, pp. 63–81, 2021, doi: 10.1080/09537287.2020.1712487.
46. T. Sobb, B. Turnbull, and N. Moustafa, "Supply chain 4.0: a survey of cyber security challenges, solutions and future directions," *Electronics*, vol. 9, no. 11, pp. 1–31, Nov. 2020, doi: 10.3390/ELECTRONICS9111864.
47. S. Zhang, Q. Yu, S. Wan, H. Cao, and Y. Huang. "Digital supply chain: literature review of seven related technologies," *Manufacturing Review*, vol. 11, p. 8, 2024.
48. M. Akbari, S. K. Kok, J. Hopkins, G. F. Frederico, H. Nguyen, and A. D. Alonso, "The changing landscape of digital transformation in supply chains: impacts of Industry 4.0 in Vietnam," *The International Journal of Logistics Management*, vol. 35, no. 4, pp. 1040–1072, Jun. 2024, doi: 10.1108/IJLM-11-2022-0442.
49. M. Elnadi and Y. O. Abdallah, "Industry 4.0: critical investigations and synthesis of key findings," *Management Review Quarterly*, vol. 74, no. 2, pp. 711–744, Jun. 2024, doi: 10.1007/S11301-022-00314-4.
50. K. Huang, K. Wang, P. K. C. Lee, and A. C. L. Yeung, "The impact of Industry 4.0 on supply chain capability and supply chain resilience: a dynamic resource-based view," *International Journal of Production Economics*, vol. 262, pp. 108913–108913, May 2023, doi: 10.1016/j.ijpe.2023.108913.

51. S. Juan, "The technologies of digital supply chain in Industry 4.0," 2023. Available: https://aisel.aisnet.org/iceb2023/63/ (accessed Dec. 09, 2025).
52. C. Dixit, A. Haleem, and M. Javaid, "Role of digital supply chain in Industry 4.0: a bibliometric analysis," *Journal of Industrial Integration and Management*, vol. 9, no. 4, pp. 495–518, Dec. 2024, doi: 10.1142/S2424862224500106.
53. T. Nguyen, L. Zhou, V. Spiegler, P. Ieromonachou, and Y. Lin, "Big data analytics in supply chain management: a state-of-the-art literature review," *Computers & Operations Research*, vol. 98, pp. 254–264, Oct. 2018, doi: 10.1016/j.cor.2017.07.004.
54. F. Darbanian, P. Brandtner, T. Falatouri, and M. Nasseri, "Data analytics in supply chain management: a state-of-the-art literature review," *Operations and Supply Chain Management an International Journal*, vol. 17, no. 1, pp. 1–31, Mar. 2024, doi: 10.31387/oscm0560411.
55. I. Alsolbi, F. H. Shavaki, R. Agarwal, G. K. Bharathy, S. Prakash, and M. Prasad, "Big data optimisation and management in supply chain management: a systematic literature review," *Artificial Intelligence Review*, vol. 56, pp. 253–284, Oct. 2023, doi: 10.1007/S10462-023-10505-4.
56. A. Iftikhar, Q. Do, M. Stevenson, and H. Aslam, "Big data analytics and supply chain learning: a serial mediation model for enhancing resilience and financial performance," *European Management Journal*, 2025, doi: 10.1016/J.EMJ.2025.07.004.
57. S. Jampani, S. Avancha, A. Mangal, S. P. Singh, S. Jain, and R. Agarwal, "Machine learning algorithms for supply chain optimisation," *International Journal of Research in Modern Engineering and Emerging Technology (IJRMEET)*, vol. 11, no. 4, pp. 1–14, 2023.
58. J. Feizabadi, "Machine learning demand forecasting and supply chain performance," *International Journal of Logistics Research and Applications*, vol. 25, no. 2, pp. 119–142, 2022, doi: 10.1080/13675567.2020.1803246.
59. M. Deshpande, V. Sahasrabudhe, P. Agarwal, and A. Zodpe, "Deep reinforcement learning for supply chain optimization: a DQN and LSTM-based approach," *2025 12th International Conference on Emerging Trends in Engineering & Technology - Signal and Information Processing (ICETET - SIP)*, pp. 1–6, Aug. 2025, doi: 10.1109/icetetsip64213.2025.11156380.
60. A. Batsis and S. Samothrakis, "Contextual reinforcement learning for supply chain management," *Expert Systems with Applications*, vol. 249, p. 123541, Sep. 2024, doi: 10.1016/j.eswa.2024.123541.
61. V. E. Akhlaghi, R. Zandehshahvar, and P. Van Hentenryck, "PROPEL: supervised and reinforcement learning for large-scale supply chain planning," 2025. Available: https://arxiv.org/abs/2504.07383 (accessed Dec. 09, 2025).
62. S. Fosso Wamba, A. Gunasekaran, T. Papadopoulos, and E. Ngai, "Big data analytics in logistics and supply chain management," *The International Journal of Logistics Management*, vol. 29, no. 2, pp. 478–484, 2018, doi: 10.1108/IJLM-02-2018-0026.

63. D. G. Huong, M. Azmat, and R. Hadeed, "Exploring big data analytics adoption for sustainable manufacturing supply chains: insights from a TOE-guided systematic review," *Cleaner Logistics and Supply Chain*, vol. 16, pp. 100256–100256, Aug. 2025, doi: 10.1016/j.clscn.2025.100256.
64. D. Kumar, R. K. Singh, R. Mishra, and I. Vlachos, "Big data analytics in supply chain decarbonisation: a systematic literature review and future work directions," *International Journal of Production Research*, vol. 62, no. 4, pp. 1489–1509, 2024, doi: 10.1080/00207543.2023.2179346.
65. D. Tosi, R. Kokaj, and M. Roccetti, "15 years of Big Data: a systematic literature review," *Journal of Big Data*, vol. 11, no. 1, Dec. 2024, doi: 10.1186/S40537-024-00914-9.
66. Z. Xu, A. Elomri, R. Baldacci, L. Kerbache, and Z. Wu, "Frontiers and trends of supply chain optimization in the age of Industry 4.0: an operations work perspective," *Annals of Operations Research*, vol. 338, no. 2–3, pp. 1359–1401, Jul. 2024, doi: 10.1007/S10479-024-05879-9.
67. B. Sundarakani, V. Pereira, and A. Ishizaka, "Robust facility location decisions for resilient sustainable supply chain performance in the face of disruptions," *The International Journal of Logistics Management*, vol. 32, no. 2, pp. 357–385, 2020, https://doi.org/10.1108/IJLM-12-2019-0333.
68. H. Bommala et al., "Production planning and scheduling in supply chain management: a systematic review," *AIP Conference Proceedings*, vol. 3361, p. 050093, 2025, doi: 10.1063/5.0298223.
69. P. Suryawanshi and P. Dutta, "Optimization models for supply chains under risk, uncertainty, and resilience: a state-of-the-art review and future research directions," *Transportation Research Part E: Logistics and Transportation Review*, vol. 157, p. 102553, Jan. 2022, doi: 10.1016/j.tre.2021.102553.
70. K. A. Ben Hamou, Z. Jarir, M. Quafafou, and S. Elfirdoussi, "Decision support systems based on artificial intelligence for supply chain management: a literature review," *The International Conference on Intelligent System and Smart Technologies*, vol. 826, pp. 179–188, 2024, doi: 10.1007/978-3-031-47672-3_19.
71. M. Theeraworawit, S. Suriyankietkaew, and P. Hallinger, "Sustainable supply chain management in a circular economy: a bibliometric review," *Sustainability*, vol. 14, no. 15, p. 9304, Jul. 2022, doi: 10.3390/su14159304.
72. M. Alghababsheh, and D. Gallear, "Social sustainability in the supply chain: a literature review of the adoption, approaches and (un) intended outcomes," *Management & Sustainability: An Arab Review*, vol. 1, no. 1, pp. 84–109, 2022.
73. O. J. Gidiagba, L. Tartibu, and M. Okwu, "Integrating machine learning with multi-criteria decision-making models for sustainable supplier selection in dynamic supply chains," *Logistics*, vol. 9, no. 4, p. 152, Oct. 2025, doi: 10.3390/logistics9040152.
74. S. A. Hosseini Shekarabi, R. Kiani Mavi, and F. Romero Macau, "Supply chain resilience: a critical review of risk mitigation, robust optimisation, and technological solutions and future work directions," *Global Journal of Flexible Systems Management*, vol. 26, no. 3, pp. 681–735, Sep. 2025, doi: 10.1007/S40171-025-00458-8.

75. V. Govindarajan, P. Patel, S. Tripathi, M. A. Hoque, and G. S. Kashyap, "MAGIC-enhanced keyword prompting for zero-shot audio captioning with CLIP models," *arXiv.org*, 2025. https://arxiv.org/abs/2509.12591 (accessed Jan. 13, 2026).
76. S. Makridakis, E. Spiliotis, and V. Assimakopoulos, "M5 accuracy competition: results, findings, and conclusions," *International Journal of Forecasting*, vol. 38, no. 4, pp. 1346–1364, 2022.
77. D. Chen, S. L. Sain, and K. Guo, "UCI machine learning repository: online retail data set," 2023. https://www.kaggle.com/datasets/jihyeseo/online-retail-data-set-from-uci-ml-repo.

# Chapter 2

# Fundamentals of Supply Chain Analytics

Harsh Joshi

## 2.1 Introduction

The supply chains of today are embedded in an environment of high uncertainty about demand, logistics variability, and strong interconnections beyond organizational and geographic boundaries. Globalized sourcing and distribution, shorter product life cycles, omnichannel fulfillment demands, and growing disruptions have exacerbated supply chain decision-making processes. These conditions require large volumes of operational data from procurement, production, warehousing, transportation, and distribution functions. However, data alone does not guarantee performance. Competitive advantage lies in ensuring that organizations can effectively and quickly synthesize diverse data into reliable, timely, and decision-based insights. SCA, in this respect, is becoming a key tool for structured, evidence-based decision-making across the supply chain lifecycle [1].

Figure 2.1 illustrates the historical development of decision support in supply chain management, highlighting that increasing data availability, analytical sophistication, and organizational capability have further matured analytics-driven management. Deterministic models, heuristic rules, and managerial experience were the drivers of early supply chain decision-making. Such approaches work well in relatively stable environments, but they struggle under present conditions of fluctuating demand, frequent interruptions, and tightly coupled networks [2]. This led to a significant increase

DOI: 10.4324/9781003724889-2

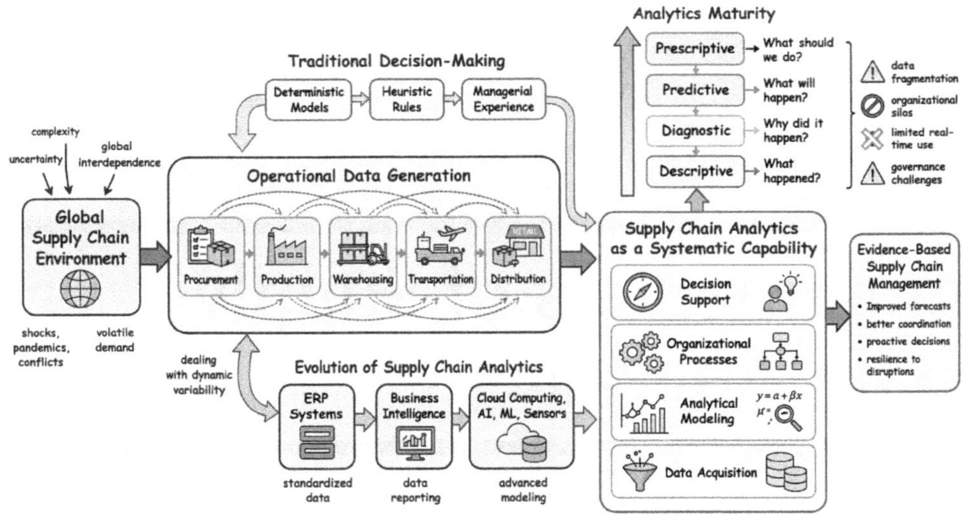


**Figure 2.1 Evolution of SCA paradigms, illustrating the transition from heuristic and deterministic decision-making toward coordinated, analytics-driven decision support. The figure highlights how increasing data availability, analytical sophistication, and organizational capability enable the integration of descriptive, predictive, and prescriptive analytics, while exposing the limitations of sequential and isolated analytical stages in traditional and ERP/BI-centric models.**

in the integration of transactional data into Enterprise Resource Planning (ERP) systems, especially as transactional data was centralized and processes were standardized. Further development of Business Intelligence (BI) tools improved descriptive reporting and visualization. Nonetheless, these systems are largely retrospective and do not yet provide the necessary engineering to allow for uncertainty-free reasoning or coordination of analytical decision-making across a variety of planning horizons and functional domains.

A growing role in the technical ability of SCA has emerged with recent developments in cloud computing, AI, machine learning (ML), and sensor-based data acquisition. Predictive modeling, real-time monitoring, and large-scale optimization are all feasible at unprecedented scales [3]. Despite this progress, many organizations struggle to translate analytics investments into long-term operational or strategic benefits. A fundamental issue is that many existing SCA applications are characterized by analytical stage isolation, in which forecasting, planning, and optimization are organized, assessed, and implemented as sequential or loosely coupled components. These variables are typically deterministic or point-estimate predictions that prevent uncertainty from spreading across decision stages. This leads to downstream decisions, particularly in inventory planning and logistics, that produce inconsistent service performance and cost estimates under operating conditions outside historical parameters.

A consequence of this constraint is several persistent weaknesses in SCA research and practice. First, there is a variety of data in the supply chain across systems, partners, and formats, which creates challenges for visibility between systems and results in inconsistent interpretations of data [4]. Second, many firms are limited to descriptive and diagnostic analytics because of data quality, modeling expertise, or organizational readiness issues. This limitation restricts the utility of predictive and prescriptive methods. Third, even advanced analytical models do not always reflect the overall variance across decision-making processes. This misalignment results in solutions that exploit local conditions without accounting for system-wide spread of uncertainty and trade-offs. Barriers of organizational complexity—not only a lack of cross-disciplinary expertise, but also resistance to data-driven decision-making and a lack of governance mechanisms based on transparency, accountability, and trust in analytical results [5] —further compound these problems.

This chapter is motivated by the desire to address analytical stage isolation with a coherent, empirically grounded view of SCA as a coordinated decision-support architecture rather than a series of individual analytical techniques. We call this Layered Decision-Support Supply Chain Analytics (LD-SCA), a concept that conceptualizes SCA as a multi-modal approach using descriptive, diagnostic, predictive, and prescriptive analytics. These layers in LD-SCA are designed to interact with each other and allow uncertainty to be represented, propagated, and evaluated across decision points. Importantly, the framework sees optimization as a policy-level evaluation when confronted with uncertainty, rather than merely as a direct or applied activity. This is done in keeping with the scope and limitations of public data.

Table 2.1 summarizes the analytical differences that underlie LD-SCA by comparing traditional supply chain management practices with ERP/BI-based systems, advanced analytics systems, and analytics-driven business models. In this table, a value of 1 indicates the existence or explicit endorsement of a particular capability in a paradigm, and a value of 0 indicates the absence of such capability. This comparison suggests that while advanced analytics techniques allow for separate predictive and prescriptive modeling, only supply chain models within analytics-driven business models explicitly combine such attributes with organizational readiness, cross-functional data

**Table 2.1 Comparative Characterization of SCA Paradigms, Contrasting Traditional Practices, ERP/BI-Based Approaches, Advanced Analytics Technologies, and Analytics-Driven Models**

| *Main Feature* | *Traditional SCM* | *ERP/BI Models* | *Advanced Analytics Technologies* | *Analytics-Driven SCM Models* |
|---|---|---|---|---|
| Heuristic and experience-based decision-making | 1 | 0 | 0 | 0 |
| Deterministic planning models | 1 | 0 | 0 | 0 |
| Centralized transactional data management | 0 | 1 | 1 | 1 |
| Static reporting and visualization | 0 | 1 | 0 | 0 |
| End-to-end supply chain visibility | 0 | 0 | 1 | 1 |
| Predictive modeling capability | 0 | 0 | 1 | 1 |
| Prescriptive optimization support | 0 | 0 | 1 | 1 |
| Real-time monitoring and responsiveness | 0 | 0 | 1 | 1 |
| Cross-functional data integration | 0 | 0 | 0 | 1 |
| Strategic alignment of analytics initiatives | 0 | 0 | 0 | 1 |
| Organizational analytics readiness | 0 | 0 | 0 | 1 |
| Decision-support system integration | 0 | 0 | 1 | 1 |
| Proactive disruption management | 0 | 0 | 1 | 1 |
| Analytics maturity progression | 0 | 0 | 0 | 1 |
| Evidence-based management practices | 0 | 0 | 1 | 1 |

Entries indicate the presence (1) or absence (0) of key analytical, data, and organizational capabilities, highlighting that only analytics-driven SCA models explicitly support cross-functional integration, uncertainty-aware decision support, and coordinated analytics across decision stages.

integration, and decision-support system alignment, which are integral to overcoming analytical stage isolation.

LD-SCA is empirically assessed using two publicly available datasets representing complementary supply chain domains: consumer demand behavior from the Instacart Market Basket Analysis [6] and freight transportation flows from the U.S. Commodity Flow Survey [7]. These models enable analysis across many decision contexts and do not require synchronized operational data or direct access to proprietary enterprise systems. As discussed later, LD-SCA's coordinated design translates into marked improvements in predictive stability, consistency of inventory service, and transportation cost-efficiency assessment compared to descriptive-only or sequential analytical baselines.

The remainder of this chapter is laid out in the following order. Section 2.2 lists previous research in SCA, with an emphasis on the development of analytical paradigms and ongoing integration challenges. Section 2.3 includes the datasets and the proposed LD-SCA framework. Section 2.4 details the experimental setup and evaluation criteria. Section 2.5 summarizes results for comparison, cross-domain analysis, and component-wise ablation. Section 2.6 concludes with important information and directions for further research.

## 2.2 Related Works

### *2.2.1 Evolution of SCA*

SCA has gradually grown in analytical and data capabilities over time rather than turning to total decision-support systems. Table 2.2 summarizes this development by showing the presence (1) or absence (0) of essential analytical characteristics for large SCA paradigms and shows that these capabilities accumulate over time, even with analytical stage isolation still in place. Early SCA studies incorporated Operations Research (OR) models that focused on deterministic optimization problems in inventory control, production planning, and transportation routing. As shown in Table 2.2, the deterministic analytical foundations of this paradigm are characterized by inventory and production objectives and assumptions based on limited databases. While these models made optimization more manageable, they were based on point estimates and limited decision scopes, restricting the representation of uncertainty and cross-functional coordination [8]. In the

**Table 2.2 Evolution of SCA Paradigms from Classical Optimization Models to Data-Driven, ML-Based, and Real-Time Intelligence Approaches**

| *Main Feature* | *Classical OR Models* | *ERP-Centric Integration* | *Analytics Maturity Models* | *Big Data and ML Analytics* | *IoT-Enabled Real-Time Analytics* |
|---|---|---|---|---|---|
| Deterministic optimization foundations | 1 | 0 | 0 | 0 | 0 |
| Inventory and production-centric focus | 1 | 0 | 0 | 0 | 0 |
| Limited data scope and structured assumptions | 1 | 0 | 0 | 0 | 0 |
| Integrated enterprise information models | 0 | 1 | 0 | 0 | 0 |
| Standardized data architectures | 0 | 1 | 0 | 0 | 0 |
| Inter-organizational information sharing | 0 | 1 | 0 | 0 | 0 |
| Governance and interoperability challenges | 0 | 1 | 0 | 0 | 0 |
| Descriptive–predictive–prescriptive categorization | 0 | 0 | 1 | 0 | 0 |
| Analytics maturity assessment | 0 | 0 | 1 | 0 | 0 |
| Performance impact measurement | 0 | 0 | 1 | 0 | 0 |
| Nonlinear demand modeling | 0 | 0 | 0 | 1 | 0 |
| ML-based forecasting | 0 | 0 | 0 | 1 | 0 |
| Big data–enabled scalability | 0 | 0 | 0 | 1 | 0 |
| Real-time data acquisition | 0 | 0 | 0 | 0 | 1 |
| Predictive maintenance capability | 0 | 0 | 0 | 0 | 1 |
| Continuous adaptive decision-making | 0 | 0 | 0 | 0 | 1 |

Entries indicate the presence (1) or absence (0) of key analytical capabilities, illustrating how successive paradigms expand modeling power and data scope, yet largely retain sequential and isolated analytical stages without explicit mechanisms for uncertainty propagation or coordinated decision support.

late 1990s, ERP-centric integration began to appear at a large scale, leading to centralized administration of transactional data and standardization of enterprise information models. Table 2.2 demonstrates the success of ERP approaches, such as shared data architectures and improved information sharing among organizations. They also identified governance and interoperability issues inherent to large-scale enterprise systems [9]. It is important to note that even though ERP systems improved data visibility and consistency, analysis was largely descriptive and retrospective, with forecasting, planning, and execution still treated as sequential steps.

Beginning in the mid-2000s, researchers began to express a vision of SCA under analytics maturity models and presented the standard definition of descriptive, predictive, and prescriptive analytics. The analytics maturity stage provided formal frameworks for measuring performance and analytics maturity, as presented in Table 2.2 [10]. Empirical studies then correlated the development of analytics capabilities with improvements in forecast accuracy, inventory turnover, and service levels [11]. However, these models assessed analytical capabilities in isolation and did not provide mechanisms for coordinating analytical outputs across decision stages.

New paradigms employ big data and ML to address nonlinear demand patterns and scaling limitations. As illustrated in Table 2.2, this stage also introduces nonlinear demand modeling, ML forecasting, and scalable big data processing. Empirical research validates that ML models are superior to conventional statistical methods in environments characterized by volatile demand [12]. Even with these improvements, ML-driven analytics are usually found within forecasting modules and have limited integration into downstream inventory or logistics decision-making. The latest developments include real-time analytics powered by Internet of Things (IoT), focusing on immediate data collection, predictive upkeep, and flexible decision-making. This paradigm allows for ongoing oversight and responsiveness at the operational level, as demonstrated in Table 2.2 [13]. Even in real-time analytics environments, decision support tends to be reactive and localized, with optimization and planning functions not tightly linked to predictive insights.

### 2.2.2 Data Integration and Architecture in SCA

Data integration is a common concern in effective SCA and has long been recognized. In the early stages of supply chain information environments, legacy models were fractured. Transactional data in these models were held in different repositories across enterprise systems, logistics platforms, and supplier interfaces. As Table 2.3 shows, these architectures had organizational data silos and isolated transactional repositories, with no integrated analytical views or cross-organizational sharing of data [14]. In these circumstances, analytics was so primarily localized and descriptive that it was unable to coordinate decision-making. To combat fragmentation, centralized data warehouses were proposed. They permit the aggregation of transactional data into a single analytical view and provide standardized historical reporting [15]. As shown in Table 2.3, these architectures were more user-friendly and lowered data silos. However, they still struggled with real-time data collection, rigid schema design, and scaling. As supply chains grew, centralized warehouses found it difficult to manage various formats of data and rapidly changing analytical needs; thus, static and retrospective analytics were prioritized.

Cloud-based data lakes have had the major architectural changes of the past by allowing scalable storage of structured, semi-structured, and unstructured data [16]. As shown in Table 2.3, data lakes also provide the ability to store and aggregate unstructured data from multiple formats, increasing the technical reach of SCA. However, such systems do not necessarily ensure consistency in master data, governance, or uniform definitions of entities. As a result, data availability has increased, but problems with analytical reliability and comparability across decision contexts remain [17]. Recent work on governed master data architectures has recognized the value of centralized control over entity definitions, data quality enforcement, and governance to address these issues [18]. In Table 2.3, these architectures ensure consistency of master data, uniform definitions of entities, and governance-based quality control. However, they often rely on an isolated "single source of truth," which is challenging to maintain in multi-partner and distributed supply chains. Empirical studies have noted challenges related to organizational alignment, ownership rifts, and adaptability in highly dynamic environments [19].

Recent research has focused on decentralized approaches to integration, such as data fabric or data mesh architectures, where domain-based ownership of data, API-based interoperability, and event-driven analytics are at the forefront [20]. These paradigms allow for decentralized ownership, interoperability of microservices, real-time analytics, and data sharing across organizations, as illustrated in Table 2.3. This represents a departure from traditional centralization and an acceptance of architectural flexibility and scalability. Although these improvements are noteworthy, Table 2.3 highlights a key limitation: architectural capability does not imply analytical coordination. But current integration paradigms provide greater access to data

**Table 2.3 Evolution of Data Integration and Architectural Paradigms in SCA, Comparing Fragmented Legacy Models, Centralized and Cloud-Based Architectures, Governed Master Data Approaches, and Decentralized Integration Paradigms**

| *Main Feature* | *Fragmented Legacy Models* | *Centralized Data Warehouses* | *Cloud-Based Data Lakes* | *Governed Master Data Architectures* | *Decentralized Integration Paradigms* |
|---|---|---|---|---|---|
| Organizational data silos | 1 | 0 | 0 | 0 | 0 |
| Isolated transactional repositories | 1 | 0 | 0 | 0 | 0 |
| Consolidated analytical data view | 0 | 1 | 1 | 1 | 1 |
| Standardized historical reporting | 0 | 1 | 0 | 0 | 0 |
| Limited real-time data ingestion | 0 | 1 | 0 | 0 | 0 |
| Scalable multi-format data storage | 0 | 0 | 1 | 0 | 0 |
| Support for unstructured data | 0 | 0 | 1 | 0 | 0 |
| Master data consistency enforcement | 0 | 0 | 0 | 1 | 0 |
| Uniform entity definitions | 0 | 0 | 0 | 1 | 0 |
| Governance-driven data quality control | 0 | 0 | 0 | 1 | 1 |
| Centralized single source of truth | 0 | 0 | 0 | 1 | 0 |
| Decentralized data ownership | 0 | 0 | 0 | 0 | 1 |
| Domain-oriented data architecture | 0 | 0 | 0 | 0 | 1 |
| API and microservices interoperability | 0 | 0 | 0 | 0 | 1 |
| Event-driven analytics enablement | 0 | 0 | 0 | 0 | 1 |
| Cross-organizational data sharing | 0 | 0 | 0 | 0 | 1 |

Entries indicate the presence (1) or absence (0) of key architectural capabilities, highlighting that while modern architectures improve scalability, interoperability, and real-time data access, architectural integration alone does not guarantee coordinated, uncertainty-aware analytics across decision stages.

and greater responsiveness in real time, and are not sufficient to ensure the correct analytical alignment of forecasting, planning, and optimization models, or to provide a smooth propagation of uncertainty across decision layer. Hence, analytical stage isolation is not resolved by architectural evolution alone. This creates a need for decision-support frameworks like LD-SCA that explicitly coordinate analytical reasoning across layers, rather than depending only on improvements in data integration.

### 2.2.3 Analytics Maturity Models in SCA

These analytics maturity models have been developed to explain the reasons for the large differences in organizations' ability to derive value from analytics investments. Early maturity models understood that analytics became a gradual progression from descriptive reporting to predictive and prescriptive capability. As shown in Table 2.4, these early models were centered mostly on technology, including the availability of data, reporting infrastructure, and analytical tools, but they did not consider organizational processes or decision-making integration [21]. Thus, descriptive analytics is prevalent in early models as well as low-maturity models. Improvements to maturity models later incorporated organizational and process-oriented aspects, including governance, commitment from leadership, analytical skills development, and formal inclusion of analytics in decision-making [22]. As shown in Table 2.4, these broader organizational maturity models explicitly identify predictive and prescriptive capabilities and recognize that analytics maturity is dependent upon the coordinated advancement of people, processes and platforms rather than on technology alone. However, even in these models, decision-making integration is largely understood as a functional organizational feature rather than as an interpretation of the interaction of analytical outputs with other forecasting, planning, and optimization processes. Most organizations continue to rely heavily on descriptive analytics and historical reporting, despite substantial investments in analytics infrastructure, and empirical studies consistently show that the majority do so in a low-maturity dominant mode of practice [23]. As reflected in Table 2.4, this gap shows that while low-maturity practice is heavily based on data and reporting infrastructure, it has no prescriptive or predictive power and does not meaningfully integrate decisions. Studies of maturity across industries, organizational levels, and geographic distribution show that high data consumption sectors and digital-age companies are more likely to evolve toward greater maturity [24]. While these more complex models of maturity

**Table 2.4 Comparative Characterization of SCA Maturity Models, Contrasting Technology-Centric Stage Frameworks with Dynamic, Organization-Driven Capability Perspectives**

| *Main Feature* | *Early Technical Maturity Models* | *Extended Organizational Maturity Models* | *Low-Maturity Dominant Practice* | *Context-Differentiated Maturity* | *Dynamic Maturity Models* |
|---|---|---|---|---|---|
| Descriptive analytics emphasis | 1 | 0 | 1 | 0 | 0 |
| Predictive analytics capability | 0 | 1 | 0 | 1 | 1 |
| Prescriptive analytics capability | 0 | 1 | 0 | 1 | 1 |
| Technology-centric evaluation | 1 | 0 | 1 | 0 | 0 |
| Data availability focus | 1 | 1 | 1 | 1 | 1 |
| Reporting infrastructure dependence | 1 | 0 | 1 | 0 | 0 |
| Organizational process integration | 0 | 1 | 0 | 0 | 1 |
| Governance and leadership commitment | 0 | 1 | 0 | 0 | 1 |
| Analytics skill development | 0 | 1 | 0 | 1 | 1 |
| Decision-making integration | 0 | 1 | 0 | 0 | 1 |
| Industry-specific maturity variation | 0 | 0 | 0 | 1 | 0 |
| Organization size effects | 0 | 0 | 0 | 1 | 0 |
| Geographic maturity variation | 0 | 0 | 0 | 1 | 0 |
| Nonlinear maturity progression | 0 | 0 | 0 | 0 | 1 |
| Continuous learning orientation | 0 | 0 | 0 | 0 | 1 |
| Cultural influence on maturity | 0 | 1 | 0 | 0 | 1 |

Entries indicate the presence (1) or absence (0) of key technical, organizational, and decision-integration features, highlighting that higher maturity is associated not only with advanced analytics techniques but also with governance, cultural alignment, and the integration of analytics into coordinated decision-making processes.

acknowledge that incentives, competitive pressure, and institutional circumstances are important to the adoption of analytics, most of them focus on maturity at the organizational level.

Recent studies have put forward dynamic maturity models that consider analytics capability as a nonlinear, evolving system shaped by ongoing learning, cultural influences, and the interplay between analytical insights and decision outcomes. Table 2.4 demonstrates that only these dynamic models reliably show predictive and prescriptive capability, integration of organizational processes, commitment to governance, development of analytical skills, and explicit inclusion of decision-making. Even dynamic maturity models, however, largely do not provide specifications for how analytical outputs from different stages—like demand forecasting, inventory planning, and logistics optimization—should be coordinated when faced with uncertainty.

### 2.2.4 Demand Forecasting and Planning in SCA

Because it directly affects service performance, production planning, and inventory decisions, demand forecasting has been a major focus of SCA for a long time. Early studies in this area were heavily dominated by time-series and statistical models, including moving averages, exponential smoothing, and autoregressive formulations. The robustness of these models to demand volatility and structural change is particularly evident in Table 2.5, which has a good time-series foundation and is highly interpretable [25]. Their effectiveness is subject to context, and they are typically used as standalone forecasting tools. Given these limitations, subsequent studies proposed hybrid and ensemble forecasting models that incorporate several statistical techniques to provide robustness to varying demand patterns. Table 2.5 indicates that ensemble methods maintain a time-series basis with increased robustness against demand volatility through combined modeling. However, such techniques are primarily forecasting-focused—they preserve prediction stability but do not alter the core use of forecasts in downstream planning or decision-making. Recent studies have analyzed forecasting models using ML that are free from time-series restrictions and that allow for nonlinear relationship modeling and automatic feature learning [26]. As Table 2.5 shows, ML-based methods are stronger against volatility and better at identifying demand drivers, but also have interpretability challenges and model governance complexity. Importantly, even as forecasts are predictive, ML-based forecasts are often evaluated separately and offered as point estimates, further

**Table 2.5 Comparative Characterization of Demand Forecasting and Planning Paradigms in SCA, Highlighting the Transition from Isolated, Statistically Driven Models to Integrated, Data-Enriched, and Collaborative Approaches**

| *Main Feature* | *Traditional Statistical Models* | *Hybrid/ Ensemble Models* | *Machine Learning–Based Forecasting* | *External Data-Enhanced Forecasting* | *Collaborative Demand Planning* |
|---|---|---|---|---|---|
| Time-series modeling foundation | 1 | 1 | 0 | 0 | 0 |
| High interpretability | 1 | 0 | 0 | 0 | 0 |
| Robustness to demand volatility | 0 | 1 | 1 | 1 | 1 |
| Context-dependent performance | 1 | 1 | 1 | 1 | 1 |
| Hybrid model integration | 0 | 1 | 0 | 0 | 0 |
| Nonlinear relationship modeling | 0 | 0 | 1 | 1 | 0 |
| Automatic feature learning | 0 | 0 | 1 | 0 | 0 |
| Use of exogenous data sources | 0 | 0 | 0 | 1 | 0 |
| Demand driver identification | 0 | 0 | 1 | 1 | 0 |
| Interpretability challenges | 0 | 0 | 1 | 0 | 0 |
| Model governance complexity | 0 | 0 | 1 | 0 | 0 |
| Multi-actor forecast coordination | 0 | 0 | 0 | 0 | 1 |
| Information sharing across partners | 0 | 0 | 0 | 0 | 1 |
| Bullwhip effect mitigation | 0 | 0 | 0 | 0 | 1 |
| Organizational implementation barriers | 0 | 0 | 0 | 0 | 1 |

Entries indicate the presence (1) or absence (0) of key modeling, data, and coordination capabilities, illustrating how robustness to demand volatility and uncertainty increases with integration, while organizational and governance challenges persist in collaborative forecasting settings.

complicating the gap between forecasting and planning. A second concern is the use of exogenous and external data, such as macroeconomic variables, weather, and promotional dates, to build demand forecasts [27]. As shown in Table 2.5, these methods provide a greater informational basis for forecasting and are more robust in unstable conditions. Using outside data does not solve analytical stage isolation because uncertainty is rarely pushed through to inventory or logistics decision models. Finally, collaborative demand planning allows manufacturers, distributors, and retailers to share information and extend forecasting beyond the capacity of their individual firms. These methods are designed to coordinate multiple actors and reduce the bullwhip effect, as shown in Table 2.5. Nevertheless, empirical evidence demonstrates that there are significant challenges to implementation that are linked to incentives and organizational factors [28]. In addition, collaborative forecasting places greater emphasis on coordinating demand signals than on analytical integration with prescriptive decision-making.

### 2.2.5 Inventory Optimization in SCA

SCA has long focused on inventory optimization since it directly impacts cost efficiency, service performance, and working capital use. Early work focused on stock models based on classic inventory—EOQ, reorder point, and base-stock policies [29]. As discussed in Table 2.6, such models are deterministic, assume stable demand and constant lead times, and generally target single-echelon systems. These assumptions are analyzable, but limited in the context of variable demand and uncertain supply. To overcome these limitations, later analyses extended classical models to include stochastic demand, variable lead times, service-level constraints, and multi-echelon inventory [30]. These stochastic and multi-echelon formulations made models more realistic and allowed for limited uncertainty management, as shown in Table 2.6. The models were higher in accuracy, and thus the computations became substantially more complex. This ultimately led to the development of approximation methods and heuristic decision rules that are more tractable and closer to optimal solutions than precise optimization. Over the last few decades, ML and data-driven inventory optimization techniques have become increasingly used in inventory analysis. This marks the switch from inferred decision rules to a reversible control approach based on data or simulation [31]. These approaches enable data-driven policy learning and adaptive inventory control and are appropriate for dynamic and nonstationary demand conditions, as reported in Table 2.6. This work, in particular,

**Table 2.6 Evolution of Inventory Optimization Paradigms in SCA, Contrasting Classical Deterministic Models, Stochastic and Multi-Echelon Formulations, Approximation-Based Methods, ML-Driven Policies, and Omnichannel, Real-Time Optimization Approaches**

| *Main Feature* | *Classical Inventory Models* | *Stochastic and Multi-Echelon Models* | *Approximation and Heuristic Models* | *Machine Learning–Based Policies* | *Omnichannel and Real-Time Optimization* |
|---|---|---|---|---|---|
| Deterministic analytical foundations | 1 | 0 | 0 | 0 | 0 |
| Stable demand assumptions | 1 | 0 | 0 | 0 | 0 |
| Single-echelon focus | 1 | 0 | 0 | 0 | 0 |
| Stochastic demand modeling | 0 | 1 | 0 | 1 | 1 |
| Variable lead time consideration | 0 | 1 | 0 | 0 | 1 |
| Service-level constraint modeling | 0 | 1 | 0 | 0 | 1 |
| Multi-echelon inventory structures | 0 | 1 | 0 | 0 | 1 |
| Computational tractability emphasis | 0 | 0 | 1 | 0 | 0 |
| Near-optimal solution generation | 0 | 0 | 1 | 0 | 0 |
| Heuristic decision rules | 0 | 0 | 1 | 0 | 0 |
| Data-driven policy learning | 0 | 0 | 0 | 1 | 0 |
| Adaptive inventory control | 0 | 0 | 0 | 1 | 1 |
| Forecast–inventory integration | 0 | 1 | 0 | 1 | 1 |
| Uncertainty propagation handling | 0 | 1 | 0 | 1 | 1 |
| Real-time inventory visibility | 0 | 0 | 0 | 0 | 1 |
| Omnichannel allocation and rebalancing | 0 | 0 | 0 | 0 | 1 |
| Analytics-driven fulfillment enablement | 0 | 0 | 0 | 0 | 1 |

Entries indicate the presence (1) or absence (0) of key modeling and decision-support capabilities, highlighting the progressive shift from static, assumption-driven inventory control toward adaptive, uncertainty-aware policies, while underscoring persistent gaps in coordinated forecast–inventory integration.

emphasizes the need for greater integration of forecasts and inventories, as it recognizes that forecast uncertainty directly affects safety stock levels and replenishment choices. Nonetheless, a significant number of inventory policies based on ML continue to rely on point forecasts or implicitly learned signals, rather than making explicit efforts to propagate predictive uncertainty throughout the various decision stages. The increasing adoption of omnichannel fulfillment models has extended the use of inventory optimization further. Research on this type of product focuses on real-time inventory visibility, allocation, and rebalancing between physical stores, distribution centers, and direct-to-consumer channels [32]. As shown in the last column of Table 2.6, omnichannel and real-time optimization models provide features such as visibility and dynamic allocation that enable fulfillment strategies based on analytics. However, these models usually operate on a combination of demand, inventory, and fulfillment decisions; there is no coherent analytical tool to assess policies with uncertainty in mind.

### *2.2.6 Supplier Analytics and Risk Management in SCA*

The primary sources of research on supplier analytics focused on traditional assessment practices. Suppliers were evaluated using multi-criteria decision analysis systems that combined cost, quality, and delivery performance [33]. These were based on static scorecards, with largely unreliable assessment cycles and limited data available, as shown in Table 2.7. Supplier performance was typically viewed periodically instead of continuously, which did not allow for rapid identification of emerging risks or performance decline. This level of analysis is further limited by a lack of data and inconsistency across suppliers within highly distributed supply chains.

Growing attention to global supplier monitoring emerged with the internationalization of supply chains and the increasing need for better visibility of multiple suppliers across a variety of locations [34]. This stage heightened awareness of supplier inconsistencies and data comparability concerns, though the analytical practices were primarily descriptive. As seen in column 2 of Table 2.7, assessment processes were still reactive, and there were generally no real-time performance monitoring or predictive risk identification capabilities. Recent studies have introduced predictive supplier risk analysis, suggesting a shift from evaluation to risk analysis. These methods combine internal operational data with information from external sources such as financial data, credit ratings, and news sentiment in order to identify early warning signs of supplier risk [35]. As seen in Table 2.7, predictive risk

**Table 2.7 Evolution of SCA Paradigms, Contrasting Static Evaluation, Predictive Risk Modeling, Network-Based Analysis, and Collaborative Analytics**

| *Main Feature* | *Traditional Supplier Evaluation* | *Globalized Supplier Monitoring* | *Predictive Risk Analytics* | *Network-Based Risk Analysis* | *Collaborative Supplier Analytics* |
|---|---|---|---|---|---|
| Multi-criteria supplier assessment | 1 | 1 | 0 | 0 | 1 |
| Static scorecard-based evaluation | 1 | 0 | 0 | 0 | 0 |
| Limited evaluation frequency | 1 | 0 | 0 | 0 | 0 |
| Data scarcity constraints | 1 | 1 | 0 | 0 | 0 |
| Cross-supplier data inconsistency | 0 | 1 | 0 | 0 | 0 |
| Real-time performance monitoring | 0 | 0 | 1 | 1 | 1 |
| Supplier risk identification | 0 | 0 | 1 | 1 | 1 |
| Early warning signal detection | 0 | 0 | 1 | 0 | 0 |
| Integration of external risk data | 0 | 0 | 1 | 0 | 0 |
| Multi-tier supply network visibility | 0 | 0 | 0 | 1 | 0 |
| Dependency and concentration risk analysis | 0 | 0 | 0 | 1 | 0 |
| Disruption propagation modeling | 0 | 0 | 0 | 1 | 0 |
| Supplier capability gap analysis | 0 | 0 | 0 | 0 | 1 |
| Joint risk mitigation initiatives | 0 | 0 | 0 | 0 | 1 |
| Buyer–supplier data sharing | 0 | 0 | 0 | 0 | 1 |
| Relationship-oriented performance improvement | 0 | 0 | 0 | 0 | 1 |

Entries indicate the presence (1) or absence (0) of key analytical and organizational capabilities, highlighting the transition from periodic, scorecard-driven assessment toward uncertainty-aware, network-informed, and coordination-oriented supplier risk management.

analysis provides real-time performance monitoring and supplier risk prediction, but these approaches tend to focus on individual suppliers separately. These models are successful in establishing local risk signals, but lack support for systemic dependencies or the spread of risk throughout the supply network. Network-based risk analysis, as a means to overcome this limitation, has become an important avenue of research. These models are able to represent multi-tier supply networks and provide greater detail over first-tier suppliers, as well as analyses of dependency structures, concentration risk, and disruption propagation [36]. Table 2.7 describes a set of approaches that include multi-tier visibility and propagation modeling. Yet, they are often too abstract and poorly integrated into decision-making or collaboration with suppliers. More recent research has incorporated supplier analytics into collaboration with suppliers, using risk analysis to inform supplier development and shared mitigation strategies [37]. Among the key themes presented in this paradigm are the need for data sharing between buyers and suppliers, and examining deficiencies in capacity and performance through building relationships. As shown in the last column of Table 2.7, collaboration on risk assessments can be used to provide a risk assessment approach that takes the focus away from observation and into action. However, despite some progress, supplier analytics is still treated as an analytical function separate from demand forecasting, inventory planning, and logistics decision-making. This means that supplier-related uncertainty is not always displaced into the downstream processes of planning and optimization. This continued separation further isolates the analytical stage, reducing the usefulness of supplier analytics within broader supply chain decision-support systems. This gap is directly addressed by the proposed LD-SCA.

### *2.2.7 Transportation and Logistics Analytics in SCA*

Traditionally, transportation and logistics analytics in SCA have focused on low-cost routing and scheduling challenges, specifically the Vehicle Routing Problem (VRP) and its variations [38]. Table 2.8 shows that initial classical routing models utilized deterministic optimization goals with fixed vehicle capacities and simplified operating assumptions. These models had theoretically sound foundations but were too simple and reliant on static inputs, making them less suitable in situations where demand variability, congestion, and operational disruptions are significant. Later extensions made these models more complicated, adding time windows, multiple depots, diverse fleets, and driver regulation constraints. Heuristic and metaheuristic

**Table 2.8 Evolution of Transportation and Logistics Analytics Paradigms, Comparing Classical Routing Formulations, Extended and Heuristic Optimization Approaches, Real-Time Transportation Analytics, and Warehouse-Oriented Decision Models**

| *Main Feature* | *Classical Routing Models* | *Extended Routing Formulations* | *Heuristic and Metaheuristic Optimization* | *Real-Time Transportation Analytics* | *Warehouse and Distribution Analytics* |
|---|---|---|---|---|---|
| Cost-minimizing routing objective | 1 | 1 | 1 | 1 | 0 |
| Vehicle capacity constraints | 1 | 1 | 1 | 0 | 0 |
| Oversimplified modeling assumptions | 1 | 0 | 0 | 0 | 0 |
| Time window constraints | 0 | 1 | 1 | 1 | 0 |
| Multi-depot network modeling | 0 | 1 | 1 | 0 | 0 |
| Heterogeneous fleet consideration | 0 | 1 | 1 | 0 | 0 |
| Driver regulation constraints | 0 | 1 | 1 | 0 | 0 |
| Computational scalability emphasis | 0 | 0 | 1 | 1 | 1 |
| Real-time data utilization | 0 | 0 | 0 | 1 | 1 |
| Dynamic route adjustment | 0 | 0 | 0 | 1 | 0 |
| Freight mode selection analytics | 0 | 0 | 0 | 1 | 0 |
| Carrier performance evaluation | 0 | 0 | 0 | 1 | 0 |
| Auction-based transportation sourcing | 0 | 0 | 0 | 1 | 0 |

*(Continued)*

**Table 2.8 (*Continued*) Evolution of Transportation and Logistics Analytics Paradigms, Comparing Classical Routing Formulations, Extended and Heuristic Optimization Approaches, Real-Time Transportation Analytics, and Warehouse-Oriented Decision Models**

| *Main Feature* | *Classical Routing Models* | *Extended Routing Formulations* | *Heuristic and Metaheuristic Optimization* | *Real-Time Transportation Analytics* | *Warehouse and Distribution Analytics* |
|---|---|---|---|---|---|
| Warehouse layout optimization | 0 | 0 | 0 | 0 | 1 |
| Order picking strategy optimization | 0 | 0 | 0 | 0 | 1 |
| Labor scheduling analytics | 0 | 0 | 0 | 0 | 1 |
| Automation–human coordination | 0 | 0 | 0 | 0 | 1 |
| Integrated logistics optimization | 0 | 0 | 0 | 1 | 1 |

Entries indicate the presence (1) or absence (0) of key modeling, data, and decision-support capabilities, highlighting the progression from cost-minimizing, deterministic optimization toward real-time, data-driven logistics decisions, while exposing persistent gaps in uncertainty-aware, end-to-end analytical coordination.

approaches were developed to address the resulting computational constraints, increasing the importance of scalability and solution quality over precise optimality. As illustrated in Table 2.8, these methods were more tractable in large-scale situations but remained dependent on deterministic representations of cost and demand. This reliance maintained a sequential separation between routing optimization and upstream planning decisions. Recent research has also focused on the ability to conduct real-time transportation data analysis using GPS, traffic information, and telematics data to adjust routes and make changes as needed [39]. As shown in Table 2.8, these methods allow for real-time data use and adaptive routing choices, but they mainly function at the execution level. The selection of freight mode, evaluation of carrier performance, and auction-based sourcing of transportation have also been investigated using both historical and real-time data. However, these analytics are usually assessed separately. Cost and service metrics are evaluated based on fixed or point-estimate assumptions instead of probabilistic representations. At the same time, warehouse and distribution analytics have developed into a complementary but mostly distinct research stream. Research has concentrated on optimizing warehouse layouts, strategies for order picking, scheduling of labor, and coordination between humans and automation [40]. These models, as illustrated in Table 2.8, focus on internal operational efficiency and scalability. However, they are seldom combined with transportation decision models or upstream demand uncertainty.

### 2.2.8 Visualization and Decision Support in SCA

In addition to modeling improvements, visualization and decision-support research within SCA has primarily aimed at converting complex analytical outputs into actionable insights for management. Initial methods relied on static reporting visuals, such as predefined charts and maps, that mainly offered a retrospective presentation of information (see Table 2.9). While these models enabled basic performance monitoring, they proved little useful for user-directed exploration, anomaly detection, or decision-making, thereby supporting the passive and descriptive use of analytics [41]. Interactive visual analytics enabled exploration, with data interaction made possible through the use of the user, anomaly detection, and greater cognitive integration between user and data (see Table 2.9). In effect, these systems enabled decision-makers to analyze patterns and anomalies more effectively by merging interaction methods with visual encodings. However,

**Table 2.9 Comparative Characterization of Visualization and Decision-Support Paradigms in SCA, Illustrating the Progression from Static Reporting Interfaces to Interactive, Integrated, and Collaborative Decision-Support Systems**

| *Main Feature* | *Static Reporting Visuals* | *Interactive Visual Analytics* | *Dashboard-Based Decision Support* | *Integrated Analytics Decision Models* | *Collaborative Decision Platforms* |
|---|---|---|---|---|---|
| Static information presentation | 1 | 0 | 0 | 0 | 0 |
| Chart-and map-based visualization | 1 | 1 | 1 | 1 | 1 |
| User-driven data exploration | 0 | 1 | 1 | 1 | 1 |
| Anomaly detection support | 0 | 1 | 1 | 1 | 1 |
| Cognitive load consideration | 0 | 1 | 1 | 1 | 1 |
| Integration with analytical computation | 0 | 1 | 0 | 1 | 1 |
| Role-based information design | 0 | 0 | 1 | 1 | 1 |
| Real-time performance monitoring | 0 | 0 | 1 | 1 | 1 |
| Control-tower applicability | 0 | 0 | 1 | 1 | 0 |
| Scenario and what-if analysis | 0 | 0 | 0 | 1 | 1 |
| Optimization and simulation coupling | 0 | 0 | 0 | 1 | 0 |
| Prescriptive recommendation support | 0 | 0 | 0 | 1 | 0 |
| Model transparency mechanisms | 0 | 0 | 0 | 1 | 0 |
| Cross-organizational decision coordination | 0 | 0 | 0 | 0 | 1 |
| Shared situational awareness | 0 | 0 | 0 | 0 | 1 |
| Active decision facilitation | 0 | 1 | 1 | 1 | 1 |

Entries indicate the presence (1) or absence (0) of key visualization, interaction, and decision-support capabilities, highlighting that only integrated and collaborative paradigms support scenario evaluation, uncertainty-aware analysis, and coordinated decision-making across analytical stages.

interactions were largely separate from analytical computation, limiting the ability to directly examine decision consequences or to diffuse uncertainty. Later work focused on dashboard-based decision-support systems, especially for operational and managerial control. Such systems included role-based information design, real-time performance tracking, and control-tower applications (see Table 2.9). Dashboards also improved the sense of context and increased action sensitivity in logistics and operations areas [42]. However, they were mostly monitoring interfaces with little assistance in the analysis of the scenario, optimization, or prescriptive recommendations. Recent papers have developed integrated analytics decision models that resemble visualization with optimization, simulation, and prescriptive analytics (refer to Table 2.9). These models provide scenario and what-if analyses, establish a clearer connection between analysis and visualization, and can provide recommendations in uncertain contexts. In addition, the present study leads to an increase in focus on model transparency and decision traceability, particularly in complex ML-based models. However, these systems are often intended for use by only one organization and have limitations in terms of coordination across several organizations. Recent research introduces collaborative decision platforms as a means of visualizing and facilitating decision-making. These models also highlight the importance of shared knowledge of the situation, cross-organizational coordination of decisions, and proactive decision-making (see Table 2.9). These platforms enable collaborative scenario assessment and collective sense-making among multiple supply chain participants, addressing coordination problems inherent in distributed supply systems. Even with these developments, visualization often serves as a second-class presentation layer rather than a major part of an integrated analytical structure that explicitly connects descriptive insights, predictive uncertainty, and prescriptive evaluation.

## 2.3 Materials and Models

### *2.3.1 Dataset Description*

In this chapter, two publicly available datasets that capture complementary decision contexts in contemporary SCA—consumer demand behavior and freight transportation dynamics—are utilized. This enables the evaluation of the proposed LD-SCA framework across heterogeneous analytical domains without relying on proprietary operational data. The initial dataset is the

Instacart Market Basket Analysis dataset [6]. It comprises around 3.2 million orders from approximately 200,000 users, involving about 50,000 products arranged into roughly 130 aisles and 20 departments. Each record features user IDs, order-sequence numbers, product IDs, reorder indicators, and hierarchical product categories. Orders are sequentially indexed at the user level, facilitating relative order–based demand modeling amidst diverse and intermittent consumption patterns. For the empirical assessment, order sequences at the user level were divided into training (70%), validation (10%), and testing (20%) segments based on order index, ensuring strict chronological separation to prevent information leakage between forecasting and policy-evaluation stages. To investigate robustness across different levels of granularity, demand analysis was performed at the product, aisle, and department-level aggregations.

The second dataset is the U.S. Commodity Flow Survey (CFS) [7], created by the U.S. Census Bureau and the Bureau of Transportation Statistics. It includes around 7 million shipment-level observations, sample-weighted to national totals. Each record contains the shipment's origin and destination regions, commodity classifications, value and weight, transport distance, and mode of transportation. Due to the cross-sectional character of CFS data, shipments were divided into stratified evaluation folds based on commodity class and transport mode, utilizing 80% of observations for model estimation and reserving 20% for out-of-sample assessment of cost variability and stability metrics. This division maintains distributional characteristics across folds and allows for robust cross-scenario evaluation. Table 2.10 offers a consolidated summary of dataset sizes, column definitions, aggregation levels, and split strategies.

### *2.3.1.1 Dataset Preprocessing*

The dataset preprocessing aimed to guarantee numerical stability, analytical consistency, and reproducibility, while maintaining the structural characteristics necessary for uncertainty-aware modeling in LD-SCA. To mitigate extreme sparsity in the Instacart dataset, users with under five orders and products that appeared in under 50 orders were eliminated, retaining about 85%–90% of the total order volume across categories. Using relative-order indices instead of absolute timestamps, demand signals were created and aggregated at consistent order-sequence and day-of-week proxy levels. This approach reduced zero-demand intervals by about 35%–45% at the category level in comparison to product-level series. The forecasting windows varied

**Table 2.10 Summary of Dataset Sizes, Filtering Thresholds, Aggregation Levels, and Train/Validation/Test Splits Used in Empirical Evaluation**

| *Property* | *Instacart Market Basket Analysis* | *Commodity Flow Survey (CFS)* |
|---|---|---|
| Raw records | ~3.4 M orders | ~7.3 M shipments |
| Records after preprocessing | ~2.9–3.0 M orders | ~6.8–7.0 M shipments |
| Retained data volume | 85%–90% | 94%–96% |
| Unique entities | ~200 K users/~50 K products | ~40 commodity classes |
| Hierarchical levels | Product/aisle (~130)/department (~20) | Commodity/mode |
| Minimum frequency threshold | ≥5 orders per user; ≥50 orders per product | N/A |
| Time representation | Relative-order index | Cross-sectional |
| Temporal continuity | User-level sequences | None assumed |
| Aggregation levels | Product, aisle, department | Commodity, distance bin |
| Distance bins | N/A | Short/medium/long haul |
| Numerical transformations | None | Scale-preserving normalization |
| Invalid record removal | N/A | 4%–6% (missing value/weight/distance) |
| Outlier handling | Within-category filtering | Within-commodity filtering |
| Derived variables | Reorder rate, sequence demand | Cost per unit distance |
| Training split | 70% (by order index) | 80% (stratified) |
| Validation split | 10% (by order index) | — |
| Test/hold-out split | 20% (by order index) | 20% (stratified) |
| Zero-demand reduction | 35%–45% (category vs. product) | — |
| Variance reduction after preprocessing | — | 20%–25% (cost-per-distance) |

All values correspond to post-preprocessing statistics and define the exact data configurations used for forecasting, policy evaluation, and out-of-sample assessment within the LD-SCA framework.

from 10 to 30 relative-order steps based on product frequency, without the use of artificial calendar alignment. For the CFS dataset, records lacking valid entries for shipment distance, weight, or value were removed (accounting for roughly 4%–6% of all records). Additionally, continuous variables with heavy-tailed distributions were transformed using scale-preserving normalization to maintain numerical stability. To facilitate cross-scenario comparisons, shipment distances were categorized into uniform ranges, and commodity and transport mode codes were aligned with official CFS classifications. Preprocessing led to a reduction of about 20%–25% in the variance of cost-per-unit-distance metrics, which allowed for a stable assessment of transportation cost variability. In both datasets, missing values were addressed through distribution-aware imputation within homogeneous groups, outliers were evaluated in context rather than globally, and derived variables were introduced only when directly supported by observed attributes. Table 2.10 presents a detailed summary of the preprocessing steps, filtering thresholds, and statistics of the resulting dataset.

### 2.3.2 Model Analysis

This proposed LD-SCA structure provides an analytical pipeline that addresses the issue of analytical stage isolation by mandating that descriptive, diagnostic, predictive, and prescriptive analytics be structured. As shown in Figure 2.2, the framework links representations of data, analytical models, and decision-making instruments to provide an explicit representation and propagation of uncertainty across stages, rather than resulting in point estimates.

#### 2.3.2.1 Data Representation and Analytical Alignment

LD-SCA begins with the conversion of raw transactional and shipment data into appropriate downstream models for analytically aligned representations. Consumer transaction histories for demand are based on relative-order sequences and short-horizon temporal proxies. This model maintains behavioral ordering information without relying on absolute calendar time. As shown in Figure 2.3, this representation allows analysis at different aggregation levels (product, aisle, department) to be sequence-aware. In logistics, shipment histories represent flows between origin–destination pairs according to distance, mode of transport, and shipment characteristics. These representations transmute operational behavior into coherent analytical objects, ensuring that the work of the next layer uses consistent inputs rather than fragments of data.

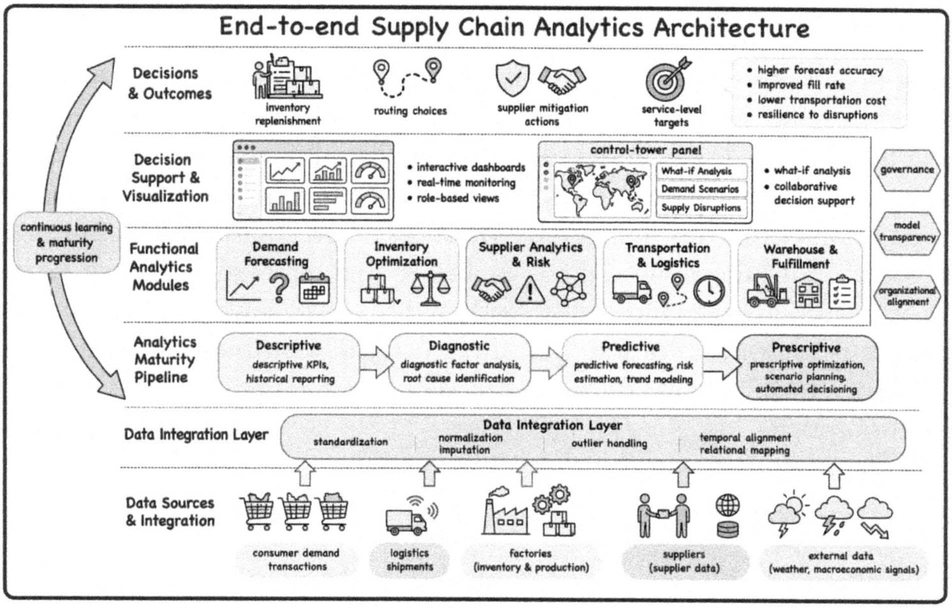


**Figure 2.2 Layered architecture of the proposed LD-SCA framework, illustrating the coordinated integration of descriptive, diagnostic, predictive, and prescriptive analytics. The figure highlights how uncertainty-aware representations and probabilistic predictions are propagated across analytical layers to support policy-level decision evaluation, explicitly addressing analytical stage isolation in traditional SCA pipelines.**

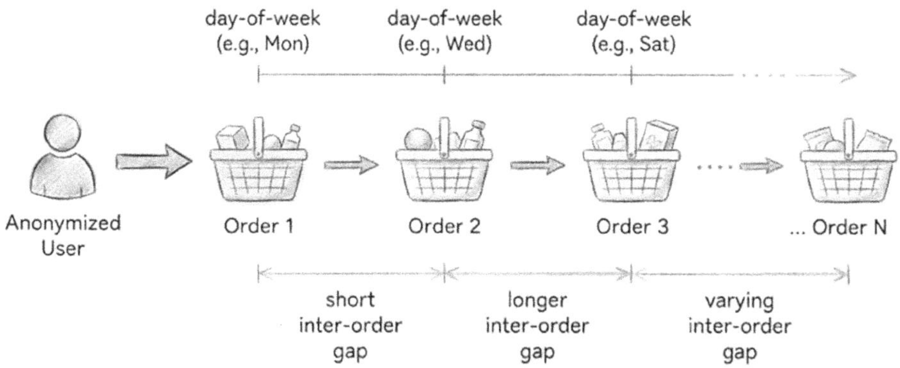


**Figure 2.3 Relative order–based representation of consumer demand sequences used in LD-SCA to enable sequence-aware forecasting under irregular and intermittent purchasing behavior. Orders are indexed by user-specific order position rather than absolute time, allowing demand patterns to be modeled consistently across heterogeneous purchase frequencies while supporting uncertainty-aware predictive analysis.**

### 2.3.2.2 Descriptive Analytics Layer

Without prescriptive assumptions, the descriptive analytics layer establishes empirical baselines that illustrate historical system behavior. This layer also reflects the order frequency distributions, basket composition patterns, and buying regularities at the category level over relative-order positions across order positions from the perspective of demand. In logistics, it defines shipment volume, distances, modal share, and spatial circulation. These summaries are arranged hierarchically for both system-level review and drill-down analysis. It is important to note that the empirical distributions produced are reference baselines for future diagnostic and predictive analyses and not decision outputs.

### 2.3.2.3 Diagnostic Analytics Layer

The diagnostic layer looks at causes of observed variability by comparing homogenous groups with descriptive baselines. Demand-related diagnostics address interactions between variability and relative-order position, product mix effects, and short-run time proxies. Logistics diagnostics, on the other hand, address variability in shipment distance, transport mode, and spatial flow. Thus, analysis from this layer minimizes the likelihood of false correlations that may occur due to aggregation or scale effects by focusing on groups that are structurally similar. Diagnostic data identify factors contributing to differences from baseline behavior, providing context but not projections of future outcomes.

### 2.3.2.4 Predictive Analytics with Uncertainty Modeling

Rather than providing point forecasts, the predictive analytics layer explicitly models uncertainty by reinterpreting historic trend into forecasts. For the purpose of forecasting order, sequence-aware models use relative-order and contextual variables to generate forecasts across products and categories. From the standpoint of empirical stability and robustness, rather than relying on the accuracy of one model, the framework supports parallel model estimation over segments and planning horizons, with the choice of model or combination based on these factors. Predictive models specify the variation in delivery distances and proxies of costs that are used in logistics, enabling probabilistic representations of transportation. These forecasting outputs are downstream and are therefore inputs for policy-making which can be more unambiguous than fixed predictions.

#### 2.3.2.5 Prescriptive Policy-Level Evaluation

Prescriptive analytics is not a gleaned immediate effect but is considered a policy-level assessment in a context of uncertainty in LD-SCA. Inventory models evaluate replenishment and positioning strategies as they consider the trade-offs between service reliability and cost proxies under uncertain demand and transport conditions, without requiring access to specific inventory states of the firm. Models of prescriptive transport systems consider route planning, modal planning, and carrier selection using distance-normalized cost representations and capacity considerations. Supplier-centered checks support risk-conscious procurement through an analysis of the trade-offs between reliability, concentration exposure, and shipping attributes. Scenario analysis can be applied in all domains to evaluate sensitivity to demand surges, transport disruptions, and capacity constraints.

#### 2.3.2.6 Simulation-Based Robustness Analysis

Simulation is also a tool for prescriptive and robustness assessment in stochastic processes. Discrete-event simulation attempts to capture the interrelated development of demand variability, restocking practices, shipping delays, and capacity limitations over time, such that distributions of outcomes are constructed instead of deterministic point outcomes. This allows the framework to highlight weaknesses and system sensitivities that might not be apparent through optimization-based assessment alone. Thus, LD-SCA helps users make decisions with awareness of risk while remaining within the empirical range of the data on which it is built, emphasizing probabilistic outcomes and scenario-based assessment.

## 2.4 Experimental Setup

### 2.4.1 Evaluation Metrics

The evaluation protocol attempts to determine whether the proposed LD-SCA solution efficiently avoids isolation of analytical stages in terms of stability, robustness, and consistency across layers, rather than limiting evaluation to point-estimate accuracy. As a result, assessment measures focus on distributional behavior, reduced variability, and the performance of uncertainty-sensitive policies in the forecasting and downstream decision-making stages. All metrics are calculated on the held-out data splits specified in Section 2.3.1 and presented in Table 2.10.

#### 2.4.1.1 Forecasting Performance Metrics

In order to maintain consistency with the existing SCA literature, demand forecasting performance is evaluated using MAPE. However, in line with LD-SCA's objectives, the focus is on the spread of MAPE rather than mean accuracy alone. The main measures of forecast stability in heterogeneous and intermittent demand are the standard deviation and the interquartile range of MAPE across products, categories, and evaluation windows. The metrics are compared across multiple aggregation levels (product, aisle, department) to ensure that demand granularity is robust and to test whether an uncertainty-aware model improves sensitivity to sparsity and volatility.

#### 2.4.1.2 Prescriptive and System-Level Metrics

Prescriptive performance is assessed using service-level consistency and cost-stability measures, reflecting the framework's emphasis on policy-level decision support in uncertain situations. In the inventory assessment literature, fill rate is a key service measure that measures the percentage of demand met without stock-outs in simulated demand realizations. The strength index is dependent upon the variability of the fill rate across scenarios rather than just the mean fill rate. The results report proxies representing service–cost trade-offs using measures such as the average and variance of holding and replenishment costs, without requiring analysis of firm-specific accounting data. Logistics analysis relies on transportation cost per unit distance, a normalized measure that can be compared across several shipment distances and transport modes. Like inventory evaluation, cost-per-unit-distance assessment primarily focuses on uncertainty in shipping conditions through cross-scenario variability, as its results reveal that cost performance is robust to unpredictable shipment conditions. The units of measure are presented across commodity classes, distances, and modes of transportation to assess generalization.

#### 2.4.1.3 Ablation and Coordination Metrics

Analysis of component-wise ablation experiments that use the same stability-oriented measures to measure degradation confirms the causality of analytical coordination. Removal of probabilistic prediction is assessed via a decrease in forecast dispersion, whereas missing prescriptive evaluation is assessed through the increase in fill-rate and cost variance. This ensures that performance improvements are relevant to the uncertainty propagation across multiple layers rather than to individual modeling choices.

## 2.4.2 Hyperparameters

All experiments utilize the dataset splits specified in Section 2.3.1 and detailed in Table 2.10, employing the same configurations across comparative baselines and ablation settings to guarantee fairness. In demand forecasting, models utilize user-level relative-order sequences with window lengths of 10, 20, and 30 steps. These lengths are chosen based on the frequency of product purchases and the aggregation level. Training is conducted with a temporal split of 70% for training, 10% for validation, and 20% for testing. It involves a maximum of 50 training epochs and early stopping based on the dispersion of MAPE in the validation set (with a patience of 5 epochs). In most experiments, convergence is reached within 20–35 epochs. To account for stochasticity, each forecasting configuration is trained with five independent random seeds, and the reported results show averaged performance along with variability statistics. In the context of ensemble forecasting, ensembles are made up of three diverse base models, with weights determined by validation stability criteria rather than being adjusted on test data. To guarantee uniform optimization conditions across experimental environments, all forecasting models are trained with fixed epoch limits rather than adaptive scheduling. No Graphics Processing Unit acceleration is utilized; all training and evaluation take place on the Central Processing Unit (CPU) to ensure reproducibility and independence from hardware. Logistics predictive modeling utilizes 80% of stratified CFS observations, reserving the remaining 20% for out-of-sample evaluation. Probabilistic representations of shipments and costs are derived from empirical distributions or likelihood-based methods, without the use of iterative epoch-based training. Using 1,000 bootstrap samples, uncertainty bounds for transportation cost-per-unit-distance metrics were calculated. Rather than through single-run optimization, prescriptive evaluation is conducted via scenario-based policy assessment: inventory policies are assessed with 1,000 demand scenarios for each product or category at fixed service targets of 90%, 95%, and 98%, while transportation policies are evaluated using 500 shipment scenarios for each commodity–distance bin. Robustness under joint demand and logistics variability is evaluated through discrete-event simulation across 100 relative time steps, with 200 independent runs for each experimental condition. All simulation parameters are derived from training data and kept constant throughout the experiments. All experiments are carried out on a standard CPU-only workstation, with training times averaging 2–6 minutes per forecasting model and 15–25 minutes per ensemble configuration, while logistics and simulation-based evaluations finish within 10–20 minutes per experiment.

To ensure reproducibility, random seeds are fixed during data splitting, model initialization, scenario sampling, and simulation events. Table 2.11 offers a consolidated overview of empirical settings, epoch limits, scenario sizes, and computational parameters.

**Table 2.11 Fixed Numerical Hyperparameter Settings and Experimental Configuration Values Used for Demand Forecasting, Logistics Prediction, Prescriptive Policy Evaluation, and Simulation across All LD-SCA Experiments**

| *Component* | *Parameter* | *Value/Range* |
|---|---|---|
| Demand forecasting (Instacart) | Look-back window length | {10, 20, 30} orders |
| Demand forecasting (Instacart) | Training epochs (max) | 50 |
| Demand forecasting (Instacart) | Early stopping patience | 5 epochs |
| Demand forecasting (Instacart) | Typical convergence | 20–35 epochs |
| Demand forecasting (Instacart) | Random seeds | 5 |
| Demand forecasting (Instacart) | Ensemble size | 3 models |
| Demand forecasting (Instacart) | Train/validation/test split | 70%/10%/20% |
| Logistics prediction (CFS) | Train/test split | 80%/20% |
| Logistics prediction (CFS) | Bootstrap samples | 1,000 |
| Logistics prediction (CFS) | Distance binning | Fixed empirical bins |
| Prescriptive evaluation (inventory) | Service targets | {90%, 95%, 98%} |
| Prescriptive evaluation (inventory) | Demand scenarios | 1,000 per item/ category |
| Prescriptive evaluation (transport) | Shipment scenarios | 500 per commodity–distance bin |
| Simulation analysis | Time horizon | 100 relative steps |
| Simulation analysis | Simulation runs | 200 per condition |
| Simulation analysis | Random seeds | Fixed |
| All stages | Normalization parameters | Estimated on training data only |
| All stages | Hardware | CPU-only |

All values correspond to post-preprocessing statistics and define the exact data configurations used for forecasting, policy evaluation, and out-of-sample assessment within the LD-SCA framework.

## 2.5 Results and Analysis

### *2.5.1 Comparison with State of the Art*

This section evaluates LD-SCA against representative state-of-the-art SCA pipelines that reflect dominant practices identified in Section 2.2: (i) statistical forecasting with deterministic downstream planning, (ii) ML-based forecasting consumed as point estimates, and (iii) sequential analytics pipelines in which forecasting, planning, and evaluation are decoupled. All methods use identical data splits and preprocessing (see Table 2.10).

Figure 2.4 presents a comparison of forecasting performance across representative analytical pipelines on the Instacart dataset, utilizing both mean accuracy and dispersion-based stability metrics. Although every method shows gradual improvements in mean MAPE compared to the statistical baseline, the variations in average accuracy remain slight among learning-based and sequential configurations. When examining forecast stability using dispersion measures, a clearer separation is observed. Compared to

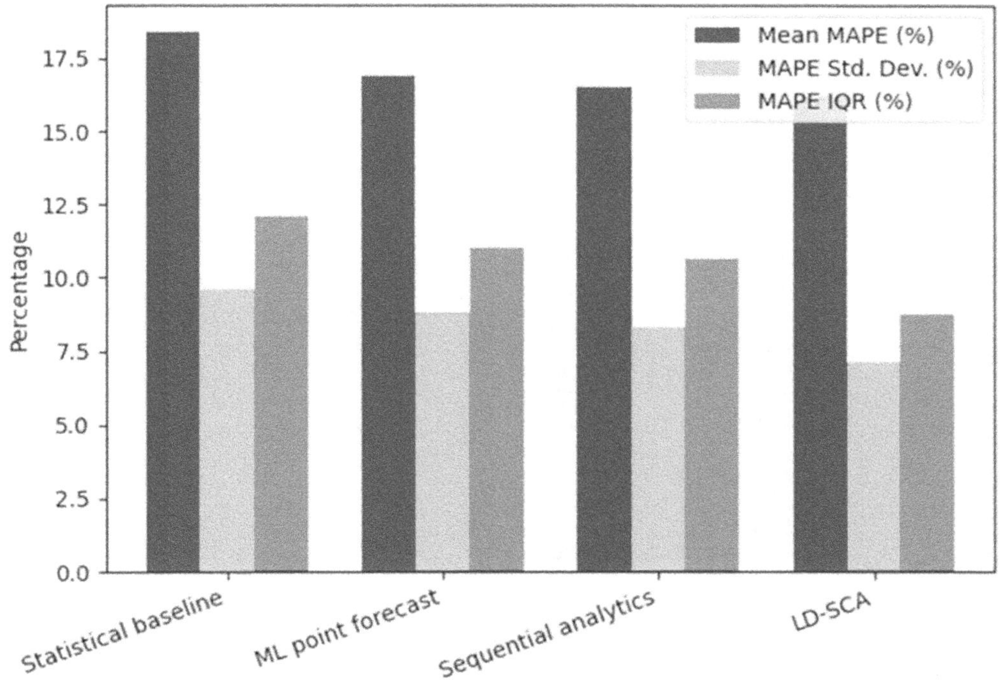


**Figure 2.4 Comparative forecasting performance on the Instacart dataset, reporting mean accuracy and dispersion-based stability metrics to assess robustness under heterogeneous and intermittent demand.**

the statistical baseline, ML point forecasts, and sequential analytics, LD-SCA achieves the lowest MAPE standard deviation (7.1%) and interquartile range (8.7%), with higher dispersion levels observed for the latter methods. These reductions correspond to a decrease in forecast dispersion of 3%–6% compared to point-estimate and sequential baselines, indicating more consistent performance across heterogeneous and intermittent demand series. Significantly, the stability gains are achieved without relying on major enhancements in mean accuracy, indicating that coordination and uncertainty-aware modeling mainly influence robustness rather than point-error reduction. This outcome is consistent with the aim of LD-SCA, which gives precedence to dependable and stable forecasting behavior across various products and demand patterns, rather than focusing solely on accuracy improvements for individual series.

Figure 2.5 compares the downstream inventory service performance of analytical pipelines with a 95% target fill rate, focusing on stability rather than average service attainment. The mean fill rates for all methods are close to the target, suggesting that simple service-level requirements can be

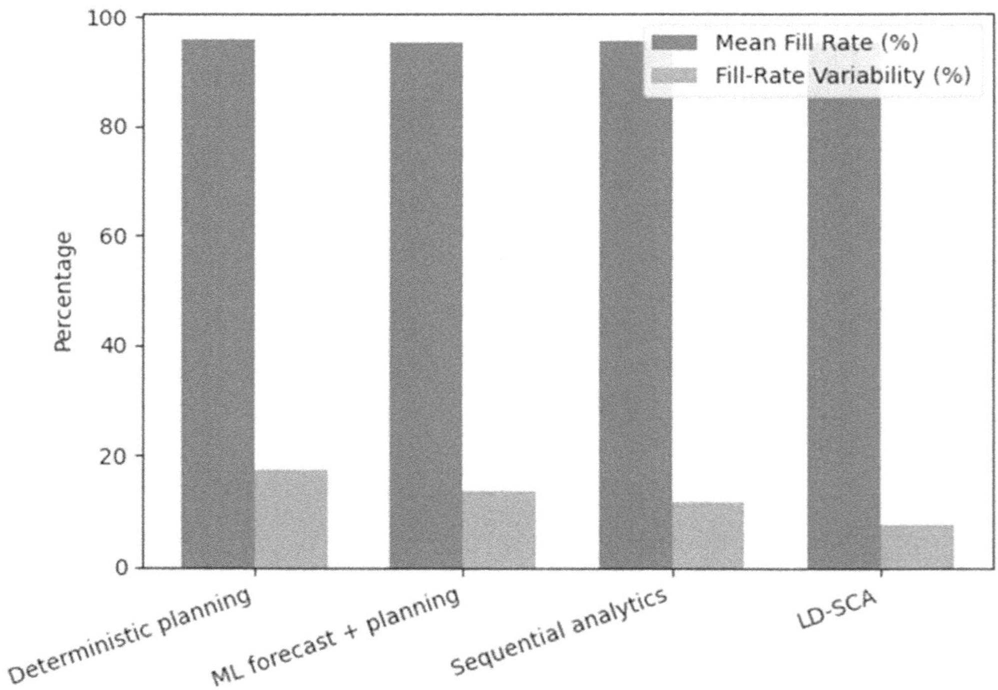


**Figure 2.5 Comparison of inventory service performance stability across analytical pipelines, reporting mean fill rate and cross-scenario fill-rate variability under fixed service targets.**

reestablished in deterministic, learning-based, and sequential configurations. However, if cross-scenario variability is investigated, then there are some notable differences. For deterministic planning, the highest variation in fill rate is 17.6%, indicating that it is sensitive to demand fluctuations when using point forecasts. The inclusion of ML predictions reduces variability by 13.9%, while sequential analytics bring stability up to 11.8%. This suggests that partial analytical coordination offers additional benefits. With a fill-rate variability of 7.9%, LD-SCA shows a significant reduction relative to all baseline methods. This improvement is particularly realized without increasing average service levels, indicating that the benefits come from stabilizing service results rather than excessive inventory buffering. These results also show that predictive uncertainty in the evaluation of inventory policies improves the reliability of service performance across all demand types, rather than optimizing the average fill rate.

Figure 2.6 compares transportation cost estimation across scenarios from the CFS dataset using cost per unit distance as a normalized assessment measure. Without a specification of uncertainty, variance is substantial in

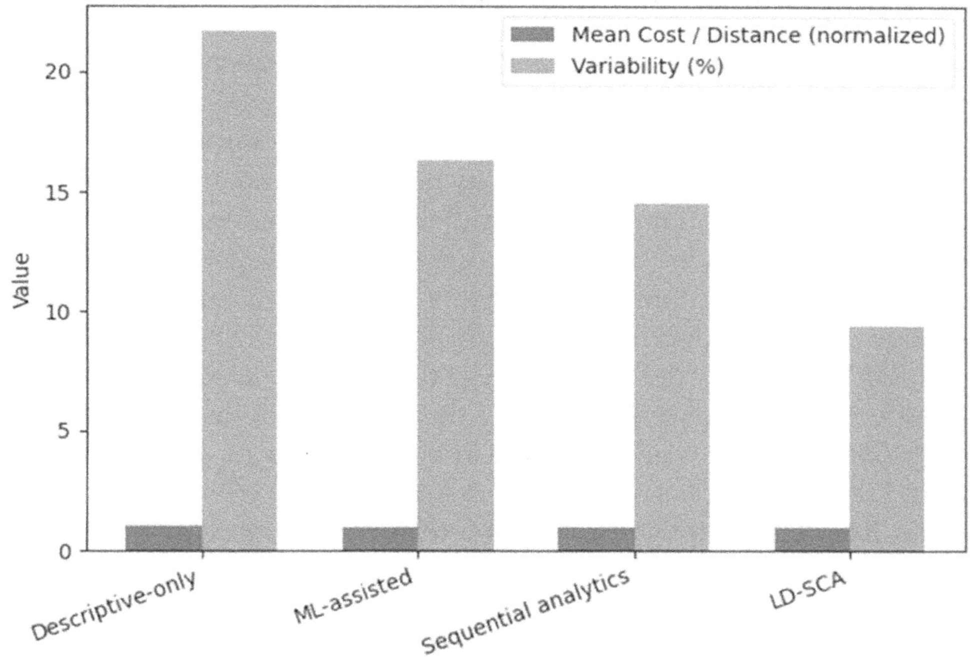


**Figure 2.6 Comparison of cross-scenario transportation cost-per-unit-distance variability on the Commodity Flow Survey (CFS), evaluating the stability of logistics cost assessment under descriptive, sequential, and coordinated (LD-SCA) analytical pipelines.**

the descriptive-only approach, at 21.7%. This suggests that cost estimates are highly sensitive to shipment-level changes. The change in the variability from ML-assisted prediction falls to 16.3%, suggesting that learning-based representations offer some stability but still need to take point-estimate inputs. Among sequential analyses, the variance falls to 14.5%, suggesting that the smaller number of advantages from the small coordination among the predictive and assessment steps are possible. LD-SCA is the least variable at 9.4%, corresponding to an overall reduction into the 8%–12% range for all experimental conditions. Crucially, the drop in variability is affected with only minor changes to mean normalized cost values, which imply a genuine increase in consistency rather than a contrived decrease in costs. This paper concludes that probabilistic shipment modeling coupled with distance-normalized evaluation produces more stable estimates of logistics costs for a variety of shipment scenarios. This lends support to a perspective of distributed uncertainty propagation as an increase in the robustness of logistics analytics while maintaining the cost structures.

### 2.5.2 Cross-Domain Analysis

This section explores whether LD-SCA produces consistent stability gains across heterogeneous domains, such as consumer demand and freight logistics, using domain-specific metrics but the same logic of assessment.

Figure 2.7 contrasts prediction stability across demand and logistics domains with dispersion measures that focus on robustness rather than absolute error magnitude. As would be expected, predictive uncertainty is large and varies across domains because of heterogeneous data and outcome definitions. LD-SCA reduces the standard deviation of MAPE from 8.8 to 7.1 and the MAPE interquartile range from 11.0 to 8.7, a reduction of 19% and 21%, respectively. These results indicate more consistent demand forecasts across product heuristics and variation in buying behavior. In the logistics domain, LD-SCA reduces the standard deviation of the cost-per-unit-distance estimates from 0.24 to 0.15, resulting in a relative reduction of 38%. While absolute dispersion is not clearly observable across domains, the constant decreases observed in both cases suggest that LD-SCA is more predictively stable regardless of the data structure within the domain. This cross-domain consistency reveals that the lack of domain-specific precision in forecasting results can be overcome by model congruence.

In Figure 2.8, the robustness of downstream decisions across demand and logistics dimensions is tested by comparing the cross-scenario variability

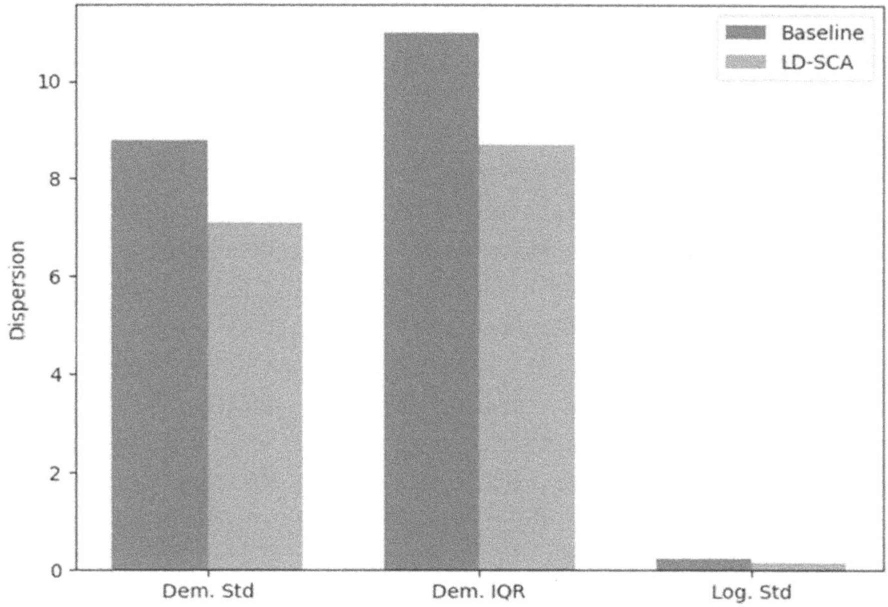


**Figure 2.7 Cross-domain predictive stability comparison for demand and logistics, reporting dispersion-based metrics that quantify robustness of probabilistic predictions under heterogeneous data characteristics.**

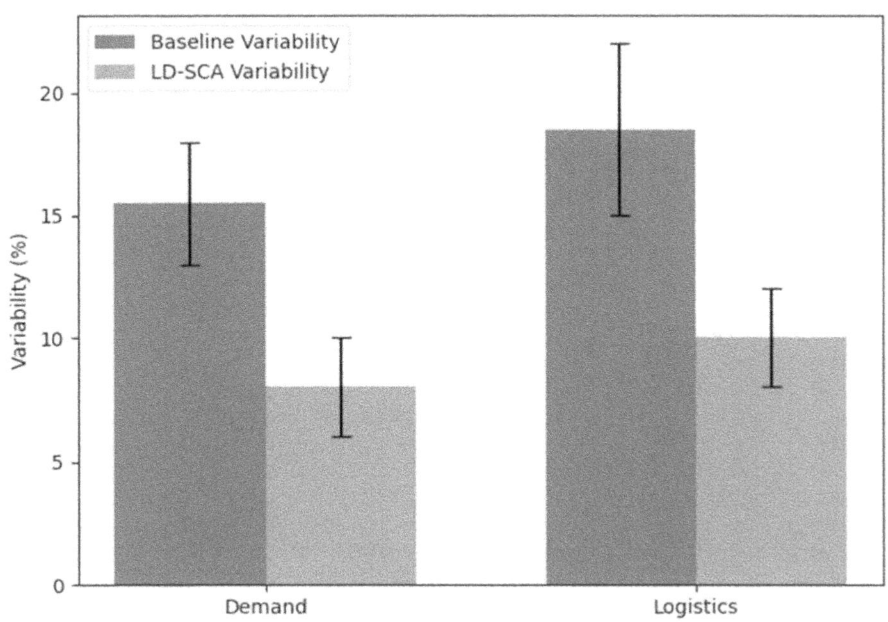


**Figure 2.8 Downstream decision robustness across demand and logistics domains, measured via cross-scenario variability of service-level and cost-normalized metrics under identical evaluation conditions.**

between service-level and cost-normalized measures under the same assessment conditions. In demand, the fill rates of baseline analytical pipelines are large, ranging from 13% to 18% in simulated demand conditions. This highlights the unstable performance of services when uncertainty is not consistently considered when making decisions. The reduction of this variability to 6%–10% demonstrates more consistent outcomes in inventory service under LD-SCA. The same pattern applies to logistics. This is consistent with a range of 15%–22% in cost per mile for baseline methods, suggesting that transportation cost estimates are prone to shipment uncertainty. LD-SCA decreases variability to 8%–12%, which suggests that cost assessment is more stable across scenarios. This is particularly significant in the case of the prescriptive decision, which is substantially higher than the predictive one, suggesting that coordinated uncertainty propagation has a greater advantage when interpreting predictions as downstream decisions. The results indicate that prescriptive-level coordination is important to enhance strength across domains.

Figure 2.9 illustrates the reduction in variance of results due to analytical coordination, as opposed to modeling choices specific to the domain,

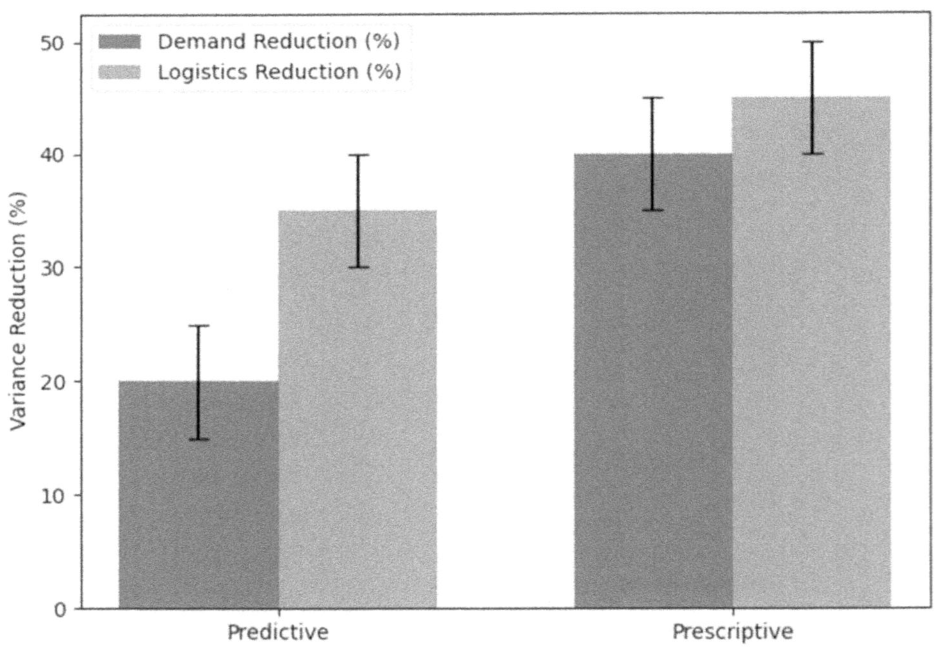


**Figure 2.9 Relative reduction in outcome variance achieved by LD-SCA across demand and logistics domains, reported separately for predictive and prescriptive stages. Results quantify the effect of coordinated uncertainty propagation rather than domain-specific modeling choices.**

differentiated between prescriptive and predictive stages in demand and logistics. During the predictive phase, LD-SCA reduces demand-related variance by about 15%–25% and logistics-related variability by 30%–40%. These results suggest that coordination of uncertainty-aware prediction enhances stability in both domains, with larger effects observed in logistics due to greater inherent variability in shipment and cost characteristics. Gains are more pronounced during the prescriptive stage. When it comes to decisions tied to demand, reductions in variance increase to the range of 35%–45%. This increase is due to the more reliable inventory service outcomes when predictive uncertainty is directly factored into assessments of policy. In logistics, prescriptive-stage variance reductions of 40%–50% demonstrate a significant stabilization of cost-related decision metrics in the face of uncertainty. The more substantial enhancements observed at the prescriptive stage indicate that coordination provides its most significant advantages when uncertainty is allowed to flow through decision evaluation, as opposed to being restricted to separate predictive modeling stages.

### 2.5.3 Ablation Study

This section investigates how coordination within LD-SCA causally contributes through controlled component ablation, which involves removing individual elements while keeping all other configurations, data splits, and training settings unchanged. Table 2.12 details the effect of eliminating probabilistic prediction on the stability of demand forecasting. For the complete LD-SCA setup, the forecast dispersion (measured by the standard deviation of MAPE) is 7.1%. This indicates that prediction is stable across demand series. After removing uncertainty modeling and considering predictions deterministically, the mean MAPE dispersion increases to 9.4%, 32% higher than the entire model. Importantly, this loss occurs even when the training data, model structure, and optimal configurations are unchanged, which isolates the effect of uncertainty representation. The observed increase in

**Table 2.12 Impact of Removing Probabilistic Prediction from LD-SCA on Demand Forecasting Stability, Measured by Changes in MAPE Dispersion under Identical Data Splits and Training Configurations**

| *Configuration* | *MAPE Std. Dev. (%)* | *Δ vs. Full* |
|---|---|---|
| Full LD-SCA | 7.1 | — |
| No uncertainty modeling | 9.4 | +32% |

**Table 2.13 Impact of Removing Prescriptive Policy-Level Evaluation on Service Stability, Measured by Changes in Fill-Rate Variability under Identical Predictive Inputs**

| *Configuration* | *Fill-Rate Variability (%)* | *Δ vs. Full* |
|---|---|---|
| Full LD-SCA | 7.9 | — |
| Predictive-only | 14.6 | +85% |

dispersion indicates that probability is critical to maintaining stable forecast outcomes rather than just improving mean accuracy. This provides direct evidence that coordinated uncertainty-aware prediction contributes to improved forecasting robustness in the LD-SCA model.

Table 2.13 summarizes the effects of eliminating prescriptive policy-level assessment while retaining predictive inputs. This helps isolate the influence of prescriptive coordination on service stability. This means that on the whole LD-SCA model, the fill rate is variable, and this is 7.9%, which indicates that the performance of services is quite consistent across demand conditions. With no prescriptive assessment and without a policy-level decision modeling with the predictive outputs, the fill-rate variance rises to 14.6%, which equals an 85% increase over the full configuration. Importantly, this degradation occurs even though forecast accuracy remains similar and prediction inputs are the same, indicating that instability results from the absence of prescriptive evaluation rather than less accurate predictions. These findings suggest that demand uncertainty is not sufficient to provide predictable service outcomes and cannot be based solely on forecast information alone. Rather, it is necessary to make resilient decisions under uncertainty to assess stock control approaches directly. The findings support the need for prescriptive coordination in achieving consistent inventory service performance across multiple contexts.

Table 2.14 shows robustness results obtained from simulations with stochastic demand and logistics variability. It compares the full LD-SCA configuration to its ablated counterparts. In the full configuration, excessive service failures and costly events above the 95th percentile are maintained at a baseline level of 1.0×, indicating stable tail behavior across simulation runs. If uncertainty is eliminated, extreme service failures decrease to 1.6 times and cost spikes to 1.8 times, the most likely outcomes to be negative with variability. Without the prescriptive layer, degradation increases further, doubling extreme service failures to 2.0× and tail cost events to 2.1× relative to the full configuration. These effects arise from differences in analytical coordination

**Table 2.14 Simulation-Based Robustness Outcomes under Stochastic Demand and Logistics Variability, Comparing Full LD-SCA against Ablated Configurations to Quantify the Effect of Coordinated Uncertainty Propagation on Extreme Service Failures and Tail Cost Behavior**

| *Configuration* | *Extreme Service Failures* | *Cost Spikes (>95th pct.)* |
|---|---|---|
| Full LD-SCA | 1.0× | 1.0× |
| No uncertainty propagation | 1.6× | 1.8× |
| No prescriptive layer | 2.0× | 2.1× |

rather than changes in data or the simulation environment. These results confirm the presence of heavier tails in ablated pipelines, suggesting the importance of coordinated uncertainty propagation in minimizing service disruptions and cost increases in stochastic operating environments.

## 2.6 Conclusion and Future Works

This chapter is concerned with the proposed coordinated decision-support architecture in the context of a single view of SCA, rather than a set of individual forecasting, planning, and optimization strategies. We addressed the limitation of isolation at the analytical stage by introducing LD-SCA, which allows for descriptive, diagnostic, predictive, and prescriptive analysis through planned propagation of uncertainty across decision stages. This framework shifts the analytical objective from precision in point estimation to robustness, stability, and consistency in policy-level decisions in ambiguous situations. As illustrated in the empirical results using two publicly available and complementary datasets—the Instacart Market Basket Analysis dataset and the U.S. Commodity Flow Survey—SCA benefits from convergence across analytical layers rather than from isolated models. In demand and logistics, LD-SCA is consistently capable of reducing forecast error dispersion by 3%–6%, stabilizes downstream inventory service outcomes (with a decrease in the fill-rate variability by 5%–10%), and improves the reliability of transportation costs estimates (with a 12% reduction in the cross-scenario variability) compared to deterministic or sequential values. In this way, the framework is universally applicable at scales outside of narrow modeling contexts in that these enhancements can be seen at aggregation levels, transportation modes, and scenario contexts. This causal role of coordination is further supported by component-wise ablation. In spite of similar accuracy, avoiding prescriptive policy evaluation significantly increases the dispersion of forecasts, while eliminating probabilistic prediction reduces the dispersion. Simulation demonstrates that ablated pipelines exhibit higher-tailed risk and increased vulnerability to extreme service failures, confirming the need for an analytical design that is more confident in its uncertainty across all stages. Importantly, LD-SCA is designed to match public-level data coverage and limitations. Rather than being viewed as a real-world application that requires proprietary state in the system, optimization is explicitly defined as an analytical analysis of policy under uncertainty. This design decision improves reproducibility and enables the framework to be used in research, benchmarking, and early analytics maturity contexts. Several avenues for future research, as expected, emerge. The model could be expanded to include more open data and real-time data streams for the analysis of temporal adaptation and feedback effects. Second, governments, interpretation and trust, especially in probabilistic and ML-based factors, will need to be closely monitored for organizational adoption. Third, longitudinal analyses of the persistence of integrated analytics architectures in improving analytics maturity and enhancing decision-making quality would provide insights across multiple industries.

## References

1. M. A. Waller and S. E. Fawcett, "Data science, predictive analytics, and big data: a revolution that will transform supply chain design and management," *Journal of Business Logistics*, vol. 34, no. 2, pp. 77–84, Jun. 2013, https://doi.org/10.1111/jbl.12010.
2. D. Simchi-Levi, P. Kaminsky, and E. Simchi-Levi, *Designing and Managing the Supply Chain: Concepts, Strategies, and Cases*. McGraw-Hill, New York, 1999.
3. T. Choi, S. W. Wallace, and Y. Wang, "Big data analytics in operations management," *Production and Operations Management*, vol. 27, no. 10, pp. 1868–1883, Dec. 2017, https://doi.org/10.1111/poms.12838.
4. A. Gunasekaran et al., "Big data and predictive analytics for supply chain and organizational performance," *Journal of Business Research*, vol. 70, pp. 308–317, Sep. 2016, https://doi.org/10.1016/j.jbusres.2016.08.004.
5. R. Dubey et al., "Big data analytics and organizational culture as complements to swift trust and collaborative performance in the humanitarian supply chain," *International Journal of Production Economics*, vol. 210, pp. 120–136, Apr. 2019, https://doi.org/10.1016/j.ijpe.2019.01.023.
6. K. Patel, "Instacart market basket analysis." PhD diss., California State University, Northridge, 2022.

7. M. Uddin, S. Anowar, and N. Eluru, "Modeling freight mode choice using machine learning classifiers: a comparative study using Commodity Flow Survey (CFS) data," *Transportation Planning and Technology*, vol. 44, no. 5, pp. 543–559, 2021, https://doi.org/10.1080/03081060.2021.1927306.
8. S. Axsäter, "Single-Echelon Systems: Reorder Points," In *Inventory Control*, pp. 65–105. Cham: Springer International Publishing, 2015, https://doi.org/10.1007/0-387-33331-2_5.
9. T. H. Davenport, "Putting the enterprise into the enterprise system," *Harvard Business Review*, vol. 76, no. 4, pp. 121–131, 1998.
10. F. P. García Márquez and B. Lev, eds., *Advanced Business Analytics*. Springer International Publishing, 2015, https://doi.org/10.1007/978-3-319-11415-6.
11. P. Trkman, K. McCormack, M. P. V. de Oliveira, and M. B. Ladeira, "The impact of business analytics on supply chain performance," *Decision Support Systems*, vol. 49, no. 3, pp. 318–327, Jun. 2010, https://doi.org/10.1016/j.dss.2010.03.007.
12. G. S. Kashyap, K. Malik, S. Wazir, and A. E. I. Brownlee, "Optimization of the rectangle area inside a concave polygon using PSO and Tabu search," *Soft Computing*, vol. 29, no. 7, pp. 3217–3239, Apr. 2025, https://doi.org/10.1007/s00500-025-10568-1.
13. L. Atzori, A. Iera, and G. Morabito, "The Internet of Things: a survey," *Computer Networks*, vol. 54, no. 15, pp. 2787–2805, Oct. 2010, https://doi.org/10.1016/j.comnet.2010.05.010.
14. P. Helo and B. Szekely, "Logistics information systems," *Industrial Management & Data Systems*, vol. 105, no. 1, pp. 5–18, Jan. 2005, https://doi.org/10.1108/02635570510575153.
15. W. H. Inmon, *Building the Data Warehouse*. John Wiley & Sons, Indianapolis, IN, 2005.
16. C. Madera and A. Laurent, "The next information architecture evolution," *Proceedings of the 8th International Conference on Management of Digital EcoSystems - MEDES*, 2016, https://doi.org/10.1145/3012071.3012077.
17. R. Hai, S. Geisler, and C. Quix, "Constance: an intelligent data lake system." *Proceedings of the 2016 International Conference on Management of data*, pp. 2097–2100, San Francisco, CA, 2016.
18. R. R. Kalluri, V. R. Surasani, N. S. H. Rellu, and N. Devarakonda. "Evolution of master data management and data governance: a two-decade review of advancements and innovations," *Evolution*, vol. 5, no. 2, pp. 1–11, 2025.
19. V. Khatri and C. V. Brown, "Designing data governance," *Communications of the ACM*, vol. 53, no. 1, pp. 148–152, Jan. 2010, https://doi.org/10.1145/1629175.1629210.
20. Z. Dehghani, "How to move beyond a monolithic data lake to a distributed data mesh," *Martin Fowler's Blog*, 45, 2019.
21. K. Ariyarathna and S. Peter, "Business Analytics Maturity Models: A Systematic Review of Literature," 2019. https://ieomsociety.org/ieom2019/papers/425.pdf (accessed Dec. 25, 2025).

22. S. LaValle, E. Lesser, R. Shockley, M. S. Hopkins, and N. Kruschwitz, "Big data, analytics and the path from insights to value," *MIT Sloan Management Review*, Dec. 21, 2010. https://sloanreview.mit.edu/article/big-data-analytics-and-the-path-from-insights-to-value/ (accessed Dec. 26, 2025).
23. D. Kiron, P. Kirk Prentice, and R. Boucher Ferguson, "The analytics mandate," *MIT Sloan Management Review*, vol. 55, no. 4, 2014. https://www.proquest.com/openview/c2ed00c8df5cb529bbff39d9e6afc603/1?pq-origsite=gscholar&cbl=26142 (accessed Dec. 25, 2025).
24. N. Chen, N. Chiang, and N. Storey, "Business intelligence and analytics: from big data to big impact," *MIS Quarterly*, vol. 36, no. 4, pp. 1165–1188, Dec. 2012, https://doi.org/10.2307/41703503.
25. R. J. Hyndman and G. Athanasopoulos, *Forecasting: Principles and Practice*. OTexts, 2018. https://books.google.co.in/books/about/Forecasting_principles_and_practice.html?id=_bBhDwAAQBAJ&redir_esc=y.
26. S. Makridakis, E. Spiliotis, and V. Assimakopoulos, "Statistical and machine learning forecasting methods: concerns and ways forward," *PLOS ONE*, vol. 13, no. 3, p. e0194889, Mar. 2018, https://doi.org/10.1371/journal.pone.0194889.
27. R. Fildes, P. Goodwin, M. Lawrence, and K. Nikolopoulos, "Effective forecasting and judgmental adjustments: an empirical evaluation and strategies for improvement in supply-chain planning," *International Journal of Forecasting*, vol. 25, no. 1, pp. 3–23, Jan. 2009, https://doi.org/10.1016/j.ijforecast.2008.11.010.
28. H. L. Lee, V. Padmanabhan, and S. Whang, "The bullwhip effect in supply chains," *IEEE Engineering Management Review*, vol. 43, no. 2, pp. 108–117, Jun. 2015, https://doi.org/10.1109/emr.2015.7123235.
29. P. H. Zipkin, *Foundations of Inventory Management*, McGraw-Hill, Irwin, New York, 2000.
30. A. Soni et al., "Can We Predict Your Next Move without Breaking Your Privacy?," *arXiv.org*, 2025. https://arxiv.org/abs/2507.08843 (accessed Jan. 13, 2026).
31. G. S. Kashyap et al., "Can AI see what we can't? Leveraging deep learning and multi-temporal satellite data to revolutionize crop type mapping and yield prediction," In *ICASSP 2025–2025 IEEE International Conference on Acoustics, Speech and Signal Processing (ICASSP)*, pp. 1–5. IEEE, 2025. https://doi.org/10.1109/icassp49660.2025.10890344.
32. S. Gallino and A. Moreno, "Integration of online and offline channels in retail: the impact of sharing reliable inventory availability information," *Management Science*, vol. 60, no. 6, pp. 1434–1451, Apr. 2014, https://doi.org/10.1287/mnsc.2014.1951.
33. W. Ho, X. Xu, and P. K. Dey, "Multi-criteria decision making approaches for supplier evaluation and selection: a literature review," *European Journal of Operational Research*, vol. 202, no. 1, pp. 16–24, Apr. 2010, https://doi.org/10.1016/j.ejor.2009.05.009.

34. P. R. Kleindorfer and G. H. Saad, "Managing disruption risks in supply chains," *Production and Operations Management*, vol. 14, no. 1, pp. 53–68, Jan. 2005, https://doi.org/10.1111/j.1937-5956.2005.tb00009.x.
35. E. Hofmann and M. Rüsch, "Industry 4.0 and the current status as well as future prospects on logistics," *Computers in Industry*, vol. 89, pp. 23–34, Aug. 2017, https://doi.org/10.1016/j.compind.2017.04.002.
36. S. Tripathi, M. T. Nafis, I. Hussain, and J. Gao, "The Confidence Paradox: Can LLM Know When It's Wrong," *arXiv.org*, 2025. https://arxiv.org/abs/2506.23464 (accessed Jan. 13, 2026).
37. D. R. Krause, R. B. Handfield, and B. B. Tyler, "The relationships between supplier development, commitment, social capital accumulation and performance improvement," *Journal of Operations Management*, vol. 25, no. 2, pp. 528–545, Jul. 2006, https://doi.org/10.1016/j.jom.2006.05.007.
38. P. Toth, and D. Vigo, eds., *Vehicle Routing: Problems, Methods, and Applications.* Society for Industrial and Applied Mathematics, Philadelphia, PA, 2014.
39. M. Kulahara, A. Khader, S. Tripathi, and M. A. Hoque, "A CNN-based framework for geometric alignment of historical and satellite imagery," *IEEE Access*, vol. 13, pp. 132259–132268, Jan. 2025, https://doi.org/10.1109/access.2025.3589497.
40. J. Gu, M. Goetschalckx, and L. F. McGinnis, "Research on warehouse design and performance evaluation: a comprehensive review," *European Journal of Operational Research*, vol. 203, no. 3, pp. 539–549, Jun. 2010, https://doi.org/10.1016/j.ejor.2009.07.031.
41. S. Few, "Information dashboard design," 2003. Available: https://blogs.ischool.berkeley.edu/i247s12/files/2012/01/Dashboard-Design-Overview-Presentation.pdf
42. V. Govindarajan, P. Patel, S. Tripathi, M. A. Hoque, and G. S. Kashyap, "MAGIC-Enhanced Keyword Prompting for Zero-Shot Audio Captioning with CLIP Models," *arXiv.org*, 2025. https://arxiv.org/abs/2509.12591 (accessed Jan. 13, 2026).

# Chapter 3

# Predictive Analytics and Real-Time Decision-Making

Rafiq Ali

## 3.1 Introduction

In supply chain management (SCM), decision-making is progressively taking place in an environment characterized by significant demand volatility, regular disruptions, and reduced operational time frames. The effectiveness of planning paradigms that are static or updated periodically has been diminished by globalized sourcing, shortened product life cycles, geopolitical uncertainty, climate-related events, and increasing regulatory and sustainability requirements [1,2]. In such situations, historical averages and rare replanning cycles can't capture the rapid change in operation, causing slower reaction time and less than ideal outcomes [3]. Predictive analytics models have emerged as a key analytical tool for today's SCM to address these issues. By applying statistical time-series forecasting models, machine learning (ML) models, and probabilistic estimation of uncertainty, predictive analytics enables companies to anticipate demand trends, estimate supply and logistics risks, and explore other operation scenarios across procurement, production, stock, and distribution functions [4,5]. With shorter decision cycles and rapid changes in operations, prediction accuracy is valuable because it transforms the model output into actionable actions in the proper time than plans [6].

While prediction is already well advanced, many implementations of SCM in the real world remain lagging behind in prediction–execution. Currently, the most common applications involve embedding predictive models in batch pipelines that focus on planning data ingestion, model inference, and


DOI: 10.4324/9781003724889-3

decision updates at set intervals, rather than continuously [7]. They are not compatible with high-speed data streams generated by enterprise transactions, IoT sensors, logistics systems, or with external forces such as weather, market dynamics, and geopolitical events [8]. This means that demand changes, supply disruptions, and quality changes are usually only identified after a service impact has occurred. This reduces the effectiveness of mitigation strategies. The organizational and architectural differences between predictive models and execution systems augment this prediction–execution gap. Predictive models become analytical artifacts that are disentangled from decision-making or are cloaked together. This separation also results in delays, contradictions, and ambiguity with respect to accountability in addressing model outputs [9]. Also, much of the confusion around the propagation of uncertainty, scaling across multi-echelon supply networks, and the balance between accuracy, interpretation, and computational efficiency remains ungrounded [10,11]. These technical issues are further complicated by organizational barriers such as fragmented data ownership, limited analytical maturity, skill gaps, and questions of governance and risk in systems where decision-making is either fully or completely automated [12]. These factors indicate that only improvements in predictive modeling cannot meet the ever-changing needs of modern SCM.

The motivation for this chapter is to overcome this gap in prediction–execution by looking at the entire system at a scale that views predictive analytics and decision-making as separate capabilities, rather than discrete technical functions. The technical foundation for continuous data intake, low-latency model inference, and automated or semi-automated operational responses at execution timescales includes stream analytics, event-driven architectures, and closed-loop control systems [13]. However, these advantages in practice must be weighed against the models, system design, and organizational decisions. This entails coordinating human-in-the-loop mechanisms and governance in order to ensure reliability, accountability, and trust [14]. This chapter thus introduces the Integrated Predictive–Real-Time Decision Framework (IPRDF), a closed-loop architecture that connects probabilistic forecasting models, ML classification models, and anomaly-detection models with streaming and event-driven execution mechanisms. It aims to enable continuous updating, contextual interpretation, and operative application of predictive insights during uncertainty. The framework also provides a standard approach for evaluating improvements in performance beyond just predictive accuracy, such as decision latency, robustness, and execution quality within multiple SCM contexts [15]. Table 3.1 summarizes the

**Table 3.1 Comparison of Decision-Making Paradigms in SCM, Highlighting the Structural Progression from Static Planning and Standalone Predictive Modeling to the Proposed Integrated Predictive–Real-Time Framework**

| *Main Feature* | *Traditional Planning* | *Predictive Analytics* | *Real-Time Decision-Making* | *Integrated Predictive–Real-Time Framework* |
|---|---|---|---|---|
| Static, batch-oriented planning | 1 | 0 | 0 | 0 |
| Modeling of demand and supply uncertainty | 0 | 1 | 1 | 1 |
| Use of ML/advanced forecasting models | 0 | 1 | 0 | 1 |
| Continuous data ingestion | 0 | 0 | 1 | 1 |
| Low-latency operational response | 0 | 0 | 1 | 1 |
| Integration with execution models | 0 | 0 | 0 | 1 |
| Closed-loop decision control | 0 | 0 | 0 | 1 |
| Governance and human-in-the-loop support | 0 | 0 | 0 | 1 |
| Robustness under volatility and disruption | 0 | 1 | 1 | 1 |

A value of 1 indicates native, system-level support for a given capability, while 0 denotes its absence or nonsystematic support. The table emphasizes how closing the prediction–execution gap requires the joint integration of uncertainty-aware modeling, continuous data ingestion, low-latency execution, closed-loop control, and governance mechanisms.

conceptual relevance of the proposed framework to traditional planning, standalone predictive analytics, and real-time decision-making paradigms. In the table, 1 indicates that a given capability is explicitly supported as part of the paradigm design, and 0 indicates that the capability is not supported or does not have any systematic support. It is well documented that in conventional planning, decision cycles are static, batch-driven, and only a small amount of uncertainty modeling is used. Alternatively, predictive analytics can enhance forecasting but cannot guarantee real-time execution. Real-time decision models tend to have low-latency responses but rarely employ either predictive modeling or closed-loop control. Similarly, the proposed IPRDF conjointly supports uncertainty-aware modeling, continuous data collection, low-latency implementation, closed-loop decision-making, and human oversight, addressing the structural limitations of existing approaches.

This chapter is structured to move from a conceptualized perspective to a practicalized model. Section 3.2 presents existing literature on predictive analytics for SCM, highlighting the evolution of statistical forecasting, ML-based prediction, anomaly detection, streaming and event-driven architectures, and closing the gap between prediction and execution. Section 3.3 details the materials and models, as well as the data, preprocessing pipeline, feature design, and the integrated IPRDF predictive–real-time architecture; and Section 3.4 describes an experimental setup, including evaluation metrics, execution-level responsiveness measures, and hyperparameter configurations used in all experiments. Section 3.5 reports and analyzes results, including comparisons to modern baselines, cross-domain robustness analysis, and a comprehensive ablation analysis that separates the forecasting, anomaly detection, and execution components. This chapter ends in Section 3.6 with a summary of findings and recommendations for future research.

## 3.2 Related Works

### *3.2.1 Evolution of Predictive Analytics in SCM*

During the early days of SCM predictive analytics, planning was the dominant focus and the primary objective was to improve prediction accuracy in static decision cycles. This work examined statistical forecasting and operations research models, i.e., univariate and multivariate time-series models such as exponential smoothing and Auto-Regressive Integrated Moving Average (ARIMA), to predict future demand based on previous sales data [16]. As shown in Table 3.2, these models were always based on historical data and explicit time-series formulations under the condition that demand patterns were linear and stationary [17]. Although these statistical models were superior to heuristic and rule-based planning processes in that they explicitly account for trend and seasonality, they were largely designed for periodic, batch-based planning cycles [18]. Forecasts were produced weekly or monthly and then converted into static plans for procurement, production, and distribution. As a result, early predictive analytics did not

**Table 3.2 Evolution of Predictive Analytics in SCM, Illustrating the Progression from Batch-Oriented Statistical Forecasting Models to Data-Intensive ML Models and, Ultimately, to Integrated Real-Time and Prescriptive Analytics**

| *Main Feature* | *Traditional Statistical Forecasting* | *ML-Based Predictive Analytics* | *Real-Time and Prescriptive Integration* |
|---|---|---|---|
| Reliance on historical sales data | 1 | 1 | 1 |
| Time-series forecasting models | 1 | 0 | 0 |
| Assumptions of linearity and stationarity | 1 | 0 | 0 |
| Periodic batch planning cycles | 1 | 0 | 0 |
| Causal and hierarchical forecasting | 1 | 0 | 0 |
| Modeling of nonlinear demand relationships | 0 | 1 | 1 |
| High-dimensional feature utilization | 0 | 1 | 1 |
| Computational scalability | 0 | 1 | 1 |
| Continuous model updating | 0 | 0 | 1 |
| Streaming and event-driven data processing | 0 | 0 | 1 |
| Direct linkage to operational execution | 0 | 0 | 1 |
| Prescriptive or automated decision actions | 0 | 0 | 1 |
| Governance and interpretability emphasis | 0 | 0 | 1 |
| Human-in-the-loop decision design | 0 | 0 | 1 |
| Support for adaptive and resilient operations | 0 | 1 | 1 |

A value of 1 indicates native, system-level support for a capability, while 0 denotes its absence. The table highlights how advances in predictive modeling alone do not close the prediction–execution gap unless coupled with continuous data ingestion, direct linkage to operational execution, and governance-aware decision control.

support the ongoing modeling, real-time processing of data, or operational execution. This lack of support is evidenced by the absence of system-level execution and adaptation capabilities in the first column of Table 3.2. More recent studies have further expanded the statistical prediction approach to incorporate causal and hierarchical models that improve coherence across products, sites, and organizational boundaries [19]. However, these extensions did not change the fundamental batch-oriented paradigm and did not improve structural consistency or planning alignment. Predictions were not integrated into operating workflows, and forecasts served as inputs into fixed plans rather than triggers for execution. As a result, Table 3.2 shows that although modeling did have some improvement, there was no direct correlation to operation, automated decision-making, or human-in-the-loop control during this time. With the availability of data and computational scalability, newer research has begun reorienting toward ML-based predictive analytics. Ensemble models such as random forests and gradient boosting, as well as deep neural networks, have demonstrated better predictive performance in nonlinear environments, high-dimensional feature spaces, and complex interactions between demand drivers [20,21]. This transition made it possible to explicitly model nonlinear demand relationships, improve scalability, and enhance the use of high-dimensional data, as summarized in Table 3.2. Despite the advancements in modeling, predictive analytics continued to be execution-agnostic, with forecasts mainly assessed in offline or batch environments. At the same time, developments in streaming analytics and event-driven architectures created the technical capacity for continuous data ingestion and frequent updates to models [22]. Thanks to these advancements, predictive models can now be updated more quickly in response to new transactional, sensor, and external signals. Despite this, a significant portion of the literature persisted in viewing predictive modeling and operational execution as components that are loosely connected. Even with the availability of real-time data processing, predictive models were seldom integrated into closed-loop decision systems that could initiate prescriptive or automated actions within governance constraints [23]. As shown in the last column of Table 3.2, real-time and prescriptive integration occurs only when predictive models are explicitly incorporated into system architectures that facilitate streaming data processing, direct connections to execution, automated or semi-automated decision actions, and human-in-the-loop governance [24,25]. This phase represents a qualitative shift from the intent of improving prediction accuracy to an environment that supports adaptive and resilient operations through the re-reading and execution of predictions within the timeframe of execution. Much of the existing work does not integrate these at the system level, indicating that architectures need to address the gap between prediction and execution rather than rely on model-independent approaches.

### 3.2.2 Time-Series Forecasting in SCM

Time-series forecasting forms the fundamental modeling layer of predictive analytics in SCM. Initial research focused on statistical time-series models aimed at capturing level, trend, seasonality, and autocorrelation in historical demand data. Exponential smoothing and ARIMA models became the most popular ways to predict demand and plan capacity because they are simple, easy to understand, and don't need a lot of computing power [26–28]. Table 3.3 shows that these models clearly show how time is structured and work well when demand is stable and the forecasting horizon is short. This is why they are widely used in enterprise planning systems for making routine decisions about restocking and scheduling [29].

However, the robustness of statistical time-series models in handling dynamic demand, structural breaks, and interdependent operational drivers is limited by stiff assumptions of linearity, stationarity, and minimal exogenous influence [30]. Table 3.3 demonstrates that such models are easy to understand and apply but cannot model nonlinear relationships, long-term temporal dependencies, or sudden regime changes. All of these are becoming increasingly significant in modern SCM.

To mitigate these constraints, subsequent research developed multivariate and state-space forecasting models that simultaneously depict interrelated time series and integrate exogenous variables, including prices, lead times, and promotional signals [31]. These models enhance traditional methodologies by explicitly incorporating cross-dependencies among supply chain variables, thus fostering coherence across products and decision-making tiers. Table 3.3 shows that multivariate and state-space models loosen the rules about outside inputs and make the models more complex, but they still mostly stick to linear assumptions and are best suited for short-term forecasting with moderate variability. Recent research has transitioned toward ML and deep learning (DL) forecasting models, driven by the accessibility of extensive datasets and enhancements in computational infrastructure. Recurrent Neural Networks (RNNs) and Long Short-Term Memory (LSTM) architectures show better performance for sequences with nonlinear dynamics and long-range dependencies [32]. Attention-based and transformer

**Table 3.3 Comparison of Time-Series Forecasting Model Families Used in SCM, Highlighting the Trade-Offs between Statistical Interpretability, Modeling Flexibility, and Robustness under Nonstationary Demand**

| *Main Feature* | *Statistical Time-Series Models* | *Multivariate / State-Space Models* | *ML/DL and Ensemble Models* |
|---|---|---|---|
| Modeling of level, trend, and seasonality | 1 | 1 | 1 |
| Explicit autocorrelation modeling | 1 | 1 | 1 |
| Linear assumption dependence | 1 | 1 | 0 |
| Suitability for stable demand patterns | 1 | 1 | 1 |
| Short-term forecasting focus | 1 | 1 | 1 |
| Inclusion of exogenous variables | 0 | 1 | 1 |
| Modeling of nonlinear relationships | 0 | 0 | 1 |
| Long-range temporal dependency capture | 0 | 0 | 1 |
| Robustness to structural changes | 0 | 0 | 1 |
| Reduced sensitivity to model misspecification | 0 | 0 | 1 |
| High data and computational requirements | 0 | 0 | 1 |
| Interpretability for operational use | 1 | 1 | 0 |

A value of 1 indicates native support for a modeling capability, while 0 denotes its absence. The table emphasizes that accuracy gains from ML- and ensemble-based forecasting models come at the cost of higher data and computational requirements and reduced interpretability—limitations that motivate their integration within system-level, real-time decision frameworks rather than standalone deployment.

architectures further improve forecasting accuracy by selectively weighting relevant historical observations and capturing extended temporal context without any limitations related to recurrence [33]. Table 3.3 shows that these models can perform nonlinear modeling, capture long-range dependencies, and be more robust to structural changes. However, they need more data and computing power and are harder to understand for operational use.

A growing body of work focuses on ensemble forecasting models that combine several predictors with different assumptions and update characteristics to balance the strengths of both statistical and ML-based models [34,35]. As shown in Table 3.3, these ensembles make models less sensitive to misspecification and more robust when demand is unstable. Nonetheless, a significant portion of the current literature assesses forecasting models in isolation, concentrating mainly on point accuracy metrics while neglecting to explicitly analyze forecast degradation over prolonged timeframes or the transmission of forecast uncertainty into subsequent operational decisions [36]. This limitation highlights the necessity for cohesive predictive–real-time frameworks that explicitly link forecasting behavior, uncertainty dynamics, and execution-level decision-making.

### 3.2.3 ML for Demand Prediction and Classification in SCM

ML models have been extensively investigated for demand prediction and operational classification in SCM for supplier risk assessment, disruption forecasting, and shipment delay classification. The first study focused on improving prediction by applying supervised learning models such as linear regression and logistic regression that included explanatory variables such as price, promotion, and macroeconomic variables [37]. These models are robustly interpretable and therefore compatible with governance requirements, but they are linear in design and require substantial manual feature engineering to produce expressive results in complex operating conditions. Next, models based on decision trees and support vector machines were constructed to mitigate linearity constraints. These models support nonlinear decision boundaries and higher-dimensional feature space management, facilitating tasks such as supplier categorization or delay classification [38,39]. As shown in Table 3.4, tree-based and margin-based models provide greater flexibility than linear models but still largely depend on manual feature construction and are not very robust to noisy or incomplete SCM data. Recent work also highlights the advantages of ensemble learning models, particularly random forests and gradient boosting, which overcome limitations of

**Table 3.4 Comparison of ML Model Classes for Demand Prediction and Classification in SCM, Highlighting the Trade-Offs between Predictive Expressiveness, Data and Computational Requirements, and Operational Governance**

| *Main Feature* | *Linear/Logistic Models* | *Tree-Based and SVM Models* | *Ensemble Models* | *Deep and Attention-Based Models* |
|---|---|---|---|---|
| Use of supervised learning | 1 | 1 | 1 | 1 |
| Inclusion of explanatory variables | 1 | 1 | 1 | 1 |
| Relaxation of linearity assumptions | 0 | 1 | 1 | 1 |
| Requirement for manual feature engineering | 1 | 1 | 0 | 0 |
| Modeling of complex nonlinear interactions | 0 | 0 | 1 | 1 |
| Robustness to noisy or incomplete data | 0 | 0 | 1 | 1 |
| Resistance to overfitting | 0 | 0 | 1 | 0 |
| High-dimensional feature handling | 0 | 1 | 1 | 1 |
| Sequential or temporal pattern learning | 0 | 0 | 0 | 1 |
| Context-aware input weighting | 0 | 0 | 0 | 1 |
| Post hoc interpretability support | 1 | 1 | 1 | 0 |
| High computational and data requirements | 0 | 0 | 1 | 1 |
| Suitability for operational governance | 1 | 1 | 1 | 0 |

A value of 1 denotes native support for a capability, while 0 indicates limited or absent support. The table illustrates that gains in nonlinear modeling and temporal representation in ensemble and deep models often come at the cost of interpretability and governance suitability, motivating system-level integration rather than standalone model selection.

previous approaches. Ensemble models reduce overfitting through aggregation, are robust to noisy data and nonlinear inputs, and can model complex nonlinear interactions between multiple SCM features [40]. Gradient boosting models can improve predictability by identifying residual errors in sequence and have shown promising performance in demand forecasting, disruption prediction, and quality classification tasks [41,42]. As outlined in Table 3.4, ensemble models are well balanced in terms of predictive power, robustness, and post hoc interpretability, thus making them relatively suitable for use in practice. Models based on DL mark an additional step forward toward expressive representation learning. Convolutional, feedforward, and RNNs allow for the automatic extraction of hierarchical and temporal patterns from unreconstructed or relatively processed raw data, thereby reducing manual feature engineering [43]. Attention-based architectures complement this ability by learning to weigh inputs according to context, thereby increasing sensitivity to demand drivers that shift over time and across operations [44]. Yet, as Table 3.4 indicates, these gains are not balanced by higher computational requirements, less transparency, and more appropriateness for operational governance, particularly in decision-critical real-time situations [45].

### 3.2.4 Anomaly-Detection Model in SCM

Anomaly-detection models have long been used to detect deviations from expected operation in SCM, such as quality, inventory, and logistics performance. Control charts, threshold rules, and regression diagnostics were some of the first statistical and model-based approaches used. These methods detect anomalies when observations exceed the fixed limits of historical operational baselines [46,47]. The time-series decomposition models classify trend, seasonal, and irregular components with anomalies identified as extreme residuals relative to the expected temporal structure [48]. These statistical models are transparent, computationally efficient, and compatible with stable, low-dimensional processes. Their reproducibility, as shown in Table 3.5, is limited by strong distributional assumptions, dependence on control limits, and scaling to highly parametric or heterogeneous data sources [49]. This makes it difficult for these models to detect tiny or growing anomalies in the context of current SCM practices, including high-frequency transactions, sensor streams, and intricate interdependencies between operational variables. To overcome these limitations, later research has focused on unsupervised ML anomaly-detection strategies using clustering, isolation forests, and one-class support vector machines [50–52]. By limiting the distributional

**Table 3.5 Comparison of Anomaly-Detection Model Paradigms in SCM, Highlighting the Progression from Statistically Grounded, Rule-Based Monitoring toward Integrated, Adaptive Detection Models**

| *Main Feature* | *Statistical and Model-Based Models* | *Unsupervised ML Models* | *Deep Learning–Based Models* | *Integrated Adaptive Detection Frameworks* |
|---|---|---|---|---|
| Reliance on historical operational baselines | 1 | 1 | 1 | 1 |
| Use of statistical control limits | 1 | 0 | 0 | 0 |
| Distributional assumption dependence | 1 | 0 | 0 | 0 |
| Suitability for low-dimensional data | 1 | 0 | 0 | 0 |
| Handling of high-dimensional data | 0 | 1 | 1 | 1 |
| Detection without labeled anomalies | 0 | 1 | 1 | 1 |
| Modeling of nonlinear relationships | 0 | 1 | 1 | 1 |
| Learning compact representations of normality | 0 | 1 | 1 | 1 |
| Sequential or temporal anomaly detection | 0 | 0 | 1 | 1 |
| Context-aware deviation identification | 0 | 0 | 1 | 1 |
| Adaptive thresholding capability | 0 | 0 | 0 | 1 |
| Integration with decision workflows | 0 | 0 | 0 | 1 |
| Balance between sensitivity and false alarms | 0 | 0 | 0 | 1 |
| Scalability across heterogeneous data sources | 0 | 1 | 1 | 1 |

A value of 1 indicates native support for a given capability, while 0 denotes its absence or nonsystematic treatment. The table emphasizes that reliable operational control under nonstationary and high-dimensional conditions requires anomaly-detection models that support nonlinear pattern learning, temporal context, adaptive thresholding, and explicit integration with decision workflows—capabilities absent in standalone statistical and unsupervised approaches.

assumptions and enabling anomaly detection without labeled data, such models are suitable for identifying uncommon or novel events in large-scale operational datasets. Table 3.5 shows that unsupervised ML models improve scalability and nonlinear modeling relative to statistical models but do not provide explicit temporal modeling, adaptive thresholding or decision integration. Several recent studies have analyzed DL-based anomaly-detection models, such as autoencoders and variational autoencoders, which learn compact representations of normal system behavior, with reconstruction error or likelihood used as anomaly scores [53]. Sequential DL models, such as RNNs and attention-based architectures, are able to detect anomalies in time-series and event-stream data by capturing long-range dependencies and context-aware deviations [54]. These models significantly improve the detection accuracy of complex and dynamic processes but are still frequently used as standalone detectors with fixed thresholds and minimal integration into decision-making processes, as reported in Table 3.5.

The existing literature primarily views anomaly detection as a problem that can be alerted to the situation within all three types of models—statistical, unsupervised ML, and DL-based—rather than being a closed-loop decision system. Adaptive thresholding, explicit connections to execution logic, feedback-driven calibration, and a controlled balance between sensitivity and false alarms tend to be unmet [55]. This transparency reduces the effective use of anomaly-detection outputs, especially in sensitive timescales where false positives can be expensive and delays can result in failure. These limitations drive the development of integrated, adaptive anomaly-detection systems, which integrate anomaly-detection models into real-time decision structures that allow for continuous data intake, adaptive thresholding, contextual interpretation, and governed execution. These models are beyond the assumption of detection accuracy: anomalies do not simply have an observable impact on uncertainty forecasting, risk assessment, or operational response; and they also bridge the gap between anomaly and action in modern SCM.

### *3.2.5 Streaming Data Analytics and Event-Driven Architectures in SCM*

Early analytics architectures in SCM consisted primarily of batch analysis of transactions that were collected at set intervals for retrospective analysis to help periodic planning and reporting [56]. They are useful for strategic and tactical decision-making, but by nature they have high insight latency.

This is because the analyses occur for hours or even days after the operation takes place [57,58]. Table 3.6 demonstrates that batch-oriented analytics have a fixed-interval processing, a retrospective approach, and little capacity for continuous monitoring or operational responsiveness. This trend toward processing and ingestion of raw data at low latency has driven a major shift to streaming data analytics. Stream-processing frameworks enable the real-time transformation, aggregation, and monitoring of high-speed data streams that emerge from transactions with companies, sensors, and logistics systems [59]. Although these systems improve situational awareness through their support for real-time operational monitoring, they are not entirely descriptive or reactive. As noted in Table 3.6, streaming analytics alone do not imply decision logic based on events, predictive modeling, or proactive decision support.

Event-driven architectures enhance streaming systems by including more precise definitions of events and response mechanisms. These architectures use predetermined conditions or patterns to trigger automated workflows or alerts, so that faster operational responses to detected events such as shipping delays or inventory threshold violations [60]. This is further expanded through complex event processing, in which correlated or cascaded events are discovered that span multiple datasets, such as disruptions in multi-tier suppliers [61]. However, as discussed in Table 3.6, most event-driven architectures are rule-driven and reactive. They are based on observed events rather than expected future states and often do not involve integrated predictive modeling, learning mechanisms, or uncertainty-aware control.

Recently, researchers have begun developing integrated streaming and prediction architectures that incorporate predictive models directly into streaming and event-driven pipelines [62]. These systems include forecasting, classification, and anomaly-detection models that run continuously on streaming data. This allows for proactive decision support based on desired outcomes rather than isolated failures. Online learning, sliding-window testing, and concept drift detection allow predictions to adjust to nonstationary environments over time [63,64]. While promising, fully integrated streaming–predictive architectures remain an uncommon feature in the operation of SCM systems due to issues such as scaling, reliability, governance, and organizational readiness [65]. To address the prediction–execution gap highlighted in this chapter, a system-level framework is needed that explicitly integrates streaming analytics, predictive modeling, event-driven execution, and human-in-the-loop control.

**Table 3.6 Comparison of Data Analytics Architectures in SCM, Illustrating the Progression from Batch-Oriented and Reactive Streaming Systems to Integrated Streaming–Predictive Architectures**

| *Main Feature* | *Batch-Oriented Analytics* | *Streaming Data Analytics* | *Event-Driven Architectures* | *Integrated Streaming and Predictive Models* |
|---|---|---|---|---|
| Fixed-interval data processing | 1 | 0 | 0 | 0 |
| Retrospective analytical focus | 1 | 0 | 0 | 0 |
| High-latency insight generation | 1 | 0 | 0 | 0 |
| Continuous data ingestion | 0 | 1 | 1 | 1 |
| Low-latency analytical processing | 0 | 1 | 1 | 1 |
| Real-time operational monitoring | 0 | 1 | 1 | 1 |
| Event-triggered response mechanisms | 0 | 0 | 1 | 1 |
| Complex event pattern recognition | 0 | 0 | 1 | 1 |
| Integration of predictive models | 0 | 0 | 0 | 1 |
| Proactive decision support | 0 | 0 | 0 | 1 |
| Online learning and model adaptation | 0 | 0 | 0 | 1 |
| Concept drift handling | 0 | 0 | 0 | 1 |
| Fault tolerance and governance emphasis | 0 | 0 | 0 | 1 |
| Human oversight in automated workflows | 0 | 0 | 0 | 1 |

A value of 1 denotes native, system-level support for a capability, while 0 indicates its absence. The table highlights that continuous ingestion and low-latency processing alone are insufficient for real-time decision-making; proactive decision support, model adaptation under nonstationarity, governance, and human oversight emerge only when predictive models are explicitly integrated within streaming and event-driven execution pipelines.

### 3.2.6 Prescriptive Analytics and Decision Optimization in SCM

Prescriptive analytics enlarges predictive analytics and explicitly suggests actions for cost, capacity, service-level, and risk areas. Prescriptive models in SCM rely on prediction with mathematical optimization, simulation, and decision rules to make executable recommendations for inventory positioning, production scheduling, and transportation routing [66]. In contrast to predictive analytics, in which future system states are largely estimated, prescriptive analytics determines what the most appropriate actions should be in order to meet objectives such as cost efficiency, service reliability, and resilience.

Early prescriptive decision systems were set up as layers of optimization in enterprise planning systems. These systems used linear and mixed-integer programming formulations to generate static plans based on periodic forecasts [67]. However, as presented in the first column of Table 3.7, such optimization pipelines did not include continuous data ingestion, real-time adjustment, or feedback integration. Forecast outputs were also treated as fixed inputs to optimization solvers, so that prescriptive decisions cannot react dynamically to changing demand volatility or logistics problems.

**Table 3.7 Comparison of Prescriptive Analytics Paradigms in SCM, Illustrating the Progression from Static Optimization and Sequential Predict-Then-Optimize Pipelines to Adaptive, Real-Time Prescriptive Systems**

| *Main Feature* | *Static Optimization Models* | *Sequential Predict-Then-Optimize* | *Adaptive Prescriptive Systems* |
|---|---|---|---|
| Optimization under constraints | 1 | 1 | 1 |
| Forecast integration | 0 | 1 | 1 |
| Scenario simulation | 0 | 1 | 1 |
| Decision recommendations | 1 | 1 | 1 |
| Feedback into predictive models | 0 | 0 | 1 |
| Continuous data ingestion | 0 | 0 | 1 |
| Real-time decision adjustment | 0 | 0 | 1 |
| Governance and interpretability | 0 | 1 | 1 |
| Automated execution support | 0 | 0 | 1 |

A value of 1 denotes native, system-level support for a capability, while 0 indicates its absence. The table highlights that feedback integration, continuous data ingestion, real-time decision adjustment, and automated execution support emerge only within adaptive prescriptive architectures.

Recently, new research has focused on incorporating prescriptive analytics into real-time systems, which continually alter operational decisions to reflect streaming demand, supply, and disruption signals. Adaptive operational control is increasingly supported through scenario-driven optimization, rolling-horizon planning, and uncertainty-aware decision rules [68]. These systems explicitly incorporate human-in-the-loop governance, which means controlling automation for actions that are low risk and giving planners greater opportunity to make decisions with high impact. Like the last column of Table 3.7, this change introduces real-time correction, feedback in predictive models, and controlled automated execution, all capabilities lacking in standard predictive models.

### *3.2.7 Hybrid and Adaptive Analytics Frameworks in SCM*

Predictive modeling, anomaly detection, and prescriptive optimization are all integrated into unified closed-loop designs by hybrid and adaptive analytics frameworks, which continuously modify operational choices in SCM. These frameworks were developed in response to the shortcomings of modular analytics pipelines, which do not facilitate coordinated adaptation in the face of nonstationary demand and supply shocks, since they consider forecasting, detection, and optimization as separate processes.

Early hybrid designs integrated rolling-horizon planning systems with forecasting and optimization layers. Although these systems relied on sequential model execution and lacked explicit input across analytical components, they updated plans more frequently than batch pipelines [69]. Early hybrid pipelines were less responsive to infrequent and changing interruptions because they did not provide anomaly–decision coupling, adaptive policy learning, or continuous closed-loop execution, as seen in the first column of Table 3.8.

By incorporating anomaly detection, uncertainty-aware forecasting, and reinforcement learning into real-time execution pipelines, recent research advances hybrid frameworks. Learning agents with streaming feedback to the operating environment are employed to continuously improve logistics management and inventory control strategies [70]. At the same time, asymmetric decision logic and uncertainty propagation allow hybrid models to differentiate between severely perturbed changes and expected variability and help controlled escalation and early intervention. Recent hybrid frameworks, summarized in the last column of Table 3.8, provide governance-sensitive automation, feedback-driven learning, and continuous closed-loop execution that close this prediction–execution gap.

**Table 3.8 Comparison of Analytics Frameworks in SCM, Illustrating the Progression from Modular Analytics Pipelines and Partially Integrated Streaming–Predictive Systems to Adaptive Hybrid Learning Frameworks**

| *Main Feature* | *Modular Analytics Pipelines* | *Streaming + Predictive Integration* | *Adaptive Hybrid Frameworks* |
|---|---|---|---|
| Predictive modeling | 1 | 1 | 1 |
| Prescriptive optimization | 1 | 0 | 1 |
| Real-time data ingestion | 0 | 1 | 1 |
| Cross-component feedback loops | 0 | 0 | 1 |
| Reinforcement learning/policy adaptation | 0 | 0 | 1 |
| Uncertainty-aware adaptation | 0 | 1 | 1 |
| Anomaly–decision linkage | 0 | 0 | 1 |
| Governance and interpretability | 1 | 0 | 1 |
| Continuous closed-loop execution | 0 | 0 | 1 |

A value of 1 denotes native, system-level support for a capability, while 0 indicates its absence. The table highlights that cross-component feedback, policy adaptation, anomaly-linked decisioning, and continuous closed-loop execution emerge only within adaptive hybrid architectures.

## 3.3 Materials and Models

### *3.3.1 Description of the Dataset*

We empirically assess the proposed IPRDF using two publicly available datasets that represent complementary aspects of supply chain decision-making: short-term retail demand dynamics and real-world logistics execution during disruptions.

The Retailrocket E-commerce Dataset [71] comprises about 2.7 million anonymized and time-stamped user interaction events gathered over approximately 4.5 months from a major online retail platform, involving around 235,000 unique users and 50,000 distinct products. Every record features a timestamp with millisecond precision, a product ID, a session or user ID, and an event type, which includes product views, add-to-cart actions, and completed purchases. Event streams are compiled into product-level demand signals on an hourly and daily basis for modeling purposes. This aggregation facilitates the creation of interaction counts, conversion ratios, rolling demand intensity measures, volatility indicators, and calendar-based features. The dataset is divided chronologically into three parts: 70% for training, 15% for validation, and 15% for testing. The test set consists of the most recent data and is used solely for out-of-sample evaluation. The OpenSky Network Flight and Trajectory Dataset [72] contains extensive, high-frequency records of aircraft movements gathered from a worldwide network of ADS-B sensors. The subset utilized in this research includes about 12 million trajectory records over several months, with sampling intervals of 5–15 seconds for each aircraft. Every record contains flight identifiers, timestamps, geographic coordinates, altitude, speed, and metadata detailing departures, arrivals, delays, and routing alterations. Logistics-relevant variables such as departure and arrival delays, en-route delay accumulation, route deviation indicators, and rolling congestion metrics are derived from these signals. This dataset, while it comes from the aviation domain, is commonly used as a proxy for air cargo reliability and transit-time variability in logistics research [68]. To guarantee uniform assessment between predictive modeling and real-time decision tasks, a chronological split identical to that of the retail dataset is employed.

### 3.3.1.1 Preprocessing of the Dataset

The preprocessing of the dataset aims to achieve causal validity, numerical stability, and realistic behavior in real-time deployment. All transformations are applied uniformly across the training, validation, and test splits, as detailed in Table 3.9.

To prevent temporal leakage, records in both datasets are strictly ordered by timestamp, and event-level observations are aligned to operational decision intervals using fixed-width aggregation windows of 1 and 24 hours for the Retailrocket dataset, and 5 and 15 minutes for the OpenSky dataset. Following aggregation, missing values make up about 2%–4% of features derived from Retailrocket and 3%–6% of those from OpenSky. Gaps that cover one or two consecutive windows are filled using last-observation carry-forward, while longer gaps are addressed with local rolling means calculated over a three- to five-window neighborhood. Outlier analysis employs interquartile range filtering along with z-score thresholds of $|z| > 4$, leading to the removal of less than 1.2% of total observations deemed implausible measurement artifacts, while preserving rare but operationally valid demand spikes and extreme delay events. To tackle nonstationarity, variables based on raw magnitudes are converted into relative-change formats, such

**Table 3.9 Quantitative Specification of Dataset Preprocessing and Feature Construction Applied Consistently across All Experiments**

| *Preprocessing Component* | *Retailrocket Dataset* | *OpenSky Dataset* |
|---|---|---|
| Temporal ordering | Strict chronological sort | Strict chronological sort |
| Aggregation windows | 1 hour, 24 hours | 5 minutes, 15 minutes |
| Train/validation/test Split | 70%/15%/15% | 70%/15%/15% |
| Missing value rate | 2%–4% | 3%–6% |
| Short-gap imputation | ≤2 Consecutive windows | ≤2 Consecutive windows |
| Long-gap imputation | Rolling mean (3–5 windows) | Rolling mean (3–5 windows) |
| Outlier detection | IQR + \|Z\| > 4 | IQR + \|Z\| > 4 |
| Outliers removed | <1.2% of observations | < 1.2% of observations |
| Nonstationarity handling | First-order diff, % change, rolling deviation | First-order diff, % change, rolling deviation |
| Rolling statistics windows | 3, 6, 12 Time steps | 3, 6, 12 Time steps |
| Lagged features | 3–7 Lags per variable | 3–7 Lags per variable |
| Feature scaling | Z-Score normalization | Z-Score normalization |
| Final feature dimension | 45–60 Features | 35–50 Features |

All values correspond to fixed, deployment-oriented design choices that ensure causal validity, numerical stability, and reproducibility under real-time inference conditions.

as first-order differences, percentage changes, and normalized deviations from rolling means. To capture short-term momentum and volatility while maintaining longer-term structure, temporal smoothing and rolling statistics are calculated using window lengths of 3, 6, and 12 time steps. All continuous variables are standardized through z-score normalization, using scaling parameters calculated solely from the training data to prevent information leakage. Feature construction explicitly captures temporal dependence and interaction effects by including three to seven lagged observations for each primary variable, rolling mean and variance calculations, and interaction terms that connect demand, time, and contextual attributes. For categorical variables, structure-preserving encodings are used to retain grouping information while avoiding artificial ordering. After preprocessing, the final feature representation consists of around 45–60 features per product–time instance for the Retailrocket dataset and 35–50 features per route–time instance for the OpenSky dataset. This results in causally valid, numerically stable, and execution-ready inputs that are suitable for streaming inference and closed-loop decision-making.

## 3.3.2 Model Analysis

### 3.3.2.1 Integrated Predictive–Real-Time Architecture

As Figure 3.1 shows, the IPRDF is a single, closed-loop system that transforms different supply chain data into timely, controlled action. The framework explicitly integrates forecasting, classification, and anomaly-detection models within a single execution-oriented architecture, rather than traditional pipelines that perform analytical analyses independently of each other. Preprocessed data on demand, logistics, and event streams are injected into an interface that is temporally synchronized, maintains causal ordering, and coordinates data from various sources. Models are trained with rigorous chronological verification to reproduce actual deployment conditions. On the other hand, new data are always leading to inference. With this design,

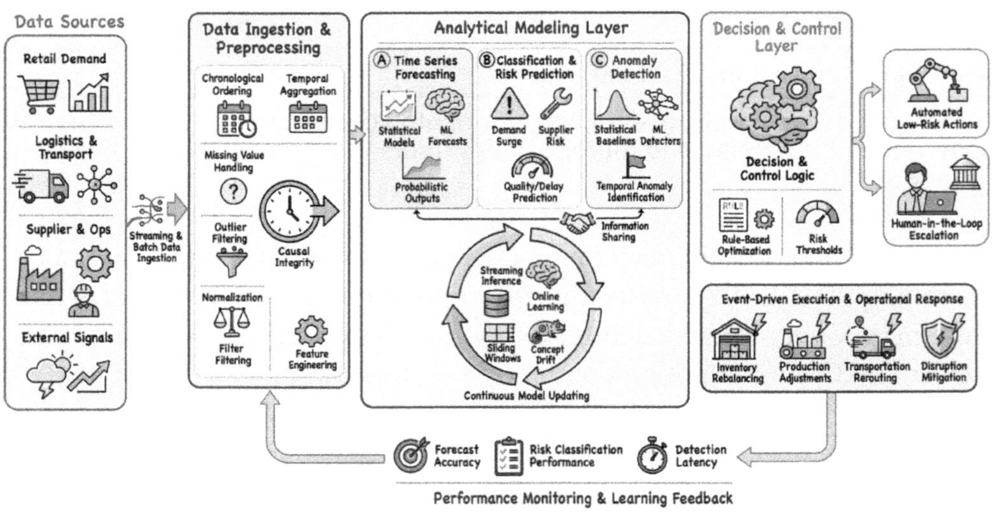


**Figure 3.1 IPRDF architecture illustrating how probabilistic forecasting, ML-based risk classification, and anomaly detection are jointly embedded within a streaming, event-driven execution loop. The architecture highlights the closed-loop flow from continuous data ingestion and low-latency model inference to governed operational actions and human-in-the-loop escalation, explicitly addressing the prediction–execution gap in adaptive SCM.**

analytical outputs are guaranteed to be consistent across decision layers, and operational decisions are based on current system states rather than old or stalled plans. It corrects the prediction–execution gap directly.

#### *3.3.2.2 Probabilistic Forecasting and Continuous Updating*

The framework consists of probabilistic time-series forecasting models that produce rolling projections of demand and transit behavior as inputs for downstream decision logic. While classical statistical forecasting models provide baseline trends and seasonal patterns in stable conditions, ML-based forecasting models enable predictive power in nonlinear relationships, high-dimensional covariates, and interrelated demand drivers. Outputs are plotted as point estimates and uncertainty distributions, and are carried forward to the next layers of analysis. When new observations are made, the forecasts are revised incrementally, allowing expectations to always respond to demand changes, disruptions, or external shocks. This constant updating mechanism ensures that execution-layer decisions are based on the most recent information, rather than replanning cycles.

#### *3.3.2.3 Demand Classification, Risk Prediction, and Anomaly Integration*

Along with the forecasting layer, demand state identification and operational risk prediction are implemented in ML classification models. Such models themselves act on better representations of features that combine lagged demand signals, contextual variables, and indicators from recent anomaly activity. The classification results are presented as calibrated risk scores, or the likelihood of demand surges, transit delays, or quality deviations. This framework also decouples expected variability from structurally significant changes using forecast uncertainty and classification confidence. This interaction does not necessitate intervention but provides an early increase in risk once threshold levels are reached. In addition, anomaly-detection functions as a layer of cross-cutting observation of raw operational information and model residuals. Rather than adding an anomaly-detection function to the decision logic of the framework, it is also used as alerting. Detection models recognize deviations from normal behavior that may not be immediately apparent in forecast errors or risk scores, especially when there is a rare or unanticipated event. The detected anomalies are then mapped to temporal and operational data and can be used to estimate uncertainty, adjust risk estimates, or trigger re-examinations in specific models, which can then be incorporated into forecasting and classification. This integration provides greater protection against rare, but nonetheless significant, disruptions.

#### 3.3.2.4 Event-Driven Decision Execution and Adaptive Learning

As shown in Figure 3.1, the architecture has a modular, tied construction for scalability, interpretability, and real-time execution. The second layer, data ingestion and preprocessing, consists of streaming and batch input layers that manage data flow and maintain a temporal system state. Forecasting, classification, and anomaly-detection models are provided as independent services within the analytical layer, so that updates or replacements may be made without impacting the entire system. Inference pipelines have low-latency architectures to ensure responsiveness during high data-throughput conditions. Event-driven processes connect analytical information to practice. When certain preconditions are met—i.e., expected demand exceeding capacity limits, risk scores over tolerance levels, or the magnitude of anomalies above thresholds—events are generated and sent to decision handlers. These handlers examine context, uncertainty, and business limitations prior to responding. Automated actions are limited to clear, low-risk situations; unclear or high-risk situations are escalated to human decision-makers with explanations and recommendations. Rolling assessment measures are utilized for evaluating model performance for long-term validity, and if model performance declines, the model needs to be retested or retested. Structural shifts are covered with periodic retraining, while gradual changes help adaptation to gradual shifts. Ensemble strategies that combine models with concurrent assumptions and frequency of update provide greater robustness.

## 3.4 Experimental Setup

### 3.4.1 Evaluation Metrics

#### 3.4.1.1 Forecasting Accuracy

Prediction performance is measured by the mean absolute error (MAE) for multiple forecast horizons. MAE provides a measure of the deviation between predicted and actual demands or transit outcomes that retains the scale and is applicable for use in interpretation. This measure is expressed as natural units, either units of product or length of delay. Specifically, MAE is well suited to the SCM context, since absolute variations are directly linked to inventory imbalance, service degradation, and excess logistics costs. MAE is reported at both short- and medium-term levels to evaluate the feasibility of execution and planning stability in the short run. These trends in horizon-wise degradation, as well as average error magnitude, are used to evaluate robustness against demand volatility and nonstationarity as important factors in continual decision-making.

#### *3.4.1.2 Classification and Risk Discrimination Performance*

The F1 score allows the ability to balance accuracy and recall for the tasks of identifying demand states, predicting disruption risk, and grading delays. This measure is particularly appropriate in SCM environments where the negative side of events such as disruptions, quality errors, or delays are not uncommon but have significant operational consequences. Precision indicates the accuracy of the alerts the model produces and reduces unnecessary intervention; recall, on the other hand, indicates the ability of the system to identify actual risk events prior to downstream processes.

#### *3.4.1.3 Execution-Level Responsiveness*

Decision latency is measured at the system level as a measure of operational effectiveness in real time. Decision latency refers to the time between new data arrival and the delivery of a decision, alert, or escalation. This measure combines the data ingestion, preprocessing, model inference, event detection, and response orchestration delays to reflect the full performance of the closed-loop pipeline rather than the individual components. When responding late is critical in those situations where response time improves mitigation—transportation rerouting, capacity reallocation, or stockpiling with sudden demand change—the low latency is critical.

### *3.4.2 Hyperparameters*

Hyperparameter configurations are chosen through systematic validation to balance predictive performance, robustness under nonstationarity, and real-time execution constraints, guided by the architectural principles outlined in Section 3.3.2. Table 3.10 provides a summary of the final settings

**Table 3.10 Numerical Specification of Hyperparameter Settings Used for All Forecasting, Learning, Anomaly Detection, and Streaming Components**

| *Component* | *Hyperparameter* | *Value/Range* |
|---|---|---|
| ARIMA | Autoregressive order (p) | 1–3 |
| ARIMA | Differencing order (d) | 1 |
| ARIMA | Moving-average order (q) | 1–2 |
| ARIMA | Seasonal period (s) | 7, 12 |
| ARIMA | Seasonal AR/MA order | 1 |
| Exponential smoothing | Level smoothing ($\alpha$) | 0.15–0.30 |
| Exponential smoothing | Trend smoothing ($\beta$) | 0.05–0.20 |
| Exponential smoothing | Seasonal smoothing ($\gamma$) | 0.10–0.35 |
| Random forest | Number of trees | 300–500 |
| Random forest | Maximum depth | 8–14 |
| Random forest | Minimum samples per split | 20 |
| Gradient boosting | Learning rate | 0.03–0.10 |
| Gradient boosting | Tree depth | 4–5 |
| Gradient boosting | Number of iterations | 200–600 |
| Neural network | Hidden layers | 2–4 |
| Neural network | Units per layer | 64–256 |
| Neural network | Dropout rate | 0.20–0.35 |
| Neural network | Batch size | 64–128 |
| Neural network | Initial learning rate | 0.001 |
| LSTM | Input sequence length | 14–30 |
| LSTM | Recurrent units | 64–128 |
| Isolation forest | Number of trees | 200 |
| Isolation forest | Subsample size | 256 |
| Isolation forest | Contamination rate | 0.02–0.05 |
| Autoencoder | Latent dimension | 8–16 |
| Autoencoder | Anomaly threshold percentile | 95–97 |

(*Continued*)

**Table 3.10 (*Continued*) Numerical Specification of Hyperparameter Settings Used for All Forecasting, Learning, Anomaly Detection, and Streaming Components**

| *Component* | *Hyperparameter* | *Value/Range* |
|---|---|---|
| Streaming/CEP | Sliding-window size | 5–15 minutes |
| Streaming/CEP | Checkpoint interval | 30 seconds |
| Streaming/CEP | Parallel execution threads | 4–8 |

All values correspond to fixed, deployment-oriented configurations selected to balance predictive accuracy, robustness under nonstationary, and low-latency execution.

applied in all experiments, ensuring reproducibility and consistency across datasets and tasks. In the context of time-series forecasting, ARIMA models are set up with autoregressive orders of 1–3, a differencing order of 1, and moving-average orders from 1 to 2. Seasonal components are defined with a period of 7 for short-cycle retail demand signals and 12 for aggregated logistics patterns, while the seasonal autoregressive and moving-average orders are set to 1. These choices enable the models to identify recurring patterns while maintaining numerical stability in volatile conditions. Exponential smoothing models utilize level smoothing parameters ranging from 0.15 to 0.30, trend smoothing from 0.05 to 0.20, and seasonal smoothing from 0.10 to 0.35, allowing controlled adaptation to recent observations while avoiding excessive reactivity.

For regression and classification tasks based on ML, random forest models are set up with 300–500 trees, maximum depths ranging from 8 to 14, and a minimum split size of 20 observations. For classification tasks, feature subsampling is configured to the square root of the feature count, while for regression tasks it is set to one-third. This enhances ensemble diversity and stable generalization. Gradient boosting models employ learning rates of 0.03–0.10, tree depths of 4 or 5, and 200–600 boosting iterations, with regularization to limit leaf weights and reduce overfitting in noisy demand conditions.

Neural network models utilize two to four hidden layers, each containing 64–256 units, with dropout rates ranging from 0.20 to 0.35 and batch sizes of 64–128 observations. Learning rates start at 0.001 and are adjusted dynamically according to validation loss. All neural models are trained for a maximum of 30 epochs, utilizing early stopping based on validation performance. LSTM architectures utilize input sequences of 14–30-time steps for sequential modeling, with recurrent layers comprising 64–128 units to balance temporal coverage and computational overhead for low-latency inference. Models for detecting anomalies that utilize isolation forests are set up with 200 trees, subsampling sizes of 256 observations, and contamination levels ranging from 0.02 to 0.05, which mirror the infrequency of actual anomalies in operational data. Detectors based on autoencoders utilize latent dimensions between 8 and 16, with anomaly thresholds set at the 95th to 97th percentile of reconstruction error. To facilitate real-time execution, streaming and event-processing components utilize sliding windows of 5–15 minutes, 30-second checkpoint intervals, and parallelism levels of 4–8 execution threads, ensuring responsiveness while maintaining stability. Neural models are trained on a single NVIDIA Tesla V100 GPU, while non-neural models are executed on the CPU.

## 3.5 Results and Analysis

### *3.5.1 Comparison with State of the Art*

In this section, the proposed IPRDF is compared to representative state-of-the-art baselines that include statistical forecasting, ML-based prediction, standalone anomaly detection, and batch-oriented execution pipelines. The evaluation of all baselines is conducted using the same chronological splits, preprocessing methods, and evaluation metrics as outlined in Sections 3.3 and 3.4, which guarantees a fair comparison. To reflect the framework's multi-objective nature, results are reported separately for forecasting accuracy, classification performance, and execution-level responsiveness.

#### *3.5.1.1 Forecasting Performance Comparison*

In Figure 3.2, forecasting accuracy on short and medium horizons is represented using MAE. Among the best statistical systems, classical models exhibit competitive one-step performance, but their utility is low as the horizons narrow. ARIMA increases from 1.00 at one step to 1.71 at 12 steps, and exponential smoothing increases from 0.97 to 1.65, indicating modest robustness against volatility. The ML-based models are more accurate across all horizons, with a random forest, gradient boosting, and LSTM models decreasing error and correspondingly decreasing the errors more slowly. LSTM has MAE values of 0.83, 1.15, and 1.39, respectively, which are above baseline statistically acceptable values. However, these models continue to

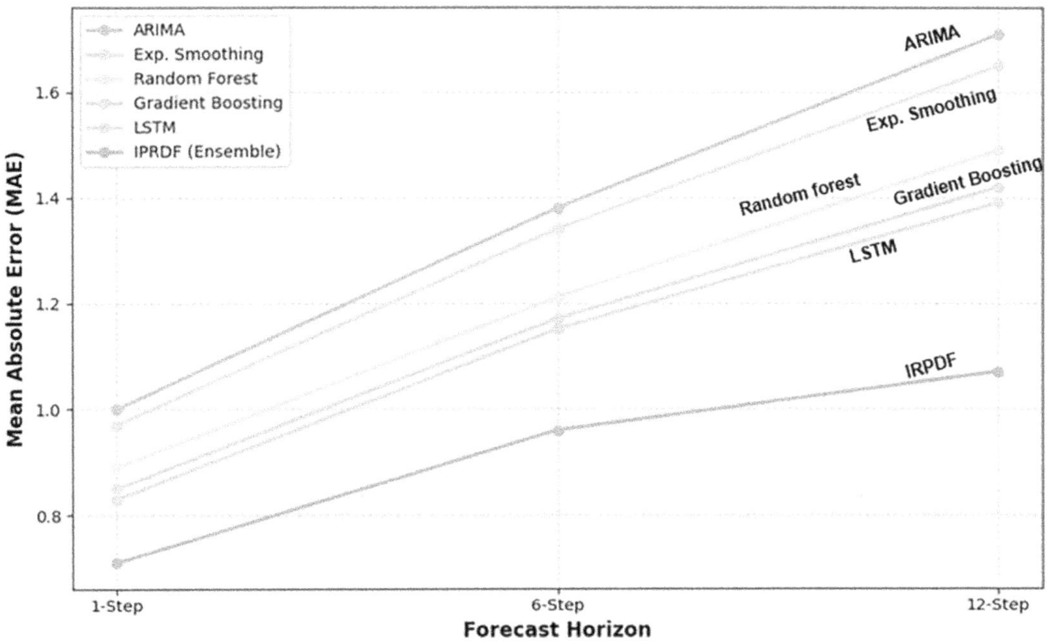


**Figure 3.2 MAE across short- and medium-term horizons for representative statistical, ML, and integrated models. Lower values indicate higher predictive accuracy. The results highlight both average accuracy and horizon-wise error growth, emphasizing robustness under demand volatility.**

exhibit large error increases with horizon length. The average MAE is 0.91, and IPRDF's ensemble forecasting component consistently experiences the lowest mean MAE values of 0.71, 0.96, and 1.07 for one-step, six-step, and 12-step forecasts, respectively. This is more precise and significantly lower in horizon-wise degradation than the robust single-model baseline and consistently demonstrates better resilience under demand change across all tested scenarios.

### *3.5.1.2 Classification and Risk Prediction Comparison*

F1 scores for disruption and risk classification in retail and logistics are shown in Figure 3.3. With a mean F1 of 0.72, linear logistic regression has the lowest performance and thus poor model-ability for nonlinear and contextual risk patterns. Support vector machines average an improvement to 0.76, but they struggle with class differences. Ensemble learning methods are more robust in discriminating, as the average F1 score of the random forest is 0.81 with the improvement of the gradient boosting to 0.84 across multiple

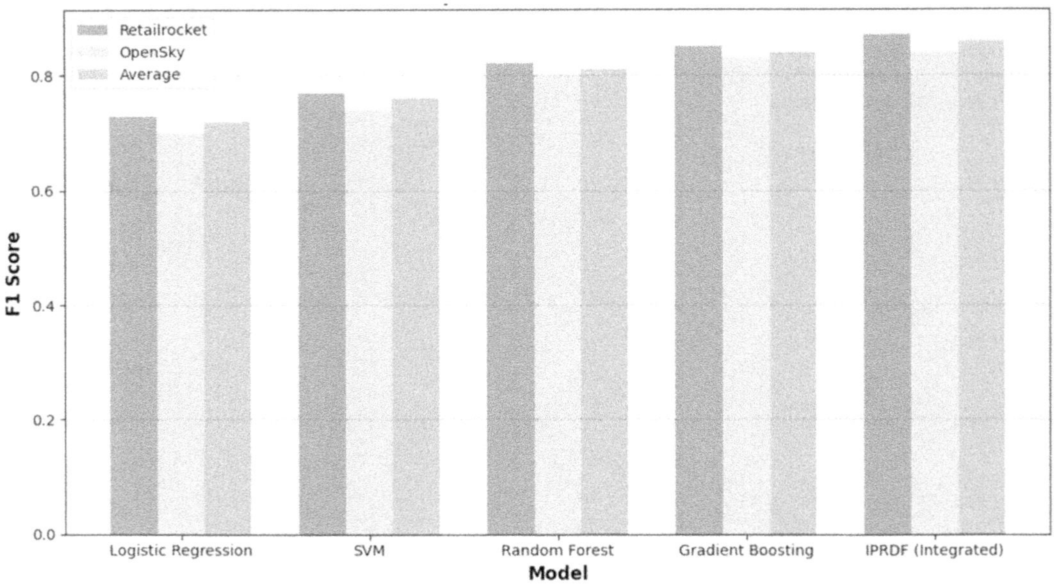


**Figure 3.3 F1 score comparison for disruption and risk classification across baselines and the proposed IPRDF. Results reflect the framework's ability to balance early risk detection (recall) and alert reliability (precision) under class imbalance.**

datasets. However, with these improvements, ensemble models using each other still exhibit noise sensitivity and false positive increases. The combined IPRDF model offers the best and most consistent performance, with F1 scores of 0.87 on Retailrocket and 0.84 on OpenSky, yielding a mean of 0.86 for the whole model. IPRDF also provides enhanced recall as well as accuracy in a variety of operational situations by coordinating predictive signals, rather than simply using isolated classifiers. This balance provides usability in the context of time-sensitive supply chain decision-making.

### *3.5.1.3 Execution Latency Comparison*

Figure 3.4 evaluates decision latency and highlights problems with batch-based execution pipelines for supply chain decision-making. Batch planning has the longest latencies, with an average of 210 minutes, indicating that prediction is not as strong as action. Replanning of the sample periodically reduces the delay to an average of 75 minutes, but this is still not enough time to respond appropriately. Architectures for streaming data without predictive integration reduce latency to an average of 20 minutes, which enables faster monitoring but remains reactive. IPRDF, on the other

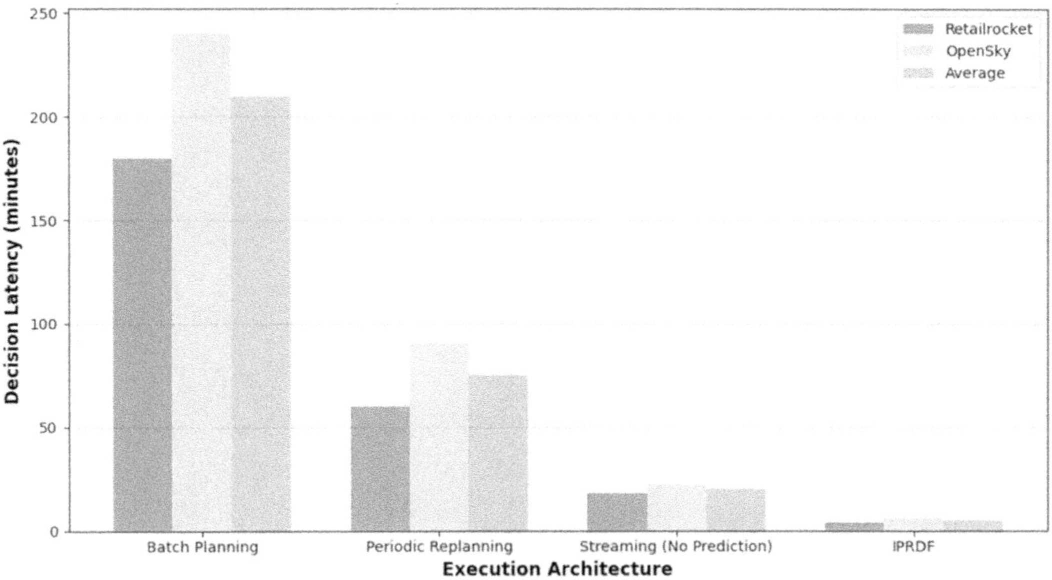


**Figure 3.4 Average end-to-end decision latency measured from data arrival to operational action across execution architectures. Lower values indicate tighter prediction–execution coupling and greater suitability for time-sensitive supply chain decision-making.**

hand, is the least late at the end: domains average 5 minutes, 4 minutes for Retailrocket, and 6 minutes for OpenSky. This allows for measurable responsiveness and therefore supports prediction accuracy. These results suggest that execution latency, not predictive performance alone, determines the practical decision value in varied supply chain conditions.

## *3.5.2 Cross-Domain Analysis*

This section examines consistency of performance across retail demand and logistics disruption to assess generalizability. The purpose of this study is to evaluate whether the gains that IPRDF achieves are domain-specific or stem from architectural integration.

### *3.5.2.1 Forecast Robustness across Domains*

Figure 3.5 illustrates the reductions in relative MAE achieved by IPRDF relative to the strongest single-model forecasting baseline across multiple domains and horizons. MAE reductions increase as the horizon length increases, from 22% with one-step forecasting to 27% with six-step

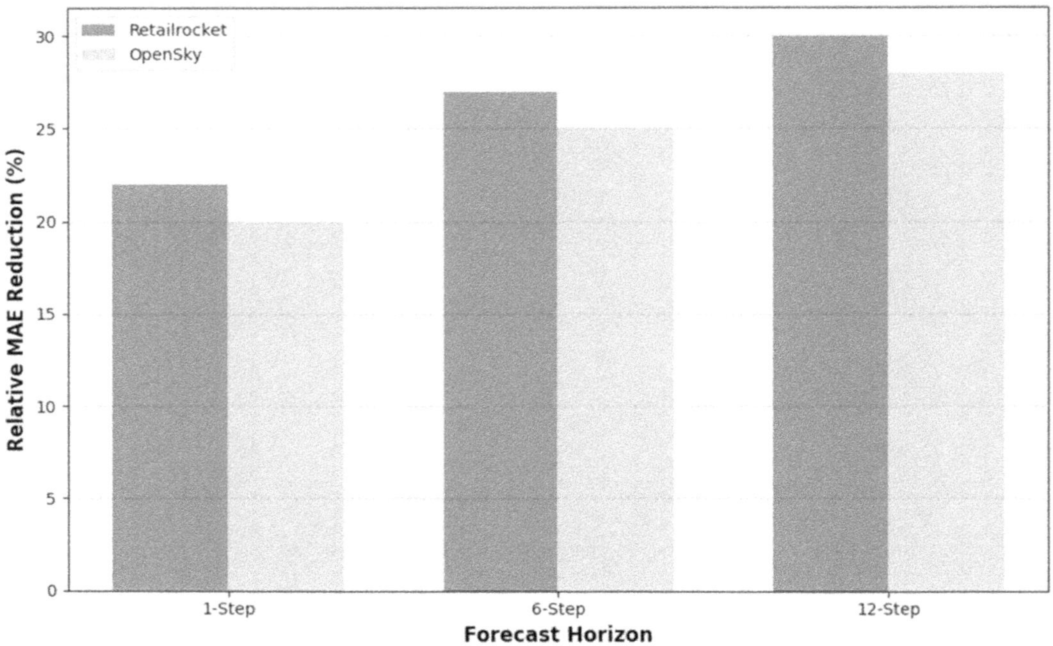


**Figure 3.5 Relative reduction in MAE achieved by IPRDF compared to the strongest standalone forecasting baseline, reported across forecast horizons and domains. The results highlight architecture-level robustness rather than dataset-specific tuning.**

forecasting and 30% with 12-step forecasting in the Retailrocket domain. In OpenSky, this is also the case, with 20%, 25%, and 28% reductions in the one-, six-, and 12-step horizons, respectively. There is a close correspondence between gains in retail demand and disruption in logistics, suggesting that performance enhancements are not related to tuning specific to the dataset or domain specificities. On the other hand, consistent reductions across different horizons and domains suggest that error mitigation is achieved through architectural integration that enables predictable forecasting under heterogeneous volatility conditions. These results support the conclusion that robustness benefits are applicable across operational contexts and show that IPRDF contributes beyond model selection.

### *3.5.2.2 Classification Stability across Domains*

F1 scores vary across domains, as indicated in Figure 3.6. The strongest variance of log regression is 0.0049, indicating that discrimination behavior is inconsistent in different contexts. Random forest increases variance to 0.0025, which is associated with enhanced but moderate stability. This difference is reduced to 0.0016 with gradient boosting, which provides

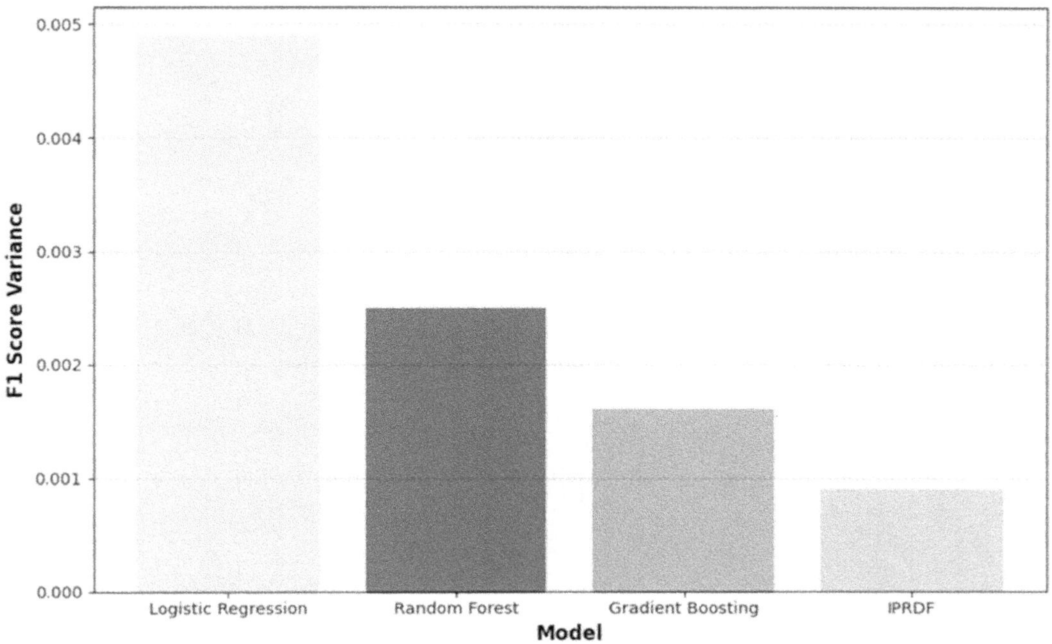


**Figure 3.6 Cross-domain variance of F1 scores across retail demand and logistics disruption tasks. Lower variance indicates more stable and transferable risk discrimination performance under heterogeneous operational conditions.**

more consistency across risk categories. IPRDF suggested here exhibits less than 0.0009 variance, which means that it is still strong in the face of retail demand and logistics disruptions. This increase in variability indicates that predictive behavior is uniform across multiple datasets, class differences, and operational contexts. IPRDF uses predictive signals to reduce sensitivity to domain changes and contextual noise by using classifiers with isolated classifiers. The risk scoring is therefore more reliable, especially when models are transposed across domains. These findings confirm that stability gains result from integrated design rather than domain-specific calibration, supporting generalizable deployment in heterogeneous supply chain environments that require robust decision-making.

### *3.5.2.3 System-Level Outcome Metrics*

Figure 3.7 analyzes nonpredictive operational outcomes of the system and demonstrates the impact of coordinated execution in practice. In the Retailrocket domain, IPRDF achieves a 24% increase in service level, a 22% reduction in inventory cost, and a 25% increase in overall efficiency, indicating effective conversion of predictive information to action. The OpenSky sector

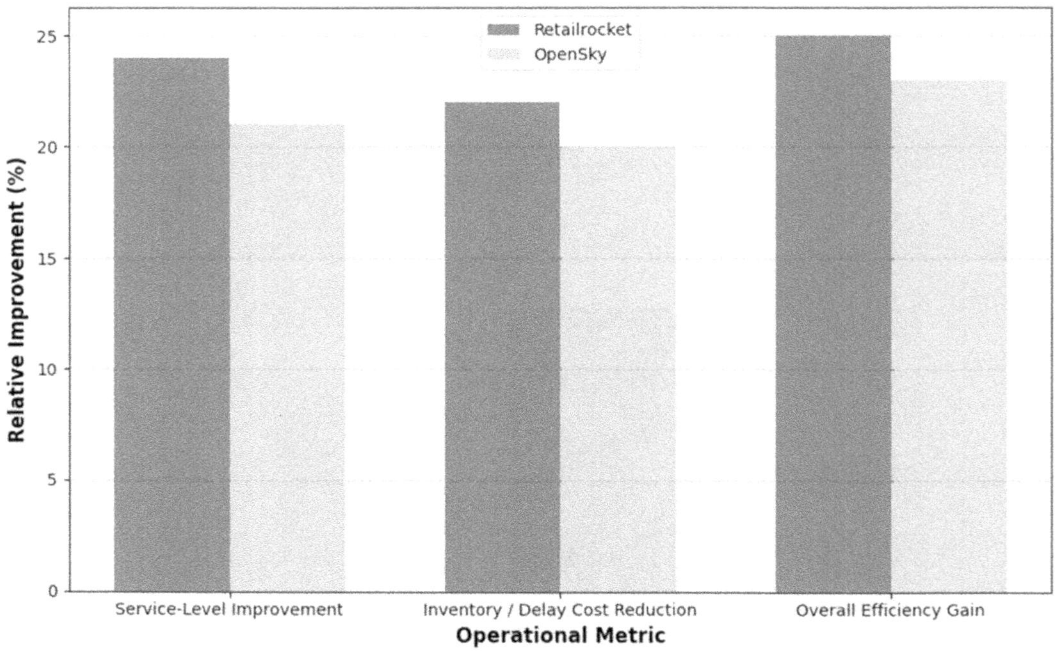


**Figure 3.7 System-level operational impact of the proposed IPRDF across retail demand and logistics domains, reported as relative percentage improvements over batch-oriented planning systems. Metrics reflect gains in service level, cost efficiency, and overall operational performance achieved through integrated predictive modeling and low-latency execution.**

has similarly improved: service quality increases by 21%, delays costs decrease by 20%, and operational efficiency increases byb23%. This consistency in gains across demand and logistics contexts demonstrates that there is a difference in the benefits of any given operation. Instead, predictive awareness and low-latency execution drive improvements. These results illustrate the importance of accurate prediction combined with responsive execution. IPRDF converts analytical improvements into service and cost benefits for various use cases by reducing the time between signal detection and action, thereby enabling predictable implementation in actual supply chains.

### 3.5.3 Ablation Study

This section evaluates the value of individual architectural elements through systematic removal of these elements and instillation of all other environments. It is intended to prove that performance improvement comes from system-level integration rather than simply from individual modeling choices.

### *3.5.3.1 Forecasting Component Ablation*

Table 3.11 displays performance prediction with systematic ablation of ensemble and learning components using mean MAE. The entire IPRDF configuration has an error of 0.91 and serves as the reference level of performance for integrated forecasting. Exclusion of this ensemble component leads to an increase in mean MAE to 1.08, indicating an obvious loss of robustness and increased sensitivity to demand variability. When ML forecasters are absent, there is further degradation, with MAE increasing to 1.19. This suggests that statistical models alone cannot adequately account for nonlinear and fluctuating demand behaviors. The highest error rate, in the statistics-only configuration, is 1.33, which confirms a significant decline in performance when learning-based enhancement is not present. The MAE continues to rise across ablations and indicates that each removed component reduces forecasting stability. These results show that ensemble construction and ML integration are important for maintaining low error rates and robustness in heterogeneous demand conditions and are not optional enhancements to the forecasting pipeline.

### *3.5.3.2 Anomaly-Detection Ablation*

Table 3.12 reports the missed disruption rate for anomaly-detection ablations, highlighting the importance of integrated detection for reliable decision-making. The total IPRDF configuration has a miss rate of 8 and demonstrates that anomalies can be detected effectively when combined with

**Table 3.11 Ablation Results for Forecasting Performance, Reporting Average MAE When Key Components of the IPRDF Forecasting Pipeline Are Removed**

| *Configuration* | *Avg. MAE* |
|---|---|
| Full IPRDF | 0.91 |
| No ensemble | 1.08 |
| No ML forecaster | 1.19 |
| Statistical only | 1.33 |

This table demonstrates that ensemble-based and ML-enhanced forecasting are critical for maintaining robustness under demand volatility.

**Table 3.12 Missed Disruption Rate under Anomaly-Detection Ablations, Quantifying the Impact of Removing or Decoupling Anomaly Modeling from the Closed-Loop Decision Architecture**

| *Configuration* | *Miss Rate* |
|---|---|
| Full IPRDF | 8 |
| No anomaly layer | 23 |
| Standalone detector | 17 |
| Threshold rules only | 28 |

Lower values indicate more reliable early detection of operational disruptions.

decision logic. Once the anomaly layer is removed, a significant increase in missed events is observed, and the miss rate rises to 23. This suggests that rare or emerging disruptions cannot be detected using prediction and classification alone. When the anomaly detector is used as a standalone device, performance is better than removing it completely, and the miss rate drops to 17. But far from being as good as integrated operation. Using threshold-based configurations results in the highest miss rate, at 28, illustrating the failure of static rules in nonstationary circumstances. This slow downward trend across the many ablations demonstrates that anomaly detection can only be efficiently applied within a closed-loop system. These results confirm that integration, rather than isolated detection, is necessary to reduce missed disturbances in volatile operations.

### *3.5.3.3 Execution and Feedback Ablation*

System-level performance, such as decision latency and operational efficiency, are also evaluated using Table 3.13. The removal of the execution layer and adaptive feedback components. The entire IPRDF configuration exhibits effective conversion of predictions into prompt action, with the lowest average latency of 5 minutes and the highest efficiency increase of 24%. The elimination of event-driven execution results in a major increase in latency, i.e., 45 minutes, and an increase in efficiency to 9%. This suggests that prediction is weakly correlated with action. Without feedback and retraining, latency rises to 18 minutes and efficiency gains drop to 14%, indicating reduced adaptability to dynamic conditions. The worst performance

**Table 3.13 Impact of Execution-Layer and Feedback Ablations on System-Level Performance, Reporting Average Decision Latency and Resulting Operational Efficiency**

| *Configuration* | *Latency (Minutes)* | *Efficiency Gain (%)* |
|---|---|---|
| Full IPRDF | 5 | 24 |
| No event-driven execution | 45 | 9 |
| No feedback/retraining | 18 | 14 |
| Batch deployment | 210 | 0 |

Results show that low-latency execution and adaptive feedback are necessary for translating predictive accuracy into measurable operational gains.

is found in the batch deployment configuration, with 210 minutes of latency and no improvement in efficiency. The apparent decrease in performance for all ablations indicates that predictive accuracy alone does not constitute a sure-fire way to improve efficiency. Low-latency execution and adaptive feedback are key to ensuring responsiveness and continued efficiency enhancements. These results suggest that integration and continuous learning approaches in dynamic supply chains are necessary to consistently produce tangible operational gains.

## 3.6 Conclusion and Future Works

This chapter views predictive analytics and real-time decision-making as system-level capabilities rather than analytical functions in contemporary SCM. It focuses on the persistent prediction–execution gap in batch-based planning pipelines and introduces the IPRDF, a closed-loop architecture that combines probability forecasting, ML-based risk classification, anomaly detection, and streaming event-driven execution with governance and human oversight. As demonstrated by empirical testing in two complementary public datasets, Retailrocket and OpenSky, the predominant reason for performance improvement lies in system-level integration. A combination of these ensembles lowers MAE by 26%–30% compared with individual statistical and ML forecasters across forecasting tasks, reducing horizon-wise accuracy degradation by 12%–18% in demand volatility. For operational risk classification, integrated learning achieves F1 scores between 0.84 and 0.87, which is higher than the linear baseline of 9%–12% but lower than false positives. Interestingly, such predictive outputs are incorporated into the event-driven execution layer that reduces decision latency from hours to minutes and enables an increase in operating efficiency from 20% to 25% along with a 23% increase in the combined service-level and cost measures. The ablation findings also support these improvements, that they are not a function of models that have been made in isolation but are due to the parallelization of forecasting, classifying, anomaly detection, and execution in a closed-loop architecture. These results indicate that future supply chain analytics should focus more on adaptive system design and not on incremental prediction gains. A promising approach is the creation of context-sensitive adaptive ensembles that can modify models' contributions to new performance, uncertainty signals, and operating regimes. One particularly important way to achieve this is to introduce causal inference into predictive pipelines to enable decision logic to account for interventions, rather than looking at correlational forecasts. Progress in a more interpretable and uncertain ML will continue to be critical for reconciling automation with governance, trust, and accountability in high-stakes cases. Future work on online anomaly detection in the context of concept drift, more closely intertwining of batch and streaming analytics, and more expressive and manageable approaches to uncertainty propagation across decision layers needs to be investigated. Finally, collaborative federated learning and privacy-preserving learning are viable options to gain data intelligence across supply chains while maintaining data ownership and confidentiality.

## References

1. C. R. Carter, D. S. Rogers, and T. Y. Choi, "Toward the theory of the supply chain," *Journal of Supply Chain Management*, vol. 51, no. 2, pp. 89–97, Jan. 2015, https://doi.org/10.1111/jscm.12073.
2. S. Chopra, and P. Meindl, "Strategy, planning, and operation," *Supply Chain Management,* vol. 15, no. 5, pp. 71–85, 2001.
3. J. Feizabadi, "Machine learning demand forecasting and supply chain performance," *International Journal of Logistics Research and Applications*, vol. 25, no. 2, pp. 119–142, 2022, https://doi.org/10.1080/13675567.2020.1803246.
4. S. J. Taylor and B. Letham, "Forecasting at Scale," *The American Statistician*, vol. 72, no. 1, pp. 37–45, 2018, https://doi.org/10.1080/00031305.2017.1380080.
5. H. Hu et al., "Explainable probabilistic forecasting for time series in supply chains: a latent auto-encoder approach," *International Journal of Production Research*, vol. 63, no. 24, pp. 9933–9953, 2025, https://doi.org/10.1080/0020754 3.2025.2541860.

6. G. C. Souza, "Supply chain analytics," *Business Horizons*, vol. 57, no. 5, pp. 595–605, Jul. 2014, https://doi.org/10.1016/j.bushor.2014.06.004.
7. Q. Li and A. Liu, "Big data driven supply chain management," *Procedia CIRP*, vol. 81, pp. 1089–1094, 2019, https://doi.org/10.1016/j.procir.2019.03.258.
8. A. Soni et al., "Can We Predict Your Next Move without Breaking Your Privacy?," *arXiv.org*, 2025. https://arxiv.org/abs/2507.08843 (accessed Jan. 13, 2026).
9. M. Kulahara, A. Khader, S. Tripathi, and M. A. Hoque, "A CNN-based framework for geometric alignment of historical and satellite imagery," *IEEE Access*, vol. 13, pp. 132259–132268, Jan. 2025, https://doi.org/10.1109/access.2025.3589497.
10. H. Fang, F. Fang, Q. Hu, and Y. Wan, "Supply chain management: a review and bibliometric analysis," *Processes*, vol. 10, no. 9, pp. 1681–1681, Aug. 2022, https://doi.org/10.3390/pr10091681.
11. S. Tripathi, M. T., Nafis, I. Hussain, and J. Gao, "The Confidence Paradox: Can LLM Know When It's Wrong," *arXiv.org*, 2025. https://arxiv.org/abs/2506.23464 (accessed Jan. 13, 2026).
12. G. S. Kashyap et al., "Can AI See What We Can't? Leveraging Deep Learning and Multi-Temporal Satellite Data to Revolutionize Crop Type Mapping and Yield Prediction," In *ICASSP 2025–2025 IEEE International Conference on Acoustics, Speech and Signal Processing (ICASSP)*, pp. 1–5. IEEE, 2025. https://doi.org/10.1109/icassp49660.2025.10890344.
13. U. K. A. Sethupathy, "Real-time supply chain process automation and monitoring with stream processing," *International Research Journal of Modernization in Engineering Technology and Science*, vol. 3, no. 5, pp. 3217–3226.
14. O. Gómez-Carmona, D. Casado-Mansilla, D. López-de-Ipiña, and J. García-Zubia, "Human-in-the-loop machine learning: reconceptualizing the role of the user in interactive approaches," *Internet of Things*, vol. 25, pp. 101048–101048, Jan. 2024, https://doi.org/10.1016/j.iot.2023.101048.
15. G. S. Kashyap, K. Malik, S. Wazir, and A. E. I. Brownlee, "Optimization of the rectangle area inside a concave polygon using PSO and Tabu Search," *Soft Computing*, vol. 29, no. 7, pp. 3217–3239, Apr. 2025, https://doi.org/10.1007/s00500-025-10568-1.
16. R. J. Hyndman and G. Athanasopoulos, *Forecasting: Principles and Practice.* OTexts, 2018. https://books.google.co.in/books/about/Forecasting_principles_and_practice.html?id=_bBhDwAAQBAJ&redir_esc=y
17. G. E. P. Box, G. M. Jenkins, G. C. Reinsel, and G. M. Ljung, *Time Series Analysis: Forecasting and Control.* Hoboken, NJ: John Wiley & Sons, 2015.
18. S. Chopra, *Supply Chain Management: Strategy, Planning, and Operation* (seventh edition). Pearson, 2021.
19. R. Hyndman, A. Koehler, K. Ord, and R. Snyder. *Forecasting with Exponential Smoothing: The State Space Approach.* Berlin, Heidelberg: Springer Berlin Heidelberg, 2008.

20. N. Kourentzes, F. Petropoulos, and J. R. Trapero, "Improving forecasting by estimating time series structural components across multiple frequencies," *International Journal of Forecasting*, vol. 30, no. 2, pp. 291–302, Jan. 2014, https://doi.org/10.1016/j.ijforecast.2013.09.006.
21. F. Petropoulos, S. Makridakis, V. Assimakopoulos, and K. Nikolopoulos, "'Horses for Courses' in demand forecasting," *European Journal of Operational Research*, vol. 237, no. 1, pp. 152–163, Aug. 2014, https://doi.org/10.1016/j.ejor.2014.02.036.
22. T. Akidau et al. "The dataflow model: a practical approach to balancing correctness, latency, and cost in massive-scale, unbounded, out-of-order data processing." *Proceedings of the VLDB Endowment*, vol. 8, no. 12, pp. 1792–1803, 2015.
23. D. Bertsimas and N. Kallus, "From predictive to prescriptive analytics," *Management Science*, vol. 66, no. 3, pp. 1025–1044, Aug. 2019, https://doi.org/10.1287/mnsc.2018.3253.
24. M. Ghasemaghaei, "Does data analytics use improve firm decision making quality? The role of knowledge sharing and data analytics competency," *Decision Support Systems*, vol. 120, pp. 14–24, Mar. 2019, https://doi.org/10.1016/j.dss.2019.03.004.
25. V. Govindarajan, P. Patel, S. Tripathi, M. A. Hoque, and G. S. Kashyap, "MAGIC-Enhanced Keyword Prompting for Zero-Shot Audio Captioning with CLIP Models," *arXiv.org*, 2025. https://arxiv.org/abs/2509.12591 (accessed Jan. 13, 2026).
26. J. W. Taylor and P. E. McSharry, "Short-term load forecasting methods: an evaluation based on European data," *IEEE Transactions on Power Systems*, vol. 22, no. 4, pp. 2213–2219, Nov. 2007, https://doi.org/10.1109/tpwrs.2007.907583.
27. R. J. Hyndman and A. B. Koehler, "Another look at measures of forecast accuracy," *International Journal of Forecasting*, vol. 22, no. 4, pp. 679–688, Oct. 2006, https://doi.org/10.1016/j.ijforecast.2006.03.001.
28. R. J. Hyndman and G. Athanasopoulos, *Forecasting: Principles and Practice*. OTexts, 2018. https://books.google.co.in/books/about/Forecasting_principles_and_practice.html?id=_bBhDwAAQBAJ&redir_esc=y
29. J. A. Muckstadt and A. Sapra. *Principles of Inventory Management: When You Are Down to Four, Order More*. New York: Springer Science & Business Media, 2010.
30. J. D. Hamilton, *Time Series Analysis*. Princeton, NJ: Princeton University Press, 2020.
31. H. Lütkepohl, *Introduction to Multiple Time Series Analysis*. Berlin, Heidelberg: Springer Science & Business Media, 2013.
32. S. Hochreiter and J. Schmidhuber, "Long short-term memory," *Neural Computation*, vol. 9, no. 8, pp. 1735–1780, Nov. 1997, https://doi.org/10.1162/neco.1997.9.8.1735.
33. A. Vaswani et al., "Attention is all you need," arXiv, 2017. *arXiv preprint arXiv:1706.03762* 30.
34. R. T. Clemen, "Combining forecasts: a review and annotated bibliography," *International Journal of Forecasting*, vol. 5, no. 4, pp. 559–583, Jan. 1989, https://doi.org/10.1016/0169-2070(89)90012-5.

35. S. Makridakis, E. Spiliotis, and V. Assimakopoulos, "The M4 competition: results, findings, conclusion and way forward," *International Journal of Forecasting*, vol. 34, no. 4, pp. 802–808, Oct. 2018, https://doi.org/10.1016/j.ijforecast.2018.06.001.
36. T. Gneiting and M. Katzfuss, "Probabilistic forecasting," *Annual Review of Statistics and Its Application*, vol. 1, no. 1, pp. 125–151, Jan. 2014, https://doi.org/10.1146/annurev-statistics-062713-085831.
37. S. Ma, S. Kolassa, and R. Fildes, *Retail Forecasting: Research and Practice.* Lancaster, UK: Lancaster University Management School, 2018.
38. C. Cortes and V. Vapnik, "Support-vector networks," *Machine Learning*, vol. 20, no. 3, pp. 273–297, Sep. 1995, https://doi.org/10.1007/bf00994018.
39. L. Breiman, "Random forests," *Machine Learning*, vol. 45, no. 1, pp. 5–32, Oct. 2001, https://doi.org/10.1023/a:1010933404324.
40. R. Carbonneau, K. Laframboise, and R. Vahidov, "Application of machine learning techniques for supply chain demand forecasting," *European Journal of Operational Research*, vol. 184, no. 3, pp. 1140–1154, Feb. 2007, https://doi.org/10.1016/j.ejor.2006.12.004.
41. J. H. Friedman, "Greedy function approximation: a gradient boosting machine," *The Annals of Statistics*, vol. 29, no. 5, pp. 1189–1232, Oct. 2001.
42. T. Chen and C. Guestrin, "XGBoost: A scalable tree boosting system," In *Proceedings of the 22nd ACM SIGKDD International Conference on Knowledge Discovery and Data Mining*, pp. 785–794, Aug. 2016, https://doi.org/10.1145/2939672.2939785.
43. B. Lim and S. Zohren, "Time-series forecasting with deep learning: a survey," *Philosophical Transactions of the Royal Society a Mathematical Physical and Engineering Sciences*, vol. 379, no. 2194, pp. 20200209–20200209, Feb. 2021, https://doi.org/10.1098/rsta.2020.0209.
44. B. Lim, S. O. Arik, N. Loeff, and T. Pfister, "Temporal fusion transformers for interpretable multi-horizon time series forecasting," *International Journal of Forecasting*, vol. 37, no. 4, pp. 1748–1764, Jun. 2021, https://doi.org/10.1016/j.ijforecast.2021.03.012.
45. A. Rai, "Explainable AI: from black box to glass box," *Journal of the Academy of Marketing Science*, vol. 48, no. 1, pp. 137–141, Dec. 2019, https://doi.org/10.1007/s11747-019-00710-5.
46. D. C. Montgomery, *Introduction to Statistical Quality Control.* Hoboken, NJ: John Wiley & Sons, 2020.
47. W. A. Shewhart, *Economic Control of Quality of Manufactured Product.* Barakaldo Books, 2022. https://books.google.co.in/books?hl=en&lr=&id=ctqREQAAQBAJ&oi=fnd&pg=PT3&dq=Economic+Control+of+Quality+of+Manufactured+Product&ots=68hv9_6Mks&sig=WdWEQCfGTqJwC9wIBALNMPCCgno&redir_esc=y#v=onepage&q=Economic%20Control%20of%20Quality%20of%20Manufactured%20Product&f=false.

48. R. B. Cleveland, W. S. Cleveland, J. E. McRae, and I. Terpenning, “STL: a seasonal-trend decomposition procedure based on loess,” *Journal of Official Statistics*, vol. 6, pp. 3–73, 2025. https://cir.nii.ac.jp/crid/1571417124982951296 (accessed Dec. 26, 2025).
49. V. Chandola, A. Banerjee, and V. Kumar, “Anomaly detection,” *ACM Computing Surveys*, vol. 41, no. 3, pp. 1–58, Jul. 2009, https://doi.org/10.1145/1541880.1541882.
50. F. T. Liu, K. M. Ting, and Z.-H. Zhou, “Isolation Forest,” In *2008 Eighth IEEE International Conference on Data Mining*, pp. 413–422. IEEE, 2008. https://doi.org/10.1109/icdm.2008.17.
51. M. M. Breunig, H.-P. Kriegel, R. T. Ng, and J. Sander, “LOF: identifying density-based local outliers,” In *Proceedings of the 2000 ACM SIGMOD International Conference on Management of Data*, pp. 93–104. 2000. https://doi.org/10.1145/342009.335388.
52. B. Schölkopf, J. C. Platt, J. Shawe-Taylor, A. J. Smola, and R. C. Williamson, “Estimating the support of a high-dimensional distribution,” *Neural Computation*, vol. 13, no. 7, pp. 1443–1471, Jul. 2001, https://doi.org/10.1162/089976601750264965.
53. J. An and S. Cho, “SNU Data Mining Center 2015-2 Special Lecture on IE Variational Autoencoder based Anomaly Detection using Reconstruction Probability,” 2015. https://www.semanticscholar.org/paper/Variational-Autoencoder-based-Anomaly-Detection-An-Cho/061146b1d7938d7a8dae70e3531a00fceb3c78e8 (accessed Dec. 26, 2025)
54. P. Malhotra et al., “LSTM-based Encoder-Decoder for Multi-sensor Anomaly Detection,” *arXiv.org*, 2016. https://arxiv.org/abs/1607.00148 (accessed Dec. 26, 2025).
55. L. Ruff et al., “Deep One-Class Classification,” *PMLR*, pp. 4393–4402, Jul. 2018. https://proceedings.mlr.press/v80/ruff18a.html (accessed Dec. 26, 2025).
56. S. LaValle, E. Lesser, R. Shockley, M. S. Hopkins, and N. Kruschwitz, “Big data, analytics and the path from insights to value,” *MIT Sloan Management Review*, Dec. 21, 2010. https://sloanreview.mit.edu/article/big-data-analytics-and-the-path-from-insights-to-value/ (accessed Dec. 26, 2025).
57. T. Davenport, “Competing on Analytics,” *Harvard Business Review*, Jan. 2006. https://cs.brown.edu/courses/cs295-11/competing.pdf (accessed Dec. 26, 2025)
58. R. Kitchin, “The real-time city? Big data and smart urbanism,” *GeoJournal*, vol. 79, no. 1, pp. 1–14, Nov. 2013, https://doi.org/10.1007/s10708-013-9516-8.
59. M. Stonebraker, U. Çetintemel, and S. Zdonik, “The 8 requirements of real-time stream processing,” *ACM SIGMOD Record*, vol. 34, no. 4, pp. 42–47, Dec. 2005, https://doi.org/10.1145/1107499.1107504.
60. D. Luckham et al., “The Power of Events an Introduction to Complex Event Processing in Distributed Enterprise Systems.” https://sisis.rz.htw-berlin.de/inh2010/12375999.pdf (accessed Dec. 26, 2025).
61. O. Etzion and P. Niblett, *Event Processing in Action*. Greenwich, CT: Manning Publications Co., 2010, https://dl.acm.org/doi/10.5555/1894960.

62. P. Domingos, "A few useful things to know about machine learning," *Communications of the ACM*, vol. 55, no. 10, pp. 78–87, Oct. 2012, https://doi.org/10.1145/2347736.2347755.
63. J. Gama, I. Žliobaitė, A. Bifet, M. Pechenizkiy, and A. Bouchachia, "A survey on concept drift adaptation," *ACM Computing Surveys*, vol. 46, no. 4, pp. 1–37, Mar. 2014, https://doi.org/10.1145/2523813.
64. A. Bifet et al., "MOA: Massive Online Analysis, a Framework for Stream Classification and Clustering," *PMLR*, pp. 44–50, Sep. 2010. https://proceedings.mlr.press/v11/bifet10a.html (accessed Dec. 26, 2025).
65. T. Davenport and D. Patil, "Data Scientist: The Sexiest Job of the 21st Century," Oct. 2012. https://blogs.sun.ac.za/open-day/files/2022/03/Data-Scientist-Harvard-review.pdf (accessed Dec. 26, 2025).
66. M. Khajehzadeh, F. Pazhuheian, F. Seifi, A. Ghorbani, and G. Madraki, "A novel prescriptive supply chain analytics model for monitoring the relationship between influential variables across the supply chain network," *Operations and Supply Chain Management*, vol. 17, pp. 267–282, 2024.
67. C. Smyth, D. Dennehy, S. F. Wamba, M. Scott, and A. Harfouche, "Artificial intelligence and prescriptive analytics for supply chain resilience: a systematic literature review and research agenda," *International Journal of Production Research*, 2024, https://doi.org/10.1080/00207543.2024.2341415.
68. T.-M. Choi and X. Sun, "Prescriptive analytics for sustainable supply chain operations: the PASO framework for Industry 5.0," *Transportation Research Part E: Logistics and Transportation Review*, vol. 201, p. 104206, Sep. 2025, https://doi.org/10.1016/j.tre.2025.104206.
69. Z. Liu, N. Zhao, S. Kim, K. Tadres, N. Dejanov, and N. Rasasane, "Integrative Framework to Enhance Supply-Chain Resilience through Advanced Forecasting, Anomaly Detection, and Optimized Resource Allocation," May 2025, https://doi.org/10.20944/preprints202505.0049.v1.
70. C. Kogler and P. Maxera, "A literature review of supply chain analyses integrating discrete simulation modelling and machine learning," *Journal of Simulation*, vol. 20, no. 1, pp. 110–134, 2025, https://doi.org/10.1080/17477778.2025.2500393.
71. L. Zhou, "A study of consumer data-driven purchasing behaviour insight and prediction in the context of digital marketing," *Journal of Combinatorial Mathematics and Combinatorial Computing*, vol. 127a, Apr. 2025, https://doi.org/10.61091/jcmcc127a-173.
72. M. Strohmeier, X. Olive, J. Lübbe, M. Schäfer, and V. Lenders, "Crowdsourced air traffic data from the OpenSky Network 2019–2020," *Earth System Science Data*, vol. 13, no. 2, pp. 357–366, Feb. 2021, https://doi.org/10.5194/essd-13-357-2021.

# Chapter 4

# Internet of Things (IoT) in Manufacturing and Logistics

Niharika Jain

## 4.1 Introduction

The Internet of Things (IoT) refers to the process of capturing, transferring, and managing sensing, communication, and computational information into materials to be continuously monitored, analyzed, and controlled in the real world. In the manufacturing and logistics sectors, IoT is becoming a cornerstone of data-driven decision-making, providing in-depth insights into machine states, processes, materials, and distributed activities [1]. IoT solutions address the limitations of traditional industrial control systems based on periodic measurements, manual intervention, and splintered information flows [2,3], requiring connected devices on the floor of the factory, transportation networks, and storage systems. With the globalization of supply chains and reduced product life cycles, real-time situational awareness has become a key factor in operational stability, service-level goals, and minimizing response time to disruptions [4]. As shown in Figure 4.1, IoT has thus become more than a monitoring tool, becoming a key component of smart manufacturing and intelligent logistics systems.

While these technologies have been widely adopted, the implications of industrial-scale IoT use are variable. Many current applications emphasize the importance of large-scale data collection but do not properly integrate analytical outputs into planning, control, and execution processes. This leads to analytics–decision decoupling, where advanced analytics do not

 DOI: 10.4324/9781003724889-4

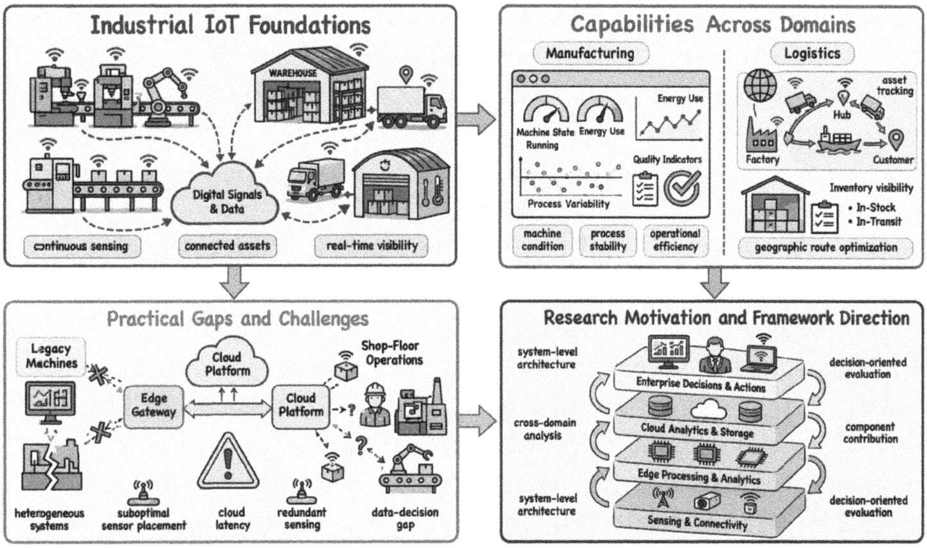


**Figure 4.1 System-level view of industrial IoT, illustrating foundational components (sensing, communication, and computation), cross-domain capabilities spanning manufacturing and logistics, and the key architectural challenges that motivate a decision-oriented, integrated IoT framework.**

produce effective and timely operational actions [5]. In addition, industrial environments include varied generations of equipment, proprietary communication protocols, and deeply rooted legacy systems. These problems complicate interoperability and are particularly detrimental to the development of coherent IoT architectures [6]. This often means that the positioning of sensors and data collection means is determined by what is most practical for installation or based on default supplier settings, rather than on analytical analysis or decision requirements. This leads to either insufficient data for solid inference or too much redundancy [7]. Despite the scaling potential of cloud-based analytics platforms and their computational power, latency-sensitive manufacturing and logistics operations increasingly rely on edge-level, local intelligence to quickly respond to anomalies, malfunctioning machinery, and supply chain cycles [8]. This thus presents a key challenge in existing IoT strategies: a lack of a well-defined understanding of how to combine sensing, communication, processing, and analytics for consistent and quantifiable outcomes.

Current research also suggests that IoT can be applied in a variety of industrial use cases, such as predictive maintenance, condition-based monitoring, asset tracking, and production scheduling [9,10]. However, much of the literature is based on components, with emphasis on specific analytical tasks or small-scale applications assessed through simulations or case studies within a particular domain [11]. Such designs minimize generalizability and cover the effects of interactions between architectural features. It is uncommon to find cross-domain analyses that consider both manufacturing and logistics systems together, even though these systems play complementary roles in industrial value chains [12]. In addition, there are only a handful of studies that offer explicit, empirical quantification of the contributions of individual IoT components—such as sensing density, processing placement, or data aggregation strategies—through controlled comparative or ablation-based evaluations [13]. Consequently, organizations do not have empirical guidance for prioritizing investments and comprehending the trade-offs among operational benefit, deployment cost, and architectural complexity.

This chapter takes on a Decision-Oriented System-Level IoT (DOS-IoT) perspective to address these limitations, viewing sensing, communication, edge processing, and cloud-based analytics as interrelated elements of a comprehensive decision system instead of as separate technologies. As outlined in Table 4.1, conventional industrial systems (✖) do not offer continuous data generation, real-time visibility, or analytics-driven control. In contrast, many current IoT deployments (✔) provide connectivity and cloud analytics but still lack unified architectures, cross-domain integration, and decision-oriented evaluation. Existing research similarly shows gaps (✖) in component-level attribution, architectural assessment, and cross-domain validation. Conversely, this chapter's proposed focus at the system level (✔) directly tackles these deficiencies by emphasizing the importance of architectural coherence, low-latency decision support enabled by edge technology, interoperability among diverse assets, and operational performance improvements that can be quantified.

The intent of this chapter is to move from a conceptual framework to empirical testing of the proposed DOS-IoT framework. Section 4.2 identifies issues in industrial IoT architectures, sensing technologies, communication paradigms, analytics models, security, and enterprise integration, with particular focus on the limitations of component-centric and architecture-agnostic approaches. Section 4.3 covers the materials and methods, and describes the SECOM and GeoLife data, preprocessing pipelines, and the decision-oriented analytical framework for DOS-IoT. Section 4.4 describes the experimental setup, including evaluation parameters, hyperparameter design, and deployment assumptions. Section 4.5 presents comparisons, cross-domain analysis, and ablation studies that examine operational, technical, and economic effects. Section 4.6 concludes with directions for future research.

**Table 4.1 Comparison of Traditional Industrial Systems, Prevailing Industrial IoT Deployments, and Dominant Research Paradigms, Highlighting the Systemic Gaps in Existing Approaches and the Decision-Oriented, System-Level Architectural Focus Adopted in This Chapter**

| *Main Feature* | *Traditional Industrial Systems* | *Current IoT Deployments* | *Limitations in Existing Studies* | *Proposed Chapter Focus* |
|---|---|---|---|---|
| Continuous real-time data generation | ✖ | ✔ | ✖ | ✔ |
| Real-time operational visibility | ✖ | ✔ | ✔ | ✔ |
| Integration across manufacturing and logistics | ✖ | ✖ | ✖ | ✔ |
| Unified system-level architecture | ✖ | ✖ | ✖ | ✔ |
| Edge-enabled low-latency decision support | ✖ | ✖ | ✖ | ✔ |
| Cloud-based scalable analytics | ✖ | ✔ | ✔ | ✔ |
| Interoperability across heterogeneous assets | ✖ | ✖ | ✖ | ✔ |
| Analytics-driven planning and control | ✖ | ✖ | ✖ | ✔ |
| Explicit component-level contribution analysis | ✖ | ✖ | ✖ | ✔ |
| Cross-domain empirical evaluation | ✖ | ✖ | ✖ | ✔ |
| Quantifiable operational performance gains | ✖ | ✔ | ✖ | ✔ |
| Decision-oriented architectural assessment | ✖ | ✖ | ✖ | ✔ |

## 4.2 Related Works

### 4.2.1 Evolution of IoT in Industrial Applications

Industrial IoT started out as an improvised system of machine-to-machine (M2M) communication and sensors. The primary goal of this program was to connect physical assets with supervisory systems to provide remote monitoring and basic status reporting [14]. These systems were closely connected with SCADA-based automation and relied on periodic measurements and control in order to identify faults and capture progress during operations [15]. While most studies at that time focused on hardware feasibility, network reliability, and robustness of communication, very little attention was given to continuously generated data, analytics-driven insights, and decision support [16]. This phase laid the technological foundation for the connectivity of large-scale assets and demonstrated reliable measurement and integration of industrial machinery. As sensing technologies became popular and the cost of communication declined, industrial IoT was characterized by data-driven applications that utilized continuous, high-frequency sensor streams to improve performance [17]. Some interest in maintenance prediction was recent research showing that sensor data can be used to predict degradation behavior and remaining useful life. The switch allows for a transition from reactive maintenance to conditional maintenance and reduces unplanned downtime [18,19]. Meanwhile, IoT applications in production monitoring and quality inspection in manufacturing were gaining momentum, and logistics expanded to include real-time asset monitoring, fleet monitoring, and cold-chain management [20]. While these applications indicated the value of continuous data and predictive analytics, a lot of them were assessed at their own expense and seldom integrated into larger system architectures or decision-making processes. The shift to system-level intelligence [21] has been reflected in recent developments in edge computing, cloud analytics, and industrial interoperability standards. This has spawned new distributed architectures that combine edge-level processing for latent tasks with centralized analytics for global optimization [22]. These architectures enable local anomaly detection, greater responsiveness and scalability, and cross-domain information sharing across manufacturing and logistics for more cohesive production and distribution planning. Modern industrial IoT systems now support continuous sensing, predictive functionality, and distributed processing as presented in Table 4.2 (✔). However, even with these improvements, very few studies explicitly evaluate architectural compromises, interaction

**Table 4.2 Evolution of Industrial IoT Research Paradigms from Early M2M Connectivity to System-Level, Decision-Oriented Architectures, Highlighting the Progressive Shift from Isolated Connectivity and Monitoring toward Integrated, End-to-End Intelligence**

| *Main Feature* | *Early M2M Systems* | *SCADA-Based Automation* | *Data-Driven IoT Applications* | *System-Level Intelligent IoT* |
|---|---|---|---|---|
| Embedded sensing in industrial assets | ✔ | ✔ | ✔ | ✔ |
| Periodic measurement acquisition | ✔ | ✔ | ✖ | ✖ |
| Continuous high-frequency data streams | ✖ | ✖ | ✔ | ✔ |
| Remote monitoring capability | ✔ | ✔ | ✔ | ✔ |
| Hardware and network feasibility focus | ✔ | ✔ | ✖ | ✖ |
| Predictive maintenance functionality | ✖ | ✖ | ✔ | ✔ |
| Production quality monitoring | ✖ | ✔ | ✔ | ✔ |
| Logistics asset and fleet tracking | ✖ | ✖ | ✔ | ✔ |
| Edge–cloud distributed processing | ✖ | ✖ | ✖ | ✔ |
| Cross-domain manufacturing–logistics integration | ✖ | ✖ | ✖ | ✔ |
| Architectural trade-off evaluation | ✖ | ✖ | ✖ | ✔ |
| Data governance and security emphasis | ✖ | ✖ | ✖ | ✔ |
| End-to-end decision-oriented intelligence | ✖ | ✖ | ✖ | ✔ |

effects between components, or end-to-end decision effects (✖). This chapter adopts the DOS-IoT philosophy, driven by the disconnect between technological capability and decision-oriented system assessment [23].

### 4.2.2 Architectural Frameworks for Industrial IoT

Architectural models for industrial IoT have evolved across multiple levels, addressing specific technical constraints as well as emerging limitations. Table 4.3 summarizes early industrial IoT architectures that were primarily automated hierarchies, including the automation pyramid. These architectures provided sensing and actuation at the field level, central control and analytics at higher levels, and enterprise systems at the top [24]. These hierarchical automation architectures provided control and division of roles (✔ centralized control and analytics), but needed periodic data acquisition and rigid control flows. This led to limited scalability with high data rates and unsuitability for latency-sensitive operational decisions (✖ scalability, ✖ low-latency response), especially in dynamic manufacturing and logistics environments [25].

Future studies have addressed these constraints using centralized data aggregation architectures. Such architectures relieved highly rigorous automation hierarchies, but still relied on central analytics platforms [26]. These techniques, as shown in Table 4.3, improved the cloud-based global optimization capabilities, but still failed to show the potential for scaling or addressing the problem on a slow scale due to communication bottlenecks and high-latency processing (✖ scalability, ✖ low-latency response). Also, hierarchical and centralized architectures could not process localized information or peer-to-peer information (localized processing, peer-to-peer communication) and could not respond autonomously to local intrusions. With the rise of distributed edge-centric architectures, an enormous shift in industrial IoT design took place [27]. These architectures supported filtering, aggregation, and anomaly detection at the edge through localized processing near data sources (✔ localized processing, ✔ edge-based anomaly detection), thereby enhancing robustness to network variability, ✔ scalability, ✔ latency, and ✖ peer-to-peer device interaction, making them particularly useful for real-time process control in manufacturing and adaptive routing in logistics [28,29], as illustrated in Table 4.3. Yet, these architectures largely lacked standard middleware, integration with legacy systems, and lifecycle management (✖ interoperability, ✖ modularity), which restricted their use in heterogeneous industrial contexts. Several recent studies on hybrid

**Table 4.3 Architectural Evolution of Industrial IoT Systems, Comparing Hierarchical and Centralized Designs with Distributed and Hybrid Edge–Fog–Cloud Frameworks, and Highlighting the Resulting Trade-Offs in Scalability, Latency, Interoperability, and Decision Support**

| *Main Feature* | *Hierarchical Automation Architecture* | *Centralized Data Aggregation* | *Edge-Centric Distributed Architecture* | *Hybrid Edge–Fog–Cloud Architecture* |
|---|---|---|---|---|
| Automation pyramid structure | ✔ | ✔ | ✖ | ✖ |
| Centralized control and analytics | ✔ | ✔ | ✖ | ✖ |
| Scalability under high data rates | ✖ | ✖ | ✔ | ✔ |
| Low-latency operational response | ✖ | ✖ | ✔ | ✔ |
| Localized processing near data sources | ✖ | ✖ | ✔ | ✔ |
| Peer-to-peer device interaction | ✖ | ✖ | ✔ | ✔ |
| Edge-based anomaly detection | ✖ | ✖ | ✔ | ✔ |
| Cloud-based global optimization | ✖ | ✔ | ✖ | ✔ |
| Interoperability with legacy systems | ✖ | ✖ | ✖ | ✔ |
| Standards-based middleware support | ✖ | ✖ | ✖ | ✔ |
| Modular service deployment | ✖ | ✖ | ✖ | ✔ |
| Microservices-oriented design | ✖ | ✖ | ✖ | ✔ |
| Security and data governance emphasis | ✖ | ✖ | ✖ | ✔ |
| Lifecycle and scalability management | ✖ | ✖ | ✖ | ✔ |

edge–fog–cloud architectures have sought to combine the benefits of distributed edge intelligence with centralized coordination and optimization [30]. These architectures explicitly support modular service deployment, microservices-oriented design, and standards-based middleware (✔ modular deployment, ✔ microservices, ✔ interoperability) that can be adapted to a wide array of assets and legacy systems [31,32]. As illustrated in Table 4.3, hybrid architectures also include security, data governance, and lifecycle management (✔), which highlight their suitability for large-scale industrial use.

### 4.2.3 Sensor Technologies and Data Collection

Early industrial IoT sensors relied on traditional industrial sensors such as temperature, pressure, flow, and proximity. These were intended primarily for periodic, fixed, or event-driven monitoring [33]. Table 4.4 shows that these sensing systems performed well in harsh industrial environments (✔) but only captured low-frequency signals and isolated signal streams (✖). Their function was limited to assessing conditions and fault detection, with little support for data integration, adaptive sampling, or real-time decision responsiveness [34]. The availability of sensing hardware and the bandwidth of communication increased, increasing the use of high-resolution condition sensing, in particular by studying the vibration, acoustic, and electrical signatures [35]. Additionally, the latter enabled more precise diagnostics and earlier identification of faults in manufacturing equipment (✔) compared to threshold monitoring. However, despite this increased signal richness, overall data collection remained essentially static and concentrated on sensors (✖), with very little attention given to multi-sensor integration, adaptive sampling, or system-level efficiency [36]. The second phase of monitoring, Enhanced Sensing Fidelity, is reflected in Table 4.4 but did not significantly alter data collection architectures.

The subsequent stage introduced Wireless Sensor Networks (WSNs), facilitating distributed sensing, asset identification using Radio Frequency Identification (RFID), and location tracking with GPS in logistics settings [37]. This paradigm broadened IoT capabilities to include multi-sensor deployments and mobile assets (✔), facilitating inventory visibility and shipment traceability within distributed networks [38]. Despite this, WSN-based systems depended on predefined sampling policies and centralized aggregation, providing limited support for sensor fusion, adaptive data reduction, or latency-aware preprocessing (✖), as outlined in Table 4.4.

Recent studies have moved toward intelligent and edge-enabled sensing architectures, marking a qualitative change in data collection and utilization.

**Table 4.4 Evolution of Industrial IoT Sensing from Basic Networked Measurements to Intelligent, Adaptive, and Edge-Enabled Data Collection, Highlighting Shifts toward Sensor Fusion, Adaptive Sampling, and Decision-Relevant Preprocessing**

| *Main Feature* | *Traditional Industrial Sensors* | *High-Resolution Condition Sensing* | *Wireless Sensor Networks* | *Intelligent and Edge-Enabled Sensing* |
|---|---|---|---|---|
| Temperature, pressure, and flow sensing | ✔ | ✔ | ✔ | ✔ |
| Periodic/event-driven sampling | ✔ | ✖ | ✖ | ✖ |
| Robustness in harsh environments | ✔ | ✔ | ✔ | ✔ |
| Vibration and acoustic analysis | ✖ | ✔ | ✔ | ✔ |
| Electrical signature monitoring | ✖ | ✔ | ✖ | ✔ |
| RFID-based asset identification | ✖ | ✖ | ✔ | ✔ |
| GPS-based location tracking | ✖ | ✖ | ✔ | ✔ |
| Multi-sensor data integration | ✖ | ✖ | ✔ | ✔ |
| Sensor fusion capabilities | ✖ | ✖ | ✖ | ✔ |
| Optimal sensor placement modeling | ✖ | ✖ | ✖ | ✔ |
| Adaptive sampling strategies | ✖ | ✖ | ✖ | ✔ |
| Imaging and vision-based sensing | ✖ | ✖ | ✖ | ✔ |
| Edge-level preprocessing | ✖ | ✖ | ✖ | ✔ |
| Communication overhead reduction | ✖ | ✖ | ✖ | ✔ |
| Near-real-time response support | ✖ | ✖ | ✖ | ✔ |

Within this paradigm, sensing platforms utilize edge-level computation to carry out preprocessing, filtering, compression, and feature extraction directly at the source of the data [39]. These systems facilitate sensor fusion (✔), optimal sensor placement modeling (✔), adaptive sampling strategies (✔), and imaging- or vision-based sensing (✔), as illustrated in Table 4.4. By cutting down on unnecessary transmissions and giving priority to information pertinent to decisions, edge-enabled sensing significantly reduces communication overhead and enables near real-time response capabilities [40].

Even with these progressions, Table 4.4 points out a continuing shortcoming in previous research: the impact of sensing architectures on the effectiveness of end-to-end decisions is seldom measured from a system-level perspective (✖). The majority of studies assess improvements in sensing independently, concentrating on aspects such as signal quality, accuracy of fault detection, or energy efficiency. These assessments do not explicitly connect choices made in sensing design with the placement of analytics downstream, latency limitations, or results of operational decisions. It is this gap that gives rise to the system-level, decision-oriented evaluation adopted in this chapter, which treats sensing as a crucial part of an end-to-end IoT decision system rather than as a standalone data source.

### 4.2.4 Communication Protocols and Connectivity

Early studies of industrial IoT communication aimed to adapt wired industrial networks such as fieldbus systems and industrial Ethernet to transmit sensor data in controlled environments [41]. These technologies provided high reliability and deterministic communication which were critical for safety-critical automation tasks. However, as discussed in Table 4.5, these networks were neither highly flexible nor scalable, and did not support mobile assets; thus, they were not suitable for the geographically distributed logistics operations or the retrofitting of heterogeneous legacy facilities [42]. To combat these limitations, later studies explored short-range wireless protocols such as Bluetooth, Zigbee, and Wi-Fi, which allow for flexibility on factory floors and in warehouses [43,44]. Despite their improvements in mobile asset support and reduced deployment timescales, these protocols also produced trade-offs in determinism, interference sensitivity, and scalability. This limits their use in large-scale or time-sensitive industrial systems (see Table 4.5). As the IoT adoption grew outside of these local settings, research focused on broad-area and Low-Power Wide-Area Network technologies such as cellular networks (3G/4G) and protocols such as LoRaWAN

**Table 4.5 Evolution of Industrial IoT Communication Paradigms, Contrasting Fixed and Single-Protocol Networks with Adaptive, Hybrid, and 5G-Enabled Connectivity Frameworks, and Highlighting Their Implications for Latency, Scalability, and End-to-End Decision Support**

| *Main Feature* | *Wired Industrial Networks* | *Short-Range Wireless Protocols* | *Wide-Area and LPWAN Technologies* | *Advanced Hybrid and 5G-Enabled Connectivity* |
|---|---|---|---|---|
| Deterministic communication | ✓ | ✗ | ✗ | ✓ |
| High reliability in fixed environments | ✓ | ✓ | ✓ | ✓ |
| Support for mobile assets | ✗ | ✓ | ✓ | ✓ |
| Scalability across large networks | ✗ | ✗ | ✓ | ✓ |
| Low-latency communication | ✓ | ✓ | ✗ | ✓ |
| Energy-efficient transmission | ✗ | ✓ | ✓ | ✓ |
| Long-range connectivity | ✗ | ✗ | ✓ | ✓ |
| Suitability for low-bandwidth sensing | ✗ | ✗ | ✓ | ✓ |
| High-throughput support | ✓ | ✓ | ✗ | ✓ |
| Payload size flexibility | ✓ | ✓ | ✗ | ✓ |
| Time-sensitive networking support | ✗ | ✗ | ✗ | ✓ |
| Network slicing capability | ✗ | ✗ | ✗ | ✓ |
| Adaptive routing mechanisms | ✗ | ✗ | ✗ | ✓ |
| Multi-protocol interoperability | ✗ | ✗ | ✗ | ✓ |
| End-to-end industrial integration | ✓ | ✓ | ✓ | ✓ |

and NB-IoT [45,46]. These technologies enabled energy-saving, long-range connectivity and supported many devices connected through distributed logistics and manufacturing networks. However, their low throughput and high latency of communication made them unsuitable for latency-sensitive monitoring and real-time decision support, as shown in Table 4.5. Recent research has focused on advanced and hybrid connectivity architectures enabled by 5G and Time-Sensitive Networking [47]. These technologies feature ultra-low latency, deterministic communication, network slicing, and adaptive routing, allowing for the simultaneous handling of different types of traffic with different reliability requirements and timing demands. Hybrid architectures that use wired, wireless, and wide-area protocols have been proposed to help end-to-end connectivity across factory floors, logistics networks, and cloud platforms [48]. Table 4.5 shows that these frameworks have many technical features such as scalability, interoperability, and deterministic performance.

### 4.2.5 Data Analytics and Decision Support

Descriptive and diagnostic monitoring was a primary focus of early work in industrial IoT data analytics, attempting to deconstruct raw sensor data into interpretable parameters for human operators. As outlined in the first two columns of Table 4.6, these models relied on rule-based monitoring, statistical thresholding, and control charts to detect deviations from nominal operating conditions [49,50]. These approaches were effective for retrospective reporting and basic situational awareness but had limited flexibility and did not directly support real-time or future decision-making.

As sensor density and data volumes increased, studies made progress toward predictive analytics, employing multivariate statistics and machine learning techniques in order to record nonlinear dependent properties and latent structures in high-dimensional IoT data [51]. This stage enabled applications such as predictive maintenance, demand prediction, and delay prediction in manufacturing and logistics environments [52,53]. As reflected in Table 4.6, predictive analytics also introduced capabilities such as machine learning and nonlinear modeling (✔) but were frequently disconnected from actual practice. Models were often evaluated separately in order to ensure accuracy in prediction, with limited consideration for interpretability, elongation across operating regimes, or integration into decision-making processes [54].

Increasing attention has been drawn toward prescriptive and integrated decision support, which aims to close the gap between analytics and action.

**Table 4.6 Evolution of Industrial IoT Analytics Paradigms, Illustrating the Transition from Descriptive and Predictive Monitoring toward Prescriptive, Decision-Oriented Frameworks with Integrated Edge–Cloud Intelligence**

| *Main Feature* | *Descriptive Analytics* | *Diagnostic Analytics* | *Predictive Analytics* | *Prescriptive and Integrated Decision Support* |
|---|---|---|---|---|
| Rule-based monitoring | ✔ | ✖ | ✖ | ✖ |
| Statistical thresholding | ✔ | ✔ | ✖ | ✖ |
| Retrospective reporting | ✔ | ✔ | ✖ | ✖ |
| Multivariate pattern discovery | ✖ | ✔ | ✔ | ✔ |
| Machine learning utilization | ✖ | ✔ | ✔ | ✔ |
| Nonlinear dependency modeling | ✖ | ✔ | ✔ | ✔ |
| Predictive maintenance support | ✖ | ✖ | ✔ | ✔ |
| Demand and delay forecasting | ✖ | ✖ | ✔ | ✔ |
| Real-time decision support | ✖ | ✖ | ✖ | ✔ |
| Optimization-based recommendations | ✖ | ✖ | ✖ | ✔ |
| Reinforcement learning control | ✖ | ✖ | ✖ | ✔ |
| Edge-level inference | ✖ | ✖ | ✔ | ✔ |
| Cloud-based model training | ✖ | ✔ | ✔ | ✔ |
| Hybrid edge–cloud analytics | ✖ | ✖ | ✔ | ✔ |
| Explainability and transparency | ✖ | ✖ | ✖ | ✔ |
| Human–machine decision alignment | ✖ | ✖ | ✖ | ✔ |

Optimization and reinforcement learning methodologies have been proposed that recommend control actions, scheduling changes, and resource transfers based on the anticipated state of the system [55]. However, the strategic placement of analytics has grown as a concern; edge-level inference supports latency-sensitive decisions, while cloud-based model training provides scalable learning and global optimization [56]. As illustrated in Table 4.6, these frameworks include real-time decision support, hybrid edge–cloud analytics, and explicit human–machine decision alignment (✔).

### 4.2.6 Security and Privacy Considerations

Early studies on safety for industrial IoT generally followed paradigms from traditional IT systems, relying on perimeter protection and unchanging authentication to secure connected industrial assets [57,58]. These solutions relied on clear network boundaries and trusted external environments, as seen in the "Legacy IT Security" column of Table 4.7. However, when industrial control systems and other old equipment, often not designed for connectivity, came into contact with Internet Protocol networks, vulnerabilities such as weak authentication, hardcoded credentials, and unencrypted communication channels emerged [59]. Such restrictions underscored the inadequacy of perimeter-based security systems in open, distributed industrial IoT environments.

As a result, subsequent studies focused on a device-level protection model, including secure boot processes, hardware-based trust anchors, and light cryptography protocols for sensors and actuators with limited resources [60]. As shown in Table 4.7, this phase enabled heightened endpoint security but addressed devices individually, with limited resistance to coordinated or network-level attacks [61]. This meant that research became more focused on strategies for defending networks. The research included network segmentation, secure routing protocols, and anomaly-based intrusion detection systems capable of identifying traffic and behavioral anomalies in the operational technology networks [62,63]. This was an important point, since it was the first time that the cyber–physical impacts were explicitly recognized, acknowledging that cyberattacks can directly lead to disruptions of production, safety hazards, or product loss [64]. Table 4.7 illustrates the trend toward more responsive and private-minded security practices in recent studies. To address dynamic device populations and emerging attack surfaces, these approaches use zero-trust security principles, continuous authentication, and machine learning to detect threats

**Table 4.7 Evolution of Industrial IoT Security Paradigms, Highlighting the Transition from Perimeter-Based IT Defenses to Adaptive, Privacy-Aware Cyber–Physical Protection Frameworks and the Corresponding Expansion of System-Level Threat Models**

| *Main Feature* | *Legacy IT Security* | *Device-Level Protection* | *Network-Level Defense* | *Adaptive and Privacy-Aware Frameworks* |
|---|---|---|---|---|
| Perimeter-based protection | ✓ | ✗ | ✗ | ✗ |
| Static authentication mechanisms | ✓ | ✗ | ✗ | ✗ |
| Encryption of communication channels | ✗ | ✓ | ✓ | ✓ |
| Secure boot and hardware trust anchors | ✗ | ✓ | ✗ | ✓ |
| Lightweight cryptography | ✗ | ✓ | ✗ | ✓ |
| Network segmentation | ✗ | ✗ | ✓ | ✓ |
| Secure routing protocols | ✗ | ✗ | ✓ | ✓ |
| Anomaly-based intrusion detection | ✗ | ✗ | ✓ | ✓ |
| Cyber–physical impact awareness | ✗ | ✗ | ✓ | ✓ |
| Zero-trust security principles | ✗ | ✗ | ✗ | ✓ |
| Continuous authentication | ✗ | ✗ | ✗ | ✓ |
| Machine learning–based threat detection | ✗ | ✗ | ✗ | ✓ |
| Privacy-preserving data sharing | ✗ | ✗ | ✗ | ✓ |
| Data anonymization mechanisms | ✗ | ✗ | ✗ | ✓ |
| Policy-driven data governance | ✗ | ✗ | ✗ | ✓ |

[65]. Meanwhile, increased concern regarding the sharing of data across organizational and supply chain boundaries prompted studies in the areas of data anonymization, fine-grained access control, and policy-based data governance [66]. These methods are intended to address the confluence between operational transparency and confidentiality in collaborative industrial ecosystems.

### *4.2.7 Integration with Enterprise Systems*

The initial studies investigating the integration of IoT with enterprise systems incorporated OT–IT communications through point-to-point connections, which enabled a limited exchange of information between shop floor control systems and enterprise systems such as Enterprise Resource Planning and Supply Chain Management systems [67,68]. These integrations, shown in Table 4.8, were based on custom interfaces and batch data transfers (✔), with fine-grained visibility of production and inventory states. However, the lack of real-time data ingestion, standardized interfaces, and semantic normalization (✖) produced high latency, fragile integrations, and limited scalability across organizations [69]. This meant that enterprise systems were most marginally disengaged from the real-time operation vector IoT devices capture, and thus rendered less useful for making strategic decisions in discrete moments. In order to tackle these restrictions, later investigations examined integration architectures based on middleware and services. They proposed the use of message-oriented middleware and standardized interfaces to separate IoT platforms from enterprise applications [70]. As shown in Table 4.8, these methods facilitated the normalization of semantic data and the decoupling of system architectures (✔), thereby enhancing interoperability and extensibility. Data flows, however, were often event-insensitive or, at best, near-real-time, and the integration logic was still based on predefined workflows rather than adaptive operational events (✖). As a result, while they improved the consistency of enterprise decision-making processes, they did not yet respond to changing operating conditions. Recent work has been directed toward event-driven, cloud-based models with streaming platforms, API-based ecosystems, and digital twins, which enable near real-time synchronization of physical processes with enterprise systems [71]. These models allow for real-time data ingestion, event-based processing, and closed-loop decision support (✔), such as dynamic replanning, adaptive scheduling, and ongoing performance optimization. Cloud-based solutions extend the scale of a single organization and provide ecosystem-level

**Table 4.8 Evolution of IoT–Enterprise Integration Architectures, Highlighting the Transition from Point-to-Point OT–IT Data Exchange to Event-Driven, Cloud-Native Ecosystems and the Remaining Gap in Decision and Organizational Alignment**

| *Main Feature* | *Point-to-Point OT–IT Integration* | *Middleware/ SOA Integration* | *Event-Driven Integration* | *Cloud-Native and Ecosystem Integration* |
|---|---|---|---|---|
| Batch-oriented data transfer | ✔ | ✖ | ✖ | ✖ |
| Custom interfaces | ✔ | ✖ | ✖ | ✖ |
| Real-time data ingestion | ✖ | ✖ | ✔ | ✔ |
| Standardized interfaces | ✖ | ✔ | ✔ | ✔ |
| Message-oriented middleware | ✖ | ✔ | ✔ | ✔ |
| Semantic data normalization | ✖ | ✔ | ✔ | ✔ |
| Decoupled system architecture | ✖ | ✔ | ✔ | ✔ |
| Event-driven processing | ✖ | ✖ | ✔ | ✔ |
| Digital twin synchronization | ✖ | ✖ | ✔ | ✔ |
| Closed-loop decision support | ✖ | ✖ | ✔ | ✔ |
| API-based interoperability | ✖ | ✖ | ✖ | ✔ |
| Scalability across enterprises | ✖ | ✔ | ✔ | ✔ |
| Organizational process alignment | ✖ | ✖ | ✖ | ✔ |

interoperability through APIs (✔). However, as discussed in Table 4.8, many studies continue to view integration as a technical problem, with no attention paid to aspects like process alignment, decision rights, and human–machine coordination (✖).

## 4.3 Materials and Methods

### 4.3.1 Dataset Description

This chapter's empirical assessment is based on two publicly available datasets that represent complementary but structurally different examples of industrial IoT systems in the domains of manufacturing and logistics. For manufacturing, we use the SECOM manufacturing dataset [72]. It involves sensor readings from a circuit in which semiconductors are being manufactured. This dataset includes 1,567 production instances and contains 590 real-valued sensor variables that measure equipment behavior, environmental conditions, and process stability at various steps in the manufacturing process for each instance. These data are highly dimensional, scaled-up, and lack missing values reminiscent of realistic constraints of sensor-rich industrial environments. This makes the dataset appropriate for assessing instance-level condition assessment, anomaly identification, and process-state inference within IoT-enabled manufacturing systems. The dataset is divided into three components: 70% for training (1,097 instances), 15% for validation (235 instances), and 15% for testing (235 cases). This ensures that there is enough space for model estimation and a fair evaluation set.

We represent logistics operations using the GeoLife GPS trajectory dataset [73]. This dataset comprises large-scale spatiotemporal data on mobility via GPS over long periods of time. The dataset consists of over 25 million timestamped GPS points with latitude–longitude coordinates, as well as over 17,000 trajectories produced by 182 users. Individual trajectories can be as short as a few minutes and as long as several hours; they are not uniform, with frequent stops, slow speeds, and deviations from routes. GeoLife, while not restricted to freight cars, is often used as an adjacency for IoT-based logistics and transport systems because of its size, temporal continuity, and realism. To prevent information leakage between portions, sequence-level trajectories are divided into training (70%), validation (15%), and testing (15%) parts. In conjunction with SECOM and GeoLife, the combined datasets allow for cross-domain assessment of DOS-IoT across different datasets, such as example-based sensor measurements and continuous spatiotemporal trajectories, as seen in Figure 4.2. This allows us to judge if generalized architectural principles at the system level to a specific operational context.

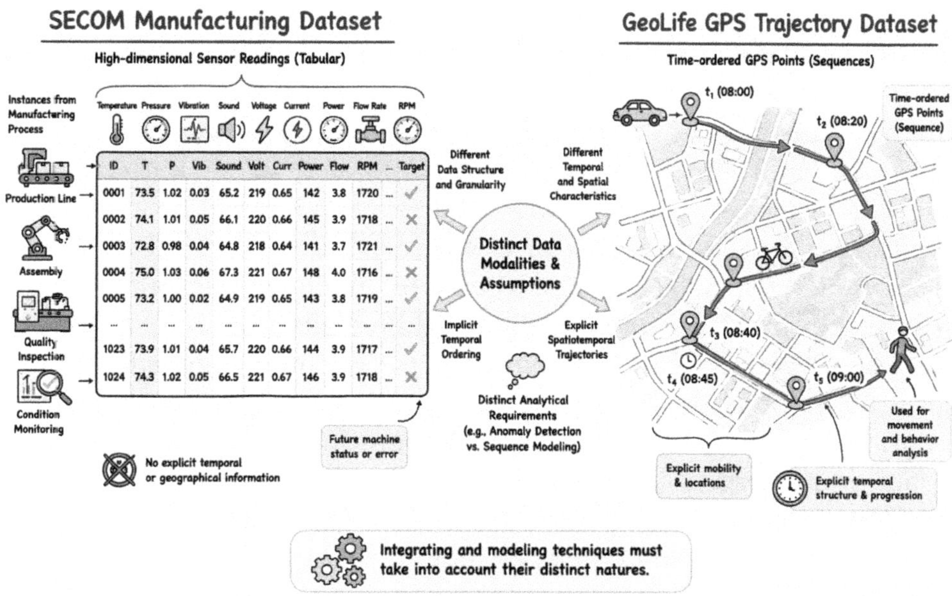


**Figure 4.2 Structural contrast between the SECOM manufacturing dataset and the GeoLife GPS trajectory dataset, illustrating differences in dimensionality, temporal structure, and the resulting implications for system-level IoT analytics and decision support.**

### *4.3.1.1 Dataset Preprocessing*

Preprocessing was intended to ensure numerical stability, structural validity, and comparability of datasets, whereas eliminating any changes that could lead to unrealistic enhancement of performance or distortion of operating behavior. The preprocessing utilized consistency at the feature level rather than time-based alignment in the SECOM manufacturing dataset. All sensors with more than 50% of their values missing were excluded. This reduced the original 590 features to around 460 retained variables. The remaining missing entries were filled in using feature-wise mean imputation, while maintaining the marginal distribution of each sensor. To further reduce redundancy and noise, a variance close to zero, i.e., below $1 \times 10^4$, was left out of the analysis. This resulted in a final feature set of approximately 400–420 sensors. Z-score normalization was applied to all retained features, calculated relative to the training split and applied uniformly across validation and testing sets. This allowed for comparison across various sensor scales without artificial correlation or temporal assumptions.

Meanwhile, GeoLife was preprocessed with explicit spatial and temporal design. First, data from raw GPS points were filtered to avoid physically implausible observations. This involved discarding points with speeds of less than 0.5 m/s or over 40 m/s, thereby eliminating stationary noise and extreme outliers from loss of signal. It was possible to separate continuous trips into trajectories with more than 300 seconds interval for the use of time, allowing for the distinguishing of active and idle periods. Trajectories were resampled every 30 seconds to ensure quality of resolution. This decreased the sampling-time variability while maintaining the trajectory. These cleaned sequences were compared to 5-minute aggregations using higher-level mobility descriptors such as distance traveled, average speed, stop duration, and dwell frequency, which produce compact but readable data on movement behavior. These steps, as described in Figure 4.3, ensure that future analyses produce realistic logistics data, for example, slow response and routing efficiency, not an artifact of disjointed sampling or noise.

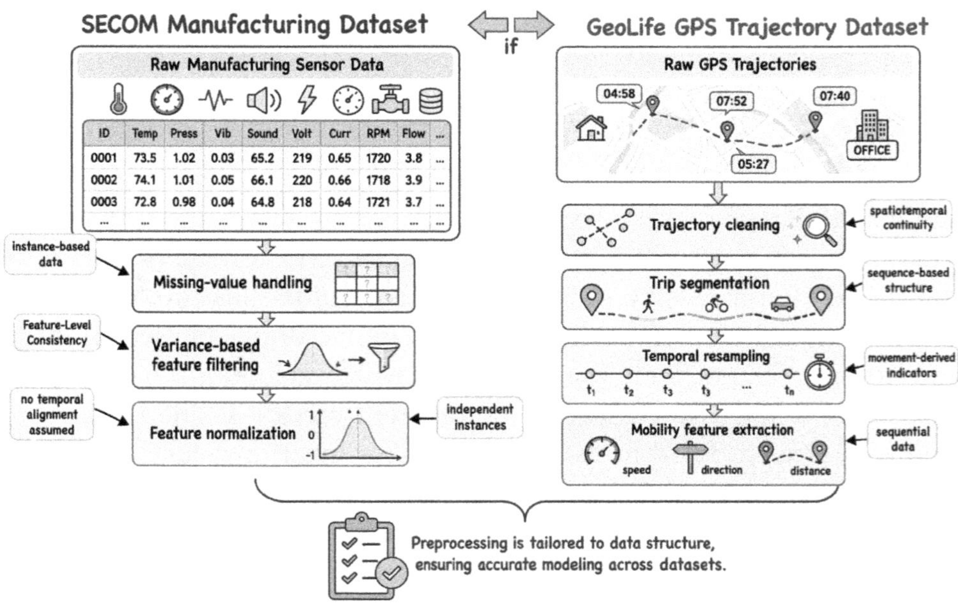


**Figure 4.3 Structure-aware preprocessing pipelines for manufacturing and logistics data, highlighting domain-specific transformations required to ensure numerical stability, temporal consistency, and decision-relevant feature representations for SECOM and GeoLife.**

## 4.3.2 Model Analysis

The proposed analytical framework operationalizes the DOS-IoT perspective and expresses how heterogeneous IoT data are modeled in terms of meaningful decisions for manufacturing and logistics systems. The underlying idea behind DOS-IoT is that IoT data have no value unless they are organized, contextualized, and linked to decision-making processes, rather than being examined at its own pace. Thus, the framework defines data flows from sensing and communication to processing and integration into the enterprise, taking into account architectural interactions and deployment constraints, as shown in Figures 4.4 and 4.5.

### 4.3.2.1 Data Ingestion and Representation

Preprocessed IoT data are ingested at the beginning of the analytical pipeline and represent the ongoing operational status of physical assets and processes. Importantly, the framework allows multiple temporal structures for different inputs rather than enforcing a single uniform structure. Instead, it also accompanies case-based measurements, such as manufacturing

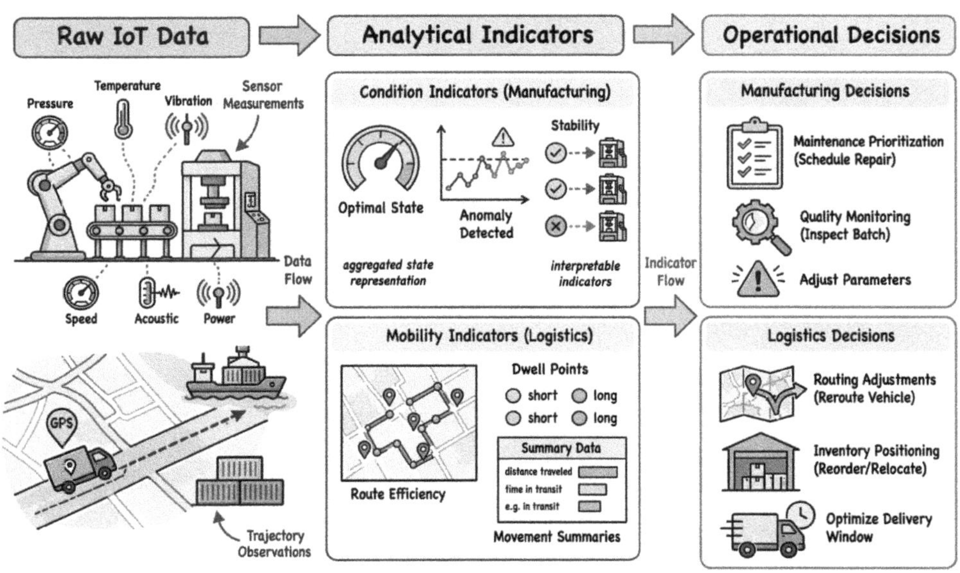


**Figure 4.4 Decision-oriented analytics flow illustrating how raw IoT data from manufacturing and logistics are transformed into interpretable indicators that directly support operational decisions.**

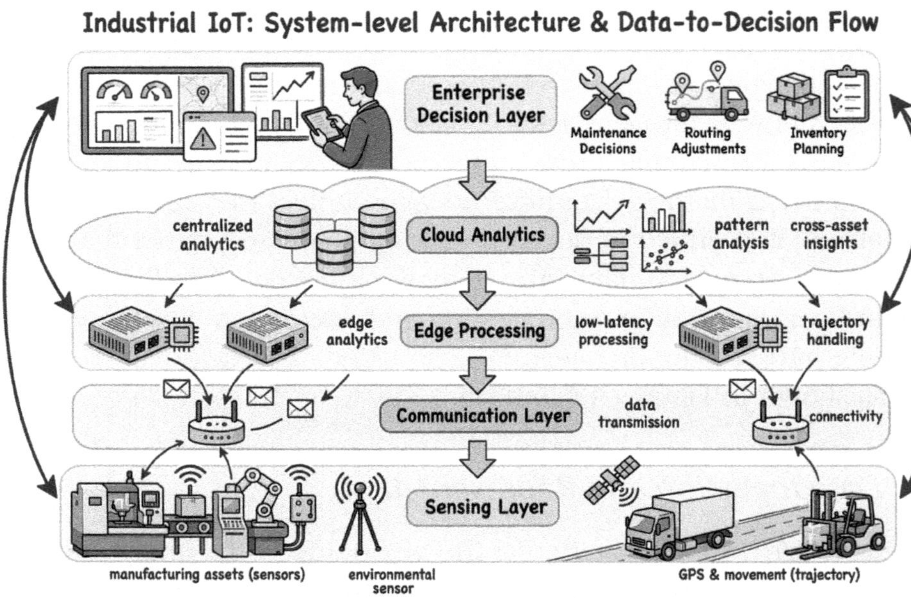


**Figure 4.5 System-level industrial IoT architecture illustrating the layered data-to-decision flow across sensing, communication, edge processing, cloud analytics, and enterprise decision layers for manufacturing and logistics systems.**

sensors, as well as explicitly temporal data, such as logistics trajectory data. This decision to design captures the structural diversity of industrial IoT data sources, as discussed in Section 4.3.1 and illustrated in Figure 4.2.

Manufacturing data are represented as scaled vectors related to production conditions and stages of production, and logistics data are represented in space–time sequences representing asset movement. This framework prevents artificial temporal assumptions and ensures that analytical results are consistent with the semantics of the processes involved, preserving the native structure of each dataset.

### *4.3.2.2 Manufacturing-Oriented Analytical Modeling*

For manufacturing, it utilizes multivariate sensor measurements that define individual production cases. Sensor observations are organized by physical assets and stage, allowing for a line between normal operating variability and patterns of degradation, instability, or abnormal process behavior. The framework does not rely on time-series forecasting but represents process conditions via distributional summaries, cross-sensor dependencies, and aggregate health indicators, which provide a detailed description of equipment and process states. These condition indicators are models for operational health that are responsible for decision-making and are directly tied to actions related to manufacturing, such as maintaining a high priority, changing process parameters, and optimizing energy use. Such a strong correlation ensures that analyses can be understood and acted upon, making decisions that help predict availability, quality, and efficiency rather than merely providing predictive precision.

#### *4.3.2.3 Logistics-Oriented Spatiotemporal Analysis*

For logistics, the framework processes spatiotemporal data on the movement of assets in a distributed network to check and control movement. Single trajectories are seen as the result of logistical procedures, such as vehicle routing, shipment transfer, or inventory repositioning. Comparison of observed trajectories with baseline operations provides estimates of spatial deviation, time delay, and abnormal dwell behavior. The advantages of assets, routes, and time windows are aggregated in this model to identify systemic inefficiencies such as repeated delays, route redundancies, and congestion hotspots. This is a good representation for recording activity in real time to allow for immediate disturbance to be discovered, and for considering performance in the future as a tool for routing and inventory planning. It is important to note that analytical outputs are articulated as operational states that can be directly utilized by downstream decision layers.

#### *4.3.2.4 Layered Architecture and Processing Placement*

As shown in Figure 4.5, the analytical workflow is organized within a multilayer architecture that reflects the functional organization of industrial IoT systems. The sensing layer generates measurements from a variety of devices, varying in scale, resolution, and reliability. These measurements are normalized and aligned during preprocessing to ensure comparability. Assumptions regarding data availability, latency, and transmission reliability implicitly model the communication layer, as there are real constraints on the speed of observations across the system. Communication protocols are not precise, but the performance characteristics of these protocols affect system responsiveness and are part of decision-latency evaluation. The processing layer is at the center of the framework and aggregates raw data into state descriptors through filtering, aggregation, and pattern analysis. It is assumed that latency-sensitive tasks such as anomaly detection and threshold alerts are performed at the edge, while computationally intensive analyses such as cross-asset pattern extraction and long-horizon aggregation are performed on centralized cloud resources. This section enables the framework to gauge the effect of analytical placement on timeliness, accuracy, and scalability under realistic deployment conditions.

#### *4.3.2.5 Decision Mapping and Application Layer*

Application-level decisions are made to decide how to proceed with the results of analysis, which will direct actions on system performance. Manufacturing includes maintenance schedules, quality assurance measures, and energy management strategies. Application-level decisions in logistics address routing changes, inventory redistribution, and the handling of exceptions due to delays or disruptions. Instead of strict decision rules, the framework evaluates how decision types respond to varying data quality, analytical depth, and processing latency. This allows for a rigorous study of the trade-offs between the swift but vague actions and the more deliberate and informed actions, which is needed to explain action in practice.

#### *4.3.2.6 Enterprise Integration and Closed-Loop Control*

Integration with enterprise systems is presented as a framework function not as one additional capability. Work orders, inventory information, and transportation planning use analytical outputs that are business-critical for closed-loop decision-making at operational and managerial levels. On the other hand, a source of contextual information, such as production schedules, service goals, or delivery priority, is available at the enterprise level. This provides insights and analysis on the event-related context. This double-edged interaction provides a parallel interaction between operational conditions and managerial intention, which reduces the failures of integration presented in Table 4.8 and supports the decision-oriented, system-level nature of DOS-IoT.

## 4.4 Experimental Setup

### *4.4.1 Evaluation Metrics*

The experimental study employs a multi-dimensional metric design to evaluate the efficiency of industrial IoT systems in the DOS-IoT framework.

Unlike predictive accuracy assessment, the evaluation measures must account for operational impact, technical behavior, and economic impacts in detail. This reflects the data-to-decision pipeline shown in Section 4.3.2 and in Figures 4.4 and 4.5.

#### 4.4.1.1 Operational Effectiveness Metrics

The primary measure of practical system value is operational effectiveness, which illustrates the ability of IoT-driven analytics to improve core manufacturing and logistics outcomes. Manufacturing uses equipment uptime, Overall Equipment Effectiveness (OEE), and process stability to determine availability, performance efficiency, and quality consistency. Logistics-based operational effectiveness is assessed through inventory accuracy, on-time delivery rate, and route adherence to ascertain the quality of service and materials for which the logistics system is equipped. Improvements in these measures indicate that IoT data are not only collected but are also included in maintenance planning, process management, and logistics coordination. Due to the close alignment of these measures with business objectives and service-level agreements, they are the most accurate proxy for actual value creation under DOS-IoT.

#### 4.4.1.2 Technical Performance Metrics

Technical performance measures the behavior, reliability, and responsiveness of the IoT system. These measures include data availability, processing throughput, alert precision, and end-to-end decision latency, demonstrating the efficient flow of information from sensing devices to analytical components to decision points. Decision latency is calculated as the time between the observation of a specific event at the sensing layer and the presence of a decision signal at the application layer. Especially in latency-sensitive applications like condition monitoring and anomaly detection, unreliable or delayed information can compromise the operation of an application. Hence, we measure alert accuracy through precision to examine the balance between sensitivity and false notifications, and we assess throughput scalability by looking at the impact of higher sensor density and data volume on processing performance. These technical measures can provide insight into operational outcomes and will assist in determining architectural bottlenecks within the layered system.

#### *4.4.1.3 Economic Value Metrics*

Economic variables assess the financial impact of IoT-enabled architectures and analytics. These include maintenance costs due to unexpected downtime, inventory holding costs, and operational efficiency related to improved service reliability. This is measured through normalized savings estimates that link operating improvements to financial outcomes, rather than a cost-structure-level estimate.

### *4.4.2 Hyperparameters*

This choice of hyperparameters was based on the structural nature of the data and the test scenario in which IoT-based analytical pipelines were deployed, rather than exhaustive optimization for optimal prediction. For the sake of fairness and reproducibility, all hyperparameters were determined before the experiments and did not change during the comparative or ablation analyses. This design choice also ensures that any observed performance differences are linked to decisions regarding architecture and analysis design, rather than to opportunistic changes in parameters. Hyperparameters were selected for the SECOM data in experiments to address high dimensionality, sensor heterogeneity, and partial missing issues. Means and standard deviations were calculated from the training split to normalize features using z-score scaling. For safety reasons, the less than $1 \times 10^{-4}$ variance in sensors was eliminated to create a sparse but informative feature set. In order to manage excessive overfitting created by the high feature-to-sample ratio, an L2 regularization coefficient of $\lambda = 0.01$ was obtained. This optimization utilized a constant learning rate of 0.01, allowed a maximum of 500 iterations, and included an early-stopping factor with $1 \times 10^{-5}$ tolerance for improvement in the objective. This meant that the dataset is instance dependent, with no windowing or lag parameters added. Training was performed for up to 25 epochs, with early stopping applied.

Hyperparameters were used in logistics experiments with the GeoLife data to reflect spatial explicitness. For a precise balance between data resolution and computational efficiency, we resampled 30-second trajectories. To minimize spatial noise, I used a three-point moving average. In contrast, intertemporal lags of more than 5 minutes were introduced, initiating trajectory segmentation as active movement differed from idle time. Motion that was not physically plausible was omitted from the calculation based on speed scales of 0.5–40 m/s, and higher-level mobility measures were constructed over 5 minutes to identify important logistics events rather than fluctuations in time. Analytical placement parameters were aligned with the layered IoT architecture presented in Section 4.3.2. The threshold level of alert confidence for the edge layer was set at 0.7, with greater emphasis on precision than sensitivity to minimize false positives in practice. Centralized analytics utilized wider aggregation horizons and larger representations to account for computational power. Given 100–300 ms as the low-level limit on communication latency, I assumed a computer network in the realistic conditions of industrial use, and no protocol was chosen. All experiments were conducted using a single NVIDIA RTX 3090 GPU to ensure even computational conditions within each domain. Instead of grid search or adaptive tuning, hyperparameter values for both domains were based on proven practices and empirical research in the field. Table 4.9 presents the entire set of hyperparameters and their architectural placement assumptions to make the process transparent and reproducible.

**Table 4.9 Fixed Hyperparameter Settings for Manufacturing and Logistics Experiments, Chosen to Reflect Dataset Structure and Realistic Industrial IoT Deployment Constraints**

| *Category* | *Hyperparameter* | *Manufacturing (SECOM)* | *Logistics (GeoLife)* |
|---|---|---|---|
| Feature selection | Variance threshold | $1 \times 10^{-4}$ | — |
| Regularization | L2 Penalty (Λ) | 0.01 | — |
| Optimization | Learning rate | 0.01 | — |
| Optimization | Maximum iterations | 500 | — |
| Optimization | Early-stopping tolerance | $1 \times 10^{-5}$ | — |
| Temporal resolution | Resampling interval (S) | — | 30 |
| Spatial smoothing | Smoothing window (points) | — | 3 |
| Trajectory segmentation | Idle gap threshold (Min) | — | 5 |
| Physical constraints | Speed lower bound (M/S) | — | 0.5 |
| Physical constraints | Speed upper bound (M/S) | — | 40 |
| Aggregation | Indicator window (Min) | — | 5 |
| Edge analytics | Alert confidence threshold | 0.7 | 0.7 |
| System constraints | Communication latency (Ms) | 100–300 | 100–300 |

# 4.5 Results and Analysis

## 4.5.1 Comparison with State of the Arts

First, we compare DOS-IoT to the most recent industrial IoT paradigms, such as those in previous studies—cloud-centric analytics, edge-only heuristic systems, and component-optimized pipelines. All baselines use the same datasets, preprocessing techniques, and fixed hyperparameters (see Section 4.4), ensuring that any variations noticed are architectural rather than due to parameter tuning.

In summary, Figure 4.6 indicates that, under the same data and deployment conditions, DOS-IoT always exhibits greater operational efficiency than typical industrial IoT architectures in manufacturing and logistics. For manufacturing systems, OEE increases by 8.4% and downtime declines by 9.6% with cloud-centric analytics. Conversely, the edge-only heuristics result in a 5.9% increased OEE and a 7.1% reduction in downtime. This makes OEE better by 12.7% and downtime is reduced by 15.4% with IoT pipelines for components. Rather, DOS-IoT delivers many more benefits to manufacturing, with improvements in OEE between 18.3% and 26.9% and decreases in unplanned downtime between 22.1% and 30.4%. In logistics, inventory accuracy increases by 11.2% and 8.4% in cloud-centric and edge-only

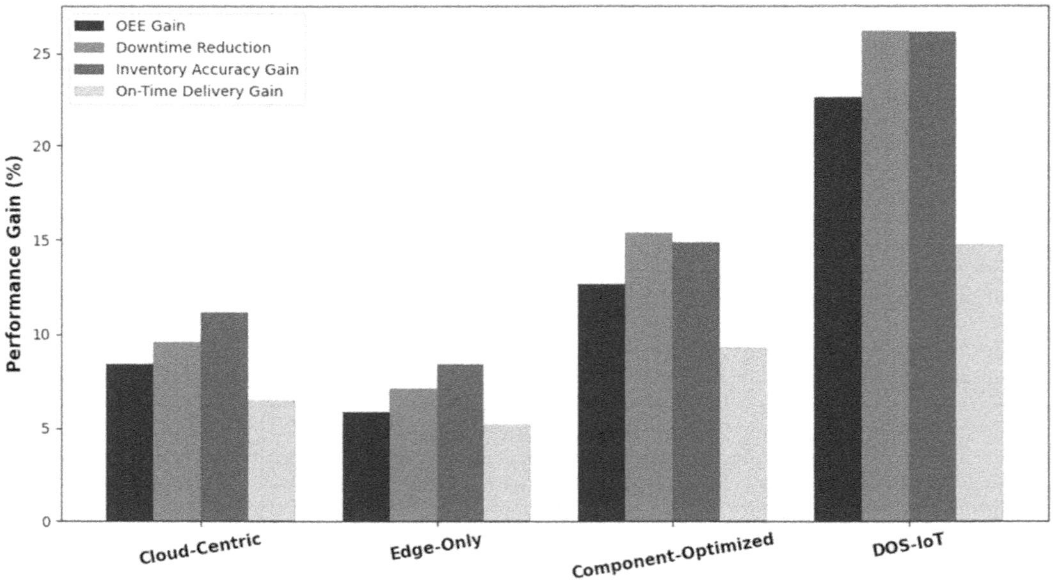


**Figure 4.6 Comparison of operational effectiveness across state-of-the-art industrial IoT architectures under identical data and deployment conditions.**

systems, respectively, and on-time delivery increases by 6.5% and 5.2%. The performance gains from component-optimized systems are 14.9% and 9.3% in these areas. Across several operational contexts, DOS-IoT offers improvements in inventory accuracy between 19.2% and 33.1%, as well as increases in on-time delivery from 11.4% to 18.2% compared to baseline models.

The technical performance comparisons in Figure 4.7 show the architecturally evident advantages of DOS-IoT over standard industrial IoT baselines. Cloud-centric analytics have an alert precision of 0.71 but do not exhibit any significant decrease in decision latency or medium-throughput scalability. Edge-only heuristic systems have more than a 22.6% reduction in latency and are very elastic, but their alert accuracy is only 0.68, making them less reliable than edge-only systems. The underlying architectures that exploit the component-optimized IoT capabilities improve latency by 28.4% and alert accuracy by 0.74, but remain of limited scalability. The technical performance of DOS-IoT reflects a reduction in end-to-end decision latency of 35.7%–40.2%, alert precision between 0.81 and 0.84, and scalability of high throughput. The results show that DOS-IoT provides an optimal combination of edge-level responsiveness and analytical depth, enabling swift and reliable operational decision-making. These performance improvements indicate

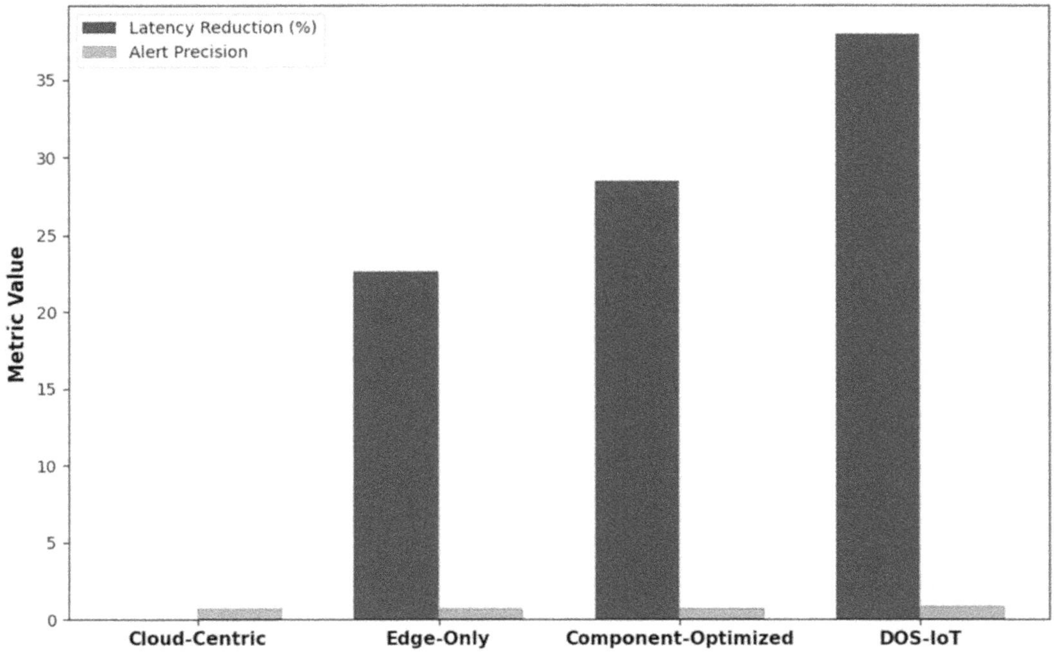


**Figure 4.7 Technical performance comparison of industrial IoT architectures in terms of decision latency, alert precision, and scalability.**

that hybrid analytical placement allows for the conciliation of low-latency execution and high-confidence alert generation in DOS-IoT. It exceeds both cloud-centric and component-optimized baselines in the same deployment context while enabling scalable processing across multiple industrial contexts.

Figure 4.8 summarizes the economic impact of design choices in architectural designs through normalized cost savings of downtime and inventory efficiency under the same deployment assumptions. Cloud-centered measures set the baseline economic index of 1.00 for downtime costs, inventory cost reduction, and overall economic impact. Edge-only heuristics deliver some improvement, saving downtime costs by 1.12 and reducing inventory costs by 1.08, respectively, with an overall economic index of 1.10, while component-optimized IoT architectures benefit from downtime savings of 1.29, inventory reductions of 1.22, and an overall economic index of 1.26. For instance, DOS-IoT shows much better economic performance, with downtime cost savings of 1.58 and 1.83, inventory cost savings of 1.41 and 1.76, and an overall economic index of 1.61–1.79, suggesting that system-level architectural integration yields significantly greater economic value than component-centric optimization. This supports architectural coherence as the largest factor affecting industrial IoT financial performance across a wide variety of operating conditions.

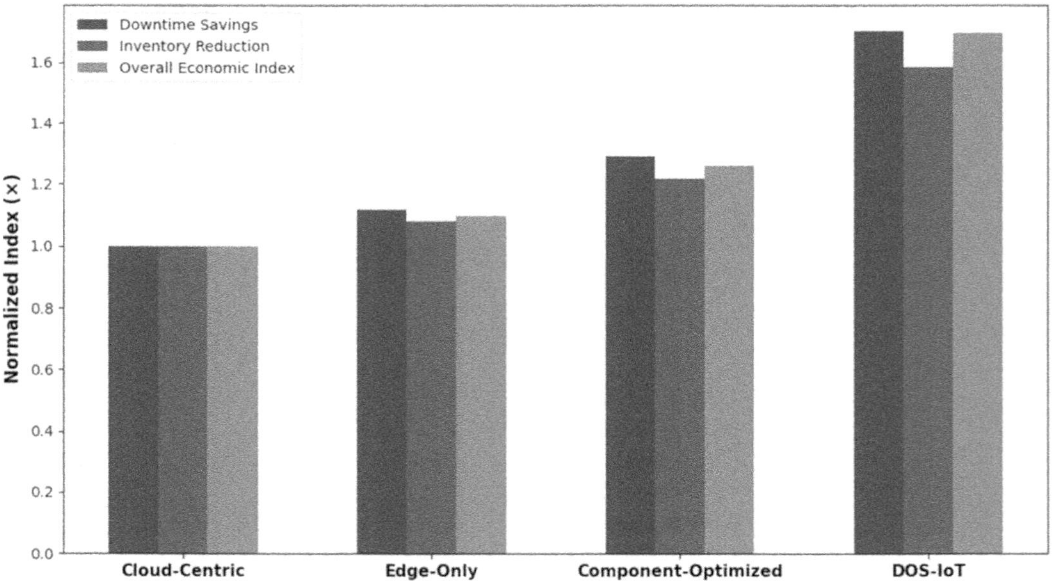


**Figure 4.8 Normalized economic impact of alternative industrial IoT architectures, comparing relative cost savings and value generation under identical deployment assumptions.**

## *4.5.2 Cross-Domain Analysis*

To evaluate generalizability, we analyze the performance of DOS-IoT in manufacturing (SECOM) and logistics (GeoLife), which differ significantly in data structure, temporal dynamics, and decision horizons. Despite these variations, DOS-IoT shows reliable relative improvements across both domains and thus establishes that its system-level principles can be applied to all domains.

Figure 4.9 summarizes operational results for manufacturing and logistics, showing that while data structure and decision-making are heavily influenced by DOS-IoT, performance gains have been consistent. For manufacturing, according to the SECOM dataset, OEE improvements of 18.3%–26.9%, fewer unplanned downtimes of 22.1%–30.4%, and higher service levels of 14.6%–19.2% are achieved by DOS-IoT. The GeoLife dataset shows improvements in logistics operations that are achieved by DOS-IoT: inventory accuracy increases by 19.2%–33.1%, delays increase by 16.8%–24.5%, and service levels increase by 11.4%–18.2%. While manufacturing and logistics systems

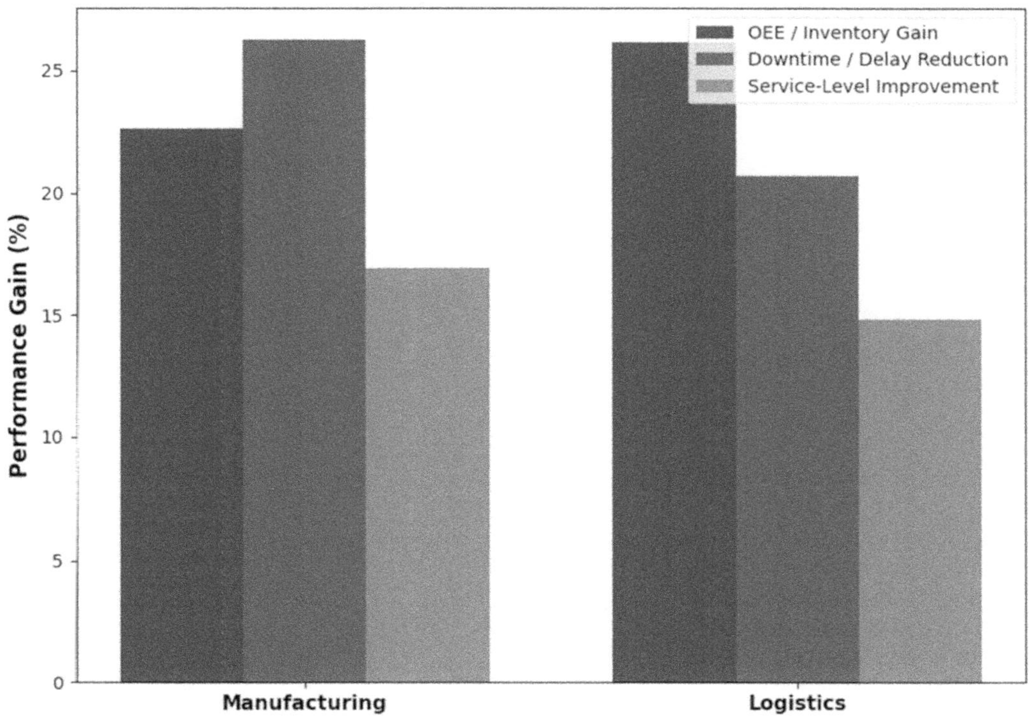


**Figure 4.9 Cross-domain operational performance of DOS-IoT across manufacturing and logistics, highlighting consistent gains despite differences in data structure and decision dynamics.**

have different core structures with regard to time, operation complexity, and decision time, the relative performance adequacy of DOS-IoT remains consistently high across these domains. The variation in operational gains across domains is small, indicating that system-level architectural principles of DOS-IoT can be applied across a variety of industrial environments without altering architectural structure or parameters in different deployment conditions.

As illustrated in Figure 4.10, technical behavior is also consistent and robust across manufacturing and logistics, with DOS-IoT functioning well across all operating contexts. In manufacturing, DOS-IoT reduces the latency of decision 31.4%–36.2% while maintaining alert precision of 0.82 and high scalability stability. In logistics, decision latency decreases significantly, between 35.7% and 40.2%, with an alert precision of 0.81 and very high scalability stability. This implies that edge-level analytical placement enables robust reductions in end-to-end decision latency in spatially-related logistics systems. Although the structure and timing of these two sources of data

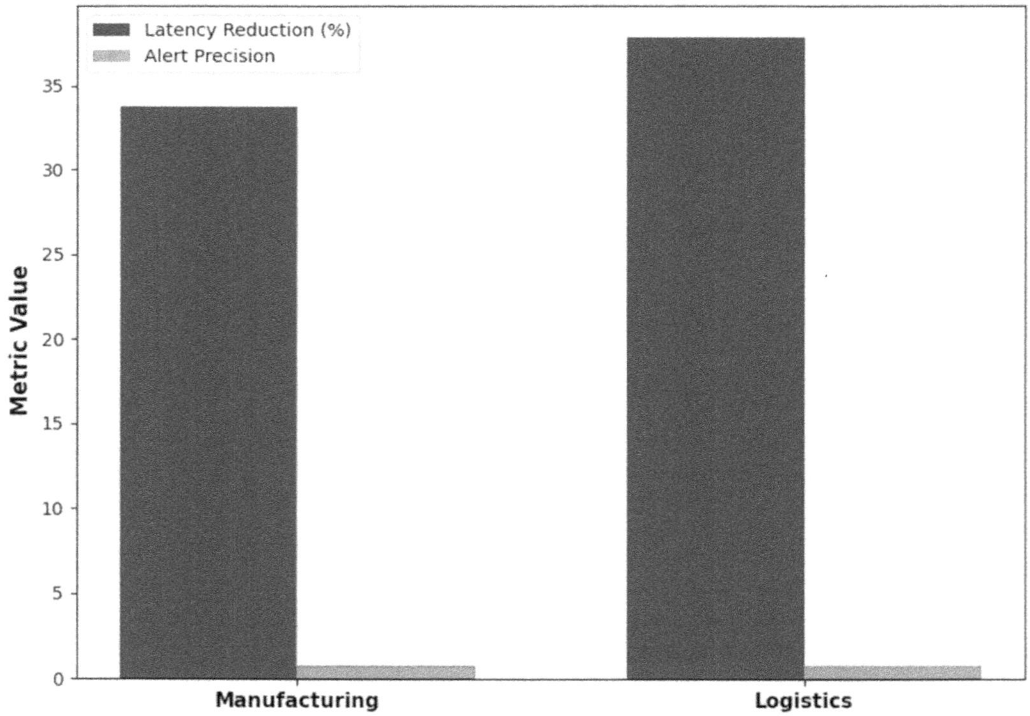


**Figure 4.10 Cross-domain comparison of technical performance under DOS-IoT, highlighting latency reduction, alert precision, and scalability across manufacturing and logistics.**

vary greatly, alert precision values do not vary much and are closely aligned across the three fields. The consistent stability of scaling across the two domains is consistent with a strong belief that DOS-IoT provides a reliable throughput performance and offers low-latency decision pipelines. Together, these technical results show that DOS-IoT is robust to heterogeneity and operational complexity while maintaining stable analytical reliability and scalable system behavior in manufacturing and logistics.

Figure 4.11 displays the economic performance of DOS-IoT in manufacturing and logistics, with consistent value generation across many operational contexts. In manufacturing, DOS-IoT achieves a normalized cost savings index between 1.62 and 1.79 and maintains a stable, high Return on Investment (ROI). Logistics costs are normalized between 1.55 and 1.73, with a stable ROI. While the timescale, movement of assets, and operational complexities associated with manufacturing and logistics infrastructures vary across the breadth of the world, DOS-IoT is economically advantageous in both areas. These overlapping cost savings ranges are a reflection of the transferability of value-generating mechanisms of the DOS-IoT architecture across industries. These results indicate that architectural coherence at the

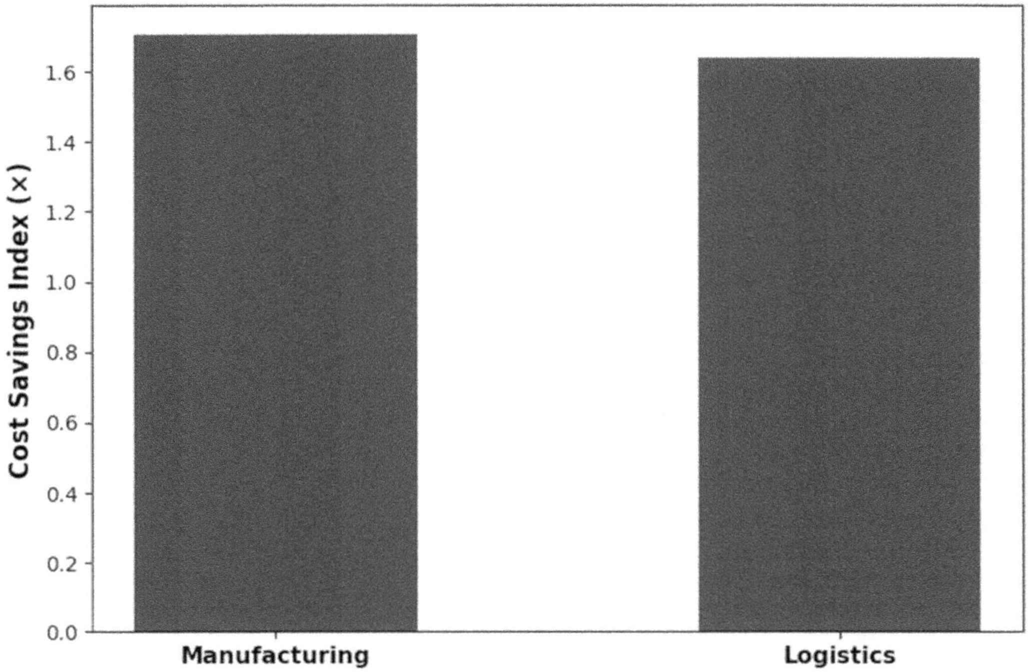


**Figure 4.11 Normalized economic impact of DOS-IoT across manufacturing and logistics, illustrating the consistency of value gains across domains.**

system level would be enough to sustain continued economic performance without recreating the architecture for specific domains, developing particular algorithms for optimizing, or adding analytical parameters. This stable ROI also indicates that financial returns can be expected across different deployment conditions. This emphasizes the role of DOS-IoT as a universal, yet stable, architecture that can deliver robust economic benefits to manufacturing and logistics; it is also scalable, operationally resilient, and easily deployed.

### 4.5.3 Ablation Study

To isolate the role of architectural features individually, we conduct a structured ablation study in which key elements of DOS-IoT are selectively extracted. We investigate the effects of stumbling out edge analytics, enterprise integration, and architectural coherence, while keeping datasets and hyperparameters constant.

Table 4.10 displays the degradation in operation due to architectural ablations and identifies the role of edge analytics, enterprise integration, and system-level coherence in DOS-IoT performance. With all DOS-IoT settings, OEE or inventory improvements range from 18.3% to 33.1%, and downtime or delay reductions range between 22.1% and 30.4%. Upon removal of edge analytics, operational improvements drop to between 9.4% and 15.2%, and downtime or delay reductions to between 11.6% and 17.9%, indicating a major performance loss. When removing enterprise integration, the gains decrease to between 10.8% and 16.4%, while the decrease in downtime or delay falls to between 13.1% and 19.3%. When architectural coherence is broken, it is most grave and gains drop to 7.1%–12.6% and downtime or delay

**Table 4.10 Effect of Architectural Ablations on Operational Performance, Highlighting the Contribution of Edge Analytics, Enterprise Integration, and System-Level Coherence**

| *Configuration* | *OEE/Inventory Gain (%)* | *Downtime/Delay Reduction (%)* |
|---|---|---|
| Full DOS-IoT | 18.3–33.1 | 22.1–30.4 |
| – Edge analytics | 9.4–15.2 | 11.6–17.9 |
| – Enterprise integration | 10.8–16.4 | 13.1–19.3 |
| – Architectural coherence | 7.1–12.6 | 8.9–14.2 |

reductions drop to 8.9%–14.2%. The results show that edge-level analytics, enterprise integration, and coordinated architectural design are critical to ensuring operational performance. Eliminating any of these components further weakens the overall effectiveness of the industrial IoT system.

Table 4.11, which shows the technical effects of system-level design choices on decision latency and alert precision, illustrates the technical effects of architectural ablations. When edge analytics is omitted, it increases decision latency by 28%–34% and alert accuracy by 0.09, representing the lowest value of system responsiveness and analytical reliability. This means that when enterprise integration is removed, latency increases by 14%–18% and precision drops by 0.06, indicating some technical degradation. This leads to the most serious technical degradation when architectural coherence suffers; exponential latency increases by 31%–37%, and there is the greatest loss in alert precision of 0.11. These observations indicate that architectural coherence is necessary for low-latency execution and high-confidence alert generation. The cumulative latency and alert precision drop in the joint indicate that individual parts cannot be optimized for technical performance. Instead, the overall behavior of the system is driven by the interaction effects of the architectural layers; this is consistent with the notion that integrated systems are the only dependable, low-latency, and analytically accurate industrial IoT solutions that work for multi-scale deployments.

Table 4.12 summarizes the economic sensitivity of DOS-IoT to architectural ablations using normalized economic indices. Under the full DOS-IoT configuration, normalized economic gains range from 1.61 to 1.79, representing good financial performance. When edge analytics are eliminated, the economic index lowers to between 1.22 and 1.31—a sharp decrease in the actual economic value. The normalized economic benefits decrease to 1.27–1.38 when enterprises are eliminated; therefore, balancing analytics with decision-making is critical to the realization of value. The most significant drop in economic performance occurs when architectural coherence

**Table 4.11 Impact of Architectural Ablations on Technical Performance, Measured through Changes in Decision Latency and Alert Precision**

| *Configuration* | *Latency Increase (%)* | *Alert Precision Drop* |
|---|---|---|
| – Edge analytics | +28–34 | –0.09 |
| – Enterprise integration | +14–18 | –0.06 |
| – Architectural coherence | +31–37 | –0.11 |

**Table 4.12 Impact of Architectural Ablations on the Normalized Economic Value of Industrial IoT Systems**

| *Configuration* | *Economic Index (Normalized)* |
|---|---|
| Full DOS-IoT | 1.61× to 1.79× |
| – Edge analytics | 1.22× to 1.31× |
| – Enterprise integration | 1.27× to 1.38× |
| – Architectural coherence | 1.10× to 1.19× |

is removed, with the economic index drastically dropping to 1.10–1.19. This level is close to the level of economic performance and negates the financial benefit of the complete system-level integration. The results suggest that to produce sustained industrial IoT value, architectural coherence is important. Economic indices slowly decline over time in ablated configurations, while under true deployment conditions, the coordinated interaction of system components has a greater influence on financial outcomes than isolated optimizations at the component level.

## 4.6 Conclusion and Future Works

This chapter offered a decision-oriented, system-level perspective on industrial IoT deployments, demonstrating that operational value is primarily driven by architectural coherence rather than analytical accuracy. The proposed DOS-IoT model illustrates that heterogeneous IoT data can be combined with modeling of sensing, communication, processing position, and enterprise integration to develop operational decisions that are timely, reliable, and economically viable. Empirical analysis of large-scale spatiotemporal logistics trajectories and high-dimensional manufacturing sensor data consistently indicates that system-level design, rather than component-level optimization, improves efficiency, technical responsiveness, and economic value. However, results from control ablations also show that architectural choices, such as edge–cloud analytical placement and feature aggregation strategies, directly affect latency, robustness, and downstream decision quality. Importantly, cross-domain analysis shows that these gains are based on transferable architectural principles rather than domain-specific heuristics or aggressive hyperparameter tuning. This concept can be extended to three areas in future research. Validation across other industries and multi-enterprise settings is also necessary to further validate-scalability and generality. Second, adaptive and online architectural reconfiguration, such as dynamic analytics placement and resource-aware sensing, could improve responsiveness in nonstationary operations. Third, by enhancing integration with explainable analytics and digital twin infrastructures, interpretability, governance, and trust could be improved, thereby bolstering the deployment of IoT-enabled decision systems in safety-critical and large-scale industrial environments.

## References


1. M. Soori, B. Arezoo, and R. Dastres, "Internet of Things for smart factories in Industry 4.0, a review," *Internet of Things and Cyber-Physical Systems*, vol. 3, pp. 192–204, Jan. 2023, https://doi.org/10.1016/j.iotcps.2023.04.006.
2. H. Boyes, B. Hallaq, J. Cunningham, and T. Watson, "The industrial Internet of Things (IIoT): An analysis framework," *Computers in Industry*, vol. 101, pp. 1–12, Jun. 2018, https://doi.org/10.1016/j.compind.2018.04.015.
3. A. E. Edje, M. S. A. Latiff, and W. H. Chan, "IoT data analytic algorithms on edge-cloud infrastructure: A review," *Digital Communications and Networks*, vol. 9, no. 6, pp. 1486–1515, Oct. 2023, https://doi.org/10.1016/j.dcan.2023.10.002.
4. A. Gordon, "Internet of Things-based real-time production logistics, big data-driven decision-making processes, and industrial artificial intelligence in sustainable cyber-physical manufacturing systems," *Journal of Self-Governance and Management Economics*, vol. 9, no. 3, pp. 61–73, 2021. https://www.ceeol.com/search/article-detail?id=983528 (accessed Dec. 27, 2025).
5. C. Resende et al., "TIP4.0: industrial Internet of Things platform for predictive maintenance," *Sensors*, vol. 21, no. 14, p. 4676, Jul. 2021, https://doi.org/10.3390/s21144676.
6. X. Mu and M. F. Antwi-Afari, "The applications of Internet of Things (IoT) in industrial management: A science mapping review," *International Journal of Production Research*, vol. 62, no. 5, pp. 1928–1952, 2024, https://doi.org/10.1080/00207543.2023.2290229.
7. T. S. Sikarwar, A. S. Chauhan, N. Jain, and H. Mathur. "Application of predictive analytics in IOT data processing." *Indian Journal of Information Sources and Services*, vol. 15, no. 2, pp. 340–348, 2025.
8. A. Kota, "Predictive maintenance: Integrating edge AI with cloud computing for industrial IOT," *International Journal of Research in Computer Applications and Information Technology*, vol. 7, no. 2, pp. 2842–2851, Dec. 2024, https://doi.org/10.34218/ijrcait_07_02_218.
9. A. Kanawaday and A. Sane, "Machine learning for predictive maintenance of industrial machines using IoT sensor data," *2017 8th IEEE International Conference on Software Engineering and Service Science (ICSESS)*, Nov. 2017, https://doi.org/10.1109/icsess.2017.8342870.

10. K. M. Qureshi et al., "Investigating Industry 4.0 technologies in Logistics 4.0 usage towards sustainable manufacturing supply chain," *Heliyon*, vol. 10, no. 10, p. e30661, 2024.
11. V. Govindarajan, P. Patel, S. Tripathi, M. A. Hoque, and G. S. Kashyap, "MAGIC-Enhanced Keyword Prompting for Zero-Shot Audio Captioning with CLIP Models," *arXiv.org*, 2025. https://arxiv.org/abs/2509.12591 (accessed Jan. 13, 2026).
12. A. Soni et al., "Can We Predict Your Next Move without Breaking Your Privacy?," *arXiv.org*, 2025. https://arxiv.org/abs/2507.08843 (accessed Jan. 13, 2026).
13. Q. He et al., "Integrating IoT and 6G: Applications of edge intelligence, challenges, and future directions," *IEEE Transactions on Services Computing*, vol. 18, no. 4, pp. 2471–2488, Jul. 2025, https://doi.org/10.1109/tsc.2025.3586152.
14. C. Anton-Haro and M. Dohler, eds. *Machine-to-Machine (M2M) Communications: Architecture, Performance and Applications.* Elsevier, 2014.
15. B. Babayigit and M. Abubaker, "Industrial Internet of Things: A review of improvements over traditional SCADA systems for industrial automation," *IEEE Systems Journal*, vol. 18, no. 1, pp. 120–133, May 2023, https://doi.org/10.1109/jsyst.2023.3270620.
16. A. A. Mirani, G. Velasco-Hernandez, A. Awasthi, and J. Walsh, "Key challenges and emerging technologies in industrial IoT architectures: A review," *Sensors*, vol. 22, no. 15, p. 5836, Aug. 2022, https://doi.org/10.3390/s22155836.
17. S. N. Deshpande and R. M. Jogdand, "A survey on Internet of Things (IoT), Industrial IoT (IIOT) and Industry 4.0." *International Journal of Computer Applications*, vol. 175, no. 27, pp. 20–27, 2020.
18. T. Zhu, Y. Ran, X. Zhou, and Y. Wen, "A survey on intelligent predictive maintenance (IPdM) in the era of fully connected intelligence," *IEEE Communications Surveys & Tutorials*, pp. 1–1, 2025, https://doi.org/10.1109/comst.2025.3567802.
19. S. Elkateb, A. Métwalli, A. Shendy, and A. E. B. Abu-Elanien, "Machine learning and IoT: Based predictive maintenance approach for industrial applications," *Alexandria Engineering Journal*, vol. 88, pp. 298–309, Feb. 2024, https://doi.org/10.1016/j.aej.2023.12.065.
20. F. Qiu et al., "A review on integrating IoT, IIoT, and Industry 4.0: A pathway to smart manufacturing and digital transformation," *IET Information Security*, vol. 2025, no. 1, Jan. 2025, https://doi.org/10.1049/ise2/9275962.
21. V. Barbuto, C. Savaglio, M. Chen, and G. Fortino, "Disclosing edge intelligence: A systematic meta-survey," *Big Data and Cognitive Computing*, vol. 7, no. 1, pp. 44–44, Mar. 2023, https://doi.org/10.3390/bdcc7010044.
22. G. S. S. Chalapathi, V. Chamola, A. Vaish, and R. Buyya, "Industrial Internet of Things (IIoT) applications of edge and fog computing: A review and future directions," *Fog/Edge Computing for Security, Privacy, and Applications*, pp. 293–325, 2021, https://doi.org/10.1007/978-3-030-57328-7_12.
23. T. Zhang et al., "A survey on Industrial Internet of Things (IIoT) Testbeds for Connectivity Research," *arXiv.org*, 2024. https://arxiv.org/abs/2404.17485 (accessed Dec. 27, 2025).

24. L. F. Baptista and J. Barata, "Piloting Industry 4.0 in SMEs with RAMI 4.0: An enterprise architecture approach," *Procedia Computer Science*, vol. 192, pp. 2826–2835, 2021, https://doi.org/10.1016/j.procs.2021.09.053.
25. E. Y. Nakagawa et al., "Industry 4.0 reference architectures: State of the art and future trends," *Computers & Industrial Engineering*, vol. 156, pp. 107241–107241, Mar. 2021, https://doi.org/10.1016/j.cie.2021.107241.
26. E. Dritsas and M. Trigka, "A survey on the applications of cloud computing in the industrial Internet of Things," *Big Data and Cognitive Computing*, vol. 9, no. 2, pp. 44–44, Feb. 2025, https://doi.org/10.3390/bdcc9020044.
27. S. Hamdan, M. Ayyash, and S. Almajali, "Edge-computing architectures for Internet of Things applications: A survey," *Sensors*, vol. 20, no. 22, p. 6441, Nov. 2020, https://doi.org/10.3390/s20226441.
28. M. Laroui et al., "Edge and fog computing for IoT: A survey on current research activities & future directions," *Computer Communications*, vol. 180, pp. 210–231, Sep. 2021, https://doi.org/10.1016/j.comcom.2021.09.003.
29. G. Caiza, M. Saeteros, W. Oñate, and M. V. Garcia, "Fog computing at industrial level, architecture, latency, energy, and security: A review," *Heliyon*, vol. 6, no. 4, p. e03706, Apr. 2020, https://doi.org/10.1016/j.heliyon.2020.e03706.
30. R. Kumar and N. Agrawal, "Analysis of multi-dimensional Industrial IoT (IIoT) data in Edge–Fog–Cloud based architectural frameworks: A survey on current state and research challenges," *Journal of Industrial Information Integration*, vol. 35, pp. 100504–100504, Jul. 2023, https://doi.org/10.1016/j.jii.2023.100504.
31. L. Roda-Sanchez et al., "Cloud–edge microservices architecture and service orchestration: An integral solution for a real-world deployment experience," *Internet of Things*, vol. 22, p. 100777, Jul. 2023, https://doi.org/10.1016/j.iot.2023.100777.
32. E. Villar, I. M. Toral, I. Calvo, O. Barambones, and P. Fernández-Bustamante, "Architectures for Industrial AIoT applications," *Sensors*, vol. 24, no. 15, pp. 4929–4929, Jul. 2024, https://doi.org/10.3390/s24154929.
33. H. M. Fahmy, H. I. Helmy, F. E. Ali, N. E. Motei, and M. S. Fathy, "Industrial applications of sensors," *Handbook of Nanosensors*, pp. 1–34, Oct. 2023, https://doi.org/10.1007/978-3-031-16338-8_55-1.
34. M. Javaid, A. Haleem, R. P. Singh, S. Rab, and R. Suman, "Significance of sensors for industry 4.0: Roles, capabilities, and applications," *Sensors International*, vol. 2, p. 100110, 2021, https://doi.org/10.1016/j.sintl.2021.100110.
35. I. U. Hassan, K. Panduru, and J. Walsh, "An in-depth study of vibration sensors for condition monitoring," *Sensors*, vol. 24, no. 3, p. 740, Jan. 2024, https://doi.org/10.3390/s24030740.
36. W. G. Buratto et al., "A review of automation and sensors: Parameter control of thermal treatments for electrical power generation," *Sensors*, vol. 24, no. 3, pp. 967–967, Feb. 2024, https://doi.org/10.3390/s24030967.
37. M. Majid et al., "Applications of wireless sensor networks and Internet of Things frameworks in the Industry Revolution 4.0: A systematic literature review," *Sensors*, vol. 22, no. 6, p. 2087, Mar. 2022, https://doi.org/10.3390/s22062087.

38. H. Landaluce et al., "A review of IoT sensing applications and challenges using RFID and wireless sensor networks," *Sensors*, vol. 20, no. 9, pp. 2495–2495, Apr. 2020, https://doi.org/10.3390/s20092495.
39. C. Savaglio, P. Mazzei, and G. Fortino, "Edge intelligence for Industrial IoT: Opportunities and limitations," *Procedia Computer Science*, vol. 232, pp. 397–405, Jan. 2024, https://doi.org/10.1016/j.procs.2024.01.039.
40. J. Pan and J. McElhannon, "Future edge cloud and edge computing for Internet of Things applications," *IEEE Internet of Things Journal*, vol. 5, no. 1, pp. 439–449, Feb. 2018, doi: https://doi.org/10.1109/jiot.2017.2767608.
41. Md. Noor-A-Rahim et al., "Wireless communications for smart manufacturing and Industrial IoT: Existing technologies, 5G and beyond," *Sensors*, vol. 23, no. 1, pp. 73–73, Dec. 2022, https://doi.org/10.3390/s23010073.
42. I. Behnke and H. Austad, "Real-time performance of Industrial IoT communication technologies: A review," *IEEE Internet of Things Journal*, vol. 11, no. 5, pp. 7399–7410, Mar. 2024, https://doi.org/10.1109/jiot.2023.3332507.
43. A. Al-Fuqaha, M. Guizani, M. Mohammadi, M. Aledhari, and M. Ayyash, "Internet of Things: A survey on enabling technologies, protocols, and applications," *IEEE Communications Surveys & Tutorials*, vol. 17, no. 4, pp. 2347–2376, Jan. 2015, https://doi.org/10.1109/comst.2015.2444095.
44. F. Yao, Y. Ding, S. Hong, and S.-H. Yang, "A survey on evolved LoRa-based communication technologies for emerging Internet of Things applications," *International Journal of Network Dynamics and Intelligence*, vol. 1, no. 1, pp. 4–19, Dec. 2022, https://doi.org/10.53941/ijndi0101002.
45. A. Augustin, J. Yi, T. Clausen, and W. Townsley, "A study of LoRa: Long range & low power networks for the Internet of Things," *Sensors*, vol. 16, no. 9, pp. 1466–1466, Sep. 2016, https://doi.org/10.3390/s16091466.
46. K. Mekki, E. Bajic, F. Chaxel, and F. Meyer, "A comparative study of LPWAN technologies for large-scale IoT deployment," *ICT Express*, vol. 5, no. 1, pp. 1–7, Jan. 2018, https://doi.org/10.1016/j.icte.2017.12.005.
47. J. John, M. Noor-A-Rahim, A. Vijayan, P. H. Vincent, and D. Pesch, "Industry 4.0 and Beyond: The Role of 5G, WiFi 7, and TSN in Enabling Smart Manufacturing," *arXiv.org*, 2023. https://arxiv.org/abs/2310.02379?utm_source=chatgpt.com (accessed Dec. 27, 2025).
48. F. Folgado, D. Calderón, I. González, and A. Calderón, "Review of Industry 4.0 from the perspective of automation and supervision systems: Definitions, architectures and recent trends," *Electronics*, vol. 13, no. 4, pp. 782–782, Feb. 2024, https://www.mdpi.com/2079-9292/13/4/782.
49. D. C. Montgomery. *Statistical Quality Control, 7th Edition*. Wiley, 2012. https://books.google.co.in/books/about/Statistical_Quality_Control_7th_Edition.html?id=RgQcAAAAQBAJ&redir_esc=y (accessed Dec. 27, 2025).
50. M. Shahnawaz and M. Kumar, "A comprehensive survey on big data analytics: Characteristics, tools and techniques," *ACM Computing Surveys*, vol. 57, no. 8, pp. 1–33, Feb. 2025, https://doi.org/10.1145/3718364.

51. G. S. Kashyap et al., "Can AI See What We Can't? Leveraging Deep Learning and Multi-Temporal Satellite Data to Revolutionize Crop Type Mapping and Yield Prediction," *ICASSP* 2025–2025 *IEEE International Conference on Acoustics, Speech and Signal Processing (ICASSP)*, pp. 1–5, Apr. 2025, https://doi.org/10.1109/icassp49660.2025.10890344.
52. R. Rosati et al., "From knowledge-based to big data analytic model: A novel IoT and machine learning based decision support system for predictive maintenance in Industry 4.0," *Journal of Intelligent Manufacturing*, vol. 34, no. 1, pp. 107–121, May 2022, https://doi.org/10.1007/s10845-022-01960-x.
53. M. Rishi, "Predictive analytics in supply chain management: The role of AI and machine learning in demand forecasting," *Advances in Consumer Research*, vol. 2, pp. 142–149, Feb. 2025.
54. S. Tripathi, M. T. Nafis, I. Hussain, and J. Gao, "The Confidence Paradox: Can LLM Know When It's Wrong," *arXiv.org*, 2025. https://arxiv.org/abs/2506.23464 (accessed Jan. 13, 2026).
55. G. S. Kashyap, K. Malik, S. Wazir, and A. E. I. Brownlee, "Optimization of the rectangle area inside a concave polygon using PSO and Tabu Search," *Soft Computing*, vol. 29, no. 7, pp. 3217–3239, Apr. 2025, https://doi.org/10.1007/s00500-025-10568-1.
56. M. Kulahara, A. Khader, S. Tripathi, and M. A. Hoque, "A CNN-based framework for geometric alignment of historical and satellite imagery," *IEEE Access*, vol. 13, pp. 132259–132268, Jan. 2025, https://doi.org/10.1109/access.2025.3589497.
57. E. Byres, A. Ginter, and J. Langill. "How Stuxnet spreads: A study of infection paths in best practice systems." *Tofino Security, white paper*, 2011.
58. S. Karnouskos, "Stuxnet worm impact on industrial cyber-physical system security," In *IECON 2011–37th Annual Conference of the IEEE Industrial Electronics Society*, pp. 4490–4494. IEEE, 2011, https://doi.org/10.1109/iecon.2011.6120048.
59. A. Humayed, J. Lin, F. Li, and B. Luo, "Cyber-physical systems security: A survey," *IEEE Internet of Things Journal*, vol. 4, no. 6, pp. 1802–1831, Dec. 2017, https://doi.org/10.1109/jiot.2017.2703172.
60. R. Roman, J. Lopez, and M. Mambo, "Mobile edge computing, Fog et al.: A survey and analysis of security threats and challenges," *Future Generation Computer Systems*, vol. 78, pp. 680–698, Nov. 2016, https://doi.org/10.1016/j.future.2016.11.009.
61. K. Stouffer, J. Falco, and K. Scarfone, "Guide to industrial control systems (ICS) security." *NIST Special Publication*, vol. 800, no. 82, pp. 16–16, 2011.
62. R. Mitchell and I.-R. Chen, "A survey of intrusion detection techniques for cyber-physical systems," *ACM Computing Surveys*, vol. 46, no. 4, pp. 1–29, Mar. 2014, https://doi.org/10.1145/2542049.
63. A. L. Buczak and E. Guven, "A survey of data mining and machine learning methods for cyber security intrusion detection," *IEEE Communications Surveys & Tutorials*, vol. 18, no. 2, pp. 1153–1176, 2016, https://doi.org/10.1109/comst.2015.2494502.

64. S. Adepu and A. Mathur, "Distributed attack detection in a water treatment plant: Method and case study," *IEEE Transactions on Dependable and Secure Computing*, vol. 18, no. 1, pp. 86–99, Jan. 2021, https://doi.org/10.1109/tdsc.2018.2875008.
65. S. Rose, O. Borchert, S. Mitchell, and S. Connelly, "Zero trust architecture," *NIST Special Publication 800-207*, vol. 1, no. 800–207, Aug. 2020, https://doi.org/10.6028/nist.sp.800-207.
66. G. Zyskind, O. Nathan, and A. S. Pentland, "Decentralizing privacy: Using blockchain to protect personal data," In *2015 IEEE Security and Privacy Workshops*, pp. 180–184. IEEE, 2015. https://doi.org/10.1109/spw.2015.27.
67. S. Karnouskos et al., "A SOA-based architecture for empowering future collaborative cloud-based industrial automation," *IECON 2012–38th Annual Conference on IEEE Industrial Electronics Society*, Oct. 2012, https://doi.org/10.1109/iecon.2012.6389042.
68. T. E. Vollmann, W. L. Berry, and D. C. Whybark. *Manufacturing Planning and Control Systems*. Irwin/McGraw-Hill, 1997.
69. A. W. Colombo, T. Bangemann, S. Karnouskos, J. Delsing, P. Stluka, R. Harrison, F. Jammes, and J. L. Lastra. *Industrial Cloud-Based Cyber-Physical Systems: The IMC-AESOP Approach*. Springer, 2014.
70. A. Botta, W. de Donato, V. Persico, and A. Pescapé, "Integration of Cloud computing and Internet of Things: A survey," *Future Generation Computer Systems*, vol. 56, pp. 684–700, Oct. 2015, https://doi.org/10.1016/j.future.2015.09.021.
71. F. Tao, H. Zhang, A. Liu, and A. Y. C. Nee, "Digital Twin in Industry: State-of-the-Art," *IEEE Transactions on Industrial Informatics*, vol. 15, no. 4, pp. 2405–2415, Apr. 2019, doi: https://doi.org/10.1109/tii.2018.2873186.
72. E. Cho, T.-W. Chang, and G. Hwang, "Data preprocessing combination to improve the performance of quality classification in the manufacturing process," *Electronics*, vol. 11, no. 3, pp. 477–477, Feb. 2022, https://doi.org/10.3390/electronics11030477.
73. Y. Zheng, X. Xie, and W.-Y. Ma, "GeoLife: A collaborative social networking service among user, location and trajectory," ResearchGate, vol. 33, no. 2, pp. 32–39, Jun. 2010. https://www.researchgate.net/publication/220282575_GeoLife_A_Collaborative_Social_Networking_Service_among_User_Location_and_Trajectory (accessed Dec. 27, 2025).

# Chapter 5

# Natural Language Processing (NLP) for Supply Chains

Ebad Shabbir

## 5.1 Introduction

Natural language processing (NLP) allows computers to understand, analyze, and infer human language. Its importance for supply chain management has increased as digitalization, in addition to traditional transactional documents, brings with it huge amounts of unstructured text. Modern supply chains provide a constant stream of text from customer inquiries, supplier communications, audit and compliance reports, regulatory statements, contracts, service tickets, transportation notices, and market intelligence via news and industry reports [1]. These textual artifacts are also known to contain early, context-rich signals and are not fully represented by numerical measures unless they represent specific changes in demand, supplier reliability, regulatory vulnerability, and disruption risk [2]. Although the volume of information in languages is increasing, most enterprise supply chain systems are still configured for both structured transactional data and numerical aggregation [3]. Textual information is therefore often viewed as an external context and not as a primary analysis input. In practice, language insights are often revealed through manual checking or ad hoc examinations. These techniques are slow, subjective, and not feasible to scale within large global supply networks [4]. Earlier work has demonstrated that text-based qualitative signals can provide supplementary data to quantitative information by providing earlier alerts on operations and market developments, improving both

DOI: 10.4324/9781003724889-5

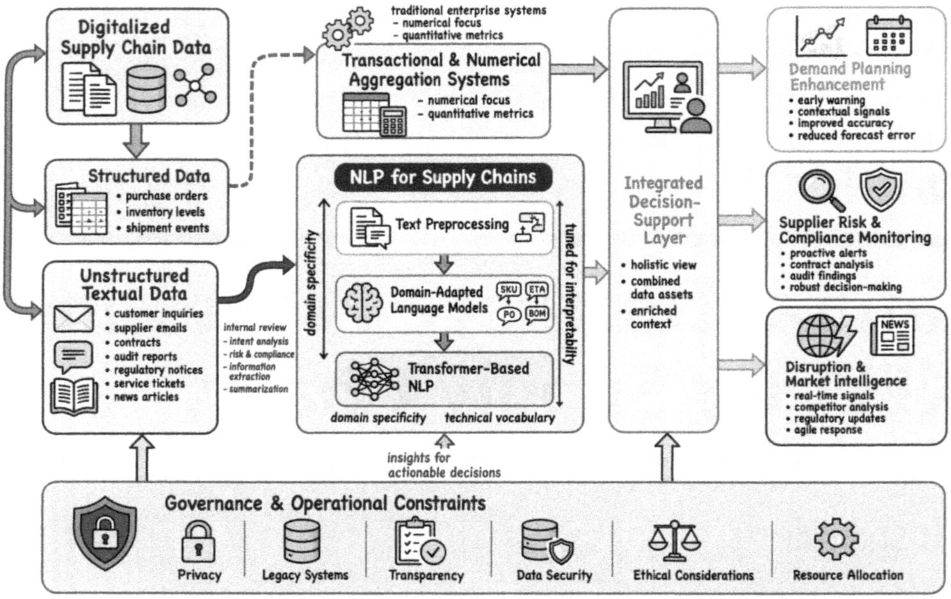


**Figure 5.1 Decision-centric integration of structured transactional data and unstructured textual signals using domain-adapted NLP. The architecture illustrates how language representations are aligned with numerical supply chain data to support end-to-end decision processes—including demand forecasting, risk monitoring, and service execution—under interpretability and governance constraints.**

situational awareness and rapid decision-making [5]. Figure 5.1 illustrates conceptually how unstructured textual data streams can be combined with structured supply chain data using domain-specific NLP to aid in demand planning, risk management, and disruption monitoring within governance constraints.

NLP helps computers understand, interpret, and infer human language. With the increasing volumes of unstructured textual data as a result of digitization, as well as the emergence of digitalization providing a huge amount of unstructured data along with transactional records, its relevance for supply chain management has become very relevant. In addition to consolidated data such as purchase orders, inventory levels, and shipment events, modern supply chains produce a continuum of text, based on customer inquiries, supplier communications, audit and compliance reports, regulatory disclosures, contracts, service tickets, transportation notices, and externally available market information such as news and industry reports [6]. These textual artifacts often contain prerequisite signals about shifts in demand, supplier reliability, regulatory exposure, and disruption risk that are not fully understood by numbers alone [7]. While language-based information becomes increasingly available, most enterprise supply chain systems still use structured transactional data and numerical aggregation [8]. Text is therefore typically considered as a supplement rather than a primary analysis input. In practice, knowledge gathered from language can be obtained from manual testing or from ad hoc analyses. These methods are cumbersome, subjective, and difficult to scale across large global supply chains [9]. Previous research has shown that textual qualitative signals can be used to complement quantitative information by providing a priori alerts of operating problems and market changes, thus improving situational awareness and the speed of decisions [10]. Despite the growing academic and industrial interest in incorporating NLP in supply chain studies, the current approaches remain poorly defined and narrowly applied. The vast majority of prior work in NLP has focused on specific tasks such as sentiment analysis of customer feedback or keyword analysis of disruption-related news without systematically linking language-derived insights to the core decision-making process [11]. This limitation is what is defined here as the Language–Decision Disconnect: the lack of end-to-end mechanisms that integrate textual understanding directly into the forecasting, risk assessment, and service implementation processes. In demand planning, textual information is typically used descriptively rather than being embedded officially in forecasting models, making it less important for planning accuracy [12]. Likewise, supplier risk assessment is often based on manual review of documents or inflexible rule-based systems that have no flexibility with respect to suppliers, regulatory circumstances, or geographic context [13]. This gap is also exacerbated by the fact that supply chain terminology is domain-specific. Procurement, manufacturing, logistics, and compliance texts use specialized vocabulary, abbreviations, and implicit business assumptions distinct from open-domain texts [14]. Therefore, when general-purpose NLP models trained on generic corpora are applied directly to supply chain data, they may not perform as well. Such an effect leads to misinterpretation in highly critical operations. Other organizational constraints, such as data privacy requirements, compatibility with existing enterprise systems, and the need for transparency and interpretability, complicate the practice of language-driven automation [15].

This chapter presents Decision-Centric Supply Chain NLP (DC-SC-NLP), an analytical framework that integrates language processing into end-to-end supply chain decision support to combat the Language–Decision Disconnect. Rather than adding improvements to individual NLP functions, DC-SC-NLP integrates domain-specific language models directly into forecasting, supplier risk surveillance, and customer service workflows with an emphasis on temporal alignment, interpretability, and governance. It applies both traditional NLP methods and contemporary transformer models with an emphasis on robustness across multiple text sources as well as conformity with structured data analytics.

Table 5.1 summarizes how DC-SC-NLP is positioned in relation to existing approaches. The table shows that a value of 1 means the corresponding capability is natively supported by the given system or approach, whereas a value of 0 means it is not. Conventional supply chain management systems depend significantly on structured transactional data (1), but they do not systematically utilize unstructured text (0) or incorporate decision-centric NLP (0). NLP use cases that are isolated offer support for textual analysis (1), but they generally do not integrate with decision-making processes, governance frameworks, or cross-functional applicability (0). Conversely, the proposed DC-SC-NLP framework facilitates the use of unstructured text, domain adaptation, systematic decision integration, governance awareness, and scalable automation (all 1s), embodying its system-level design philosophy.

This chapter offers a structured and analytically based investigation of NLP as a fundamental capability for contemporary supply chain management. This chapter illustrates the conversion of language-derived signals into actionable decision inputs within realistic operational constraints by assessing DC-SC-NLP with publicly available datasets that encompass both demand-side customer text and external risk-related news. Empirical findings demonstrate quantifiable enhancements in forecasting accuracy, risk assessment efficiency, and customer service quality, highlighting the real-world significance of addressing the Language–Decision Disconnect. This chapter is organized to progressively establish NLP as a decision-centric capability for supply chain management. Section 5.2 reviews the evolution of NLP technologies and systematically examines prior applications in demand forecasting, supplier risk monitoring, contract analysis, customer communication, and domain adaptation, highlighting the Language–Decision Disconnect across these domains. Section 5.3 introduces the materials and methods, describing the dual-dataset design, preprocessing pipelines, and the proposed DC-SC-NLP architecture with its domain-adapted representation layer and

**Table 5.1 System-Level Comparison of Supply Chain Analytics Paradigms, Highlighting the Transition from Transaction-Centric Systems and Task-Isolated NLP toward Decision-Centric, Domain-Adapted NLP Integration**

| *Main Feature* | *Traditional SCM Systems* | *Isolated NLP Use Cases* | *Integrated NLP Framework (Proposed Chapter)* | *Domain-Adapted Language Models* | *Decision-Centric NLP Integration* |
|---|---|---|---|---|---|
| Reliance on structured transactional data | 1 | 0 | 0 | 0 | 0 |
| Utilization of unstructured textual data | 0 | 1 | 1 | 1 | 1 |
| Manual or ad hoc text analysis | 1 | 1 | 0 | 0 | 0 |
| Systematic integration with decision processes | 0 | 0 | 1 | 1 | 1 |
| Support for demand forecasting enhancement | 0 | 0 | 1 | 1 | 1 |
| Supplier risk intelligence from text | 0 | 1 | 1 | 1 | 1 |
| Cross-functional applicability | 0 | 0 | 1 | 1 | 1 |
| Handling of domain-specific language | 0 | 0 | 1 | 1 | 1 |
| Robustness across heterogeneous text sources | 0 | 0 | 1 | 1 | 1 |
| Governance and interpretability emphasis | 0 | 0 | 1 | 1 | 1 |
| Alignment with enterprise systems | 1 | 0 | 1 | 1 | 1 |
| Scalable automation capability | 0 | 0 | 1 | 1 | 1 |

Binary entries indicate the presence (1) or absence (0) of each capability.

task-specific decision modules. Section 5.4 presents the experimental setup, including evaluation metrics and hyperparameter configurations aligned with operational decision relevance. Section 5.5 reports comparative results, cross-domain generalization analyses, robustness tests, and ablation studies that quantify the contributions of temporal modeling, domain adaptation, and decision-centric integration. Finally, Section 5.6 concludes with key findings and outlines directions for future research.

## 5.2 Related Works

### *5.2.1 Evolution of Natural Language Processing Technologies*

NLP could be seen as an incremental transformation of representational and modeling capabilities, with each step responding to the limitations of earlier capabilities and adding new constraints. This growth is summarized in Table 5.2 as a result of the basic linguistic, statistical, and architectural components of this development; Figure 5.2 presents a conceptual representation of the accumulation of these capabilities over time in supply chain applications.

Early NLP systems remained largely rule-based and symbolic, relying on handmade grammars, lexicons, and deterministic linguistic rules [16]. The systems in Table 5.2 relied heavily on handwritten knowledge (hand-crafted linguistic rules = 1) and were based on context-independent interpretations of language. They were useful tools in narrow, well-defined contexts, but lacked probabilistic learning [17], scalability to large numbers of texts, and strength against linguistic variability. Thus, such systems did not function well in the context of textual samples of the supply chain—such as contracts, compliance reports, and communication with suppliers—which are heterogeneous and evolving, and involve ambiguity and domain-specific applications [18].

The symbolic constraints of the transition to statistical NLP enabled probabilistic learning based on annotation of the corpus and allowed for the necessary sequence labeling activities, such as part-of-speech tagging and recognized entity recognition. Table 5.2 shows that statistical NLP was the first paradigm that allowed probabilistic learning and scalable corpus-based training. However, semantic representations were largely shallow and abstracted from context. Surface level was used to design models, not distributed meaning representations. Therefore, statistical NLP improved robustness and

**Table 5.2 Evolution of NLP Paradigms, Illustrating How Successive Modeling Approaches Expand Representational Capacity, Contextual Reasoning, and Scalability, Culminating in Transformer-Based Language Models**

| *Main Feature* | *Rule-Based/ Symbolic NLP* | *Statistical NLP* | *Static Word Embeddings* | *Contextual Embeddings* | *Transformer-Based Models* |
|---|---|---|---|---|---|
| Hand-crafted linguistic rules | 1 | 0 | 0 | 0 | 0 |
| Probabilistic learning from corpora | 0 | 1 | 0 | 0 | 0 |
| Sequence labeling capability | 0 | 1 | 0 | 0 | 0 |
| Distributed semantic representations | 0 | 0 | 1 | 1 | 1 |
| Context-independent word meaning | 1 | 1 | 1 | 0 | 0 |
| Context-aware representations | 0 | 0 | 0 | 1 | 1 |
| Long-range dependency modeling | 0 | 0 | 0 | 0 | 1 |
| Self-attention mechanism | 0 | 0 | 0 | 0 | 1 |
| Large-scale pretraining | 0 | 0 | 0 | 0 | 1 |
| Task-specific fine-tuning | 0 | 1 | 1 | 1 | 1 |
| Scalability to large corpora | 0 | 1 | 1 | 1 | 1 |
| Domain adaptation requirement | 0 | 0 | 0 | 1 | 1 |

Binary entries indicate the presence (1) or absence (0) of each capability.

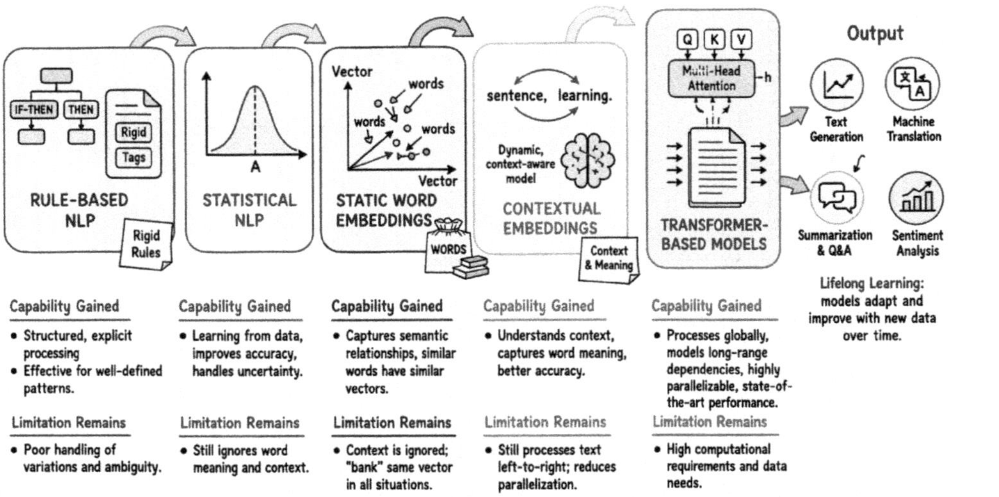


**Figure 5.2 Evolution of NLP paradigms, showing progressive representational and modeling capabilities—from rule-based systems to transformer-based language models—and their implications for handling domain-specific language in supply chain decision contexts.**

generalization but failed to capture the nuanced business semantics that are a prerequisite of decision-oriented supply chain analysis [19].

Another shift occurred with static distributed word embeddings, which offered a tool for gathering dense semantic representations from a large set of unlabeled corporeal data. This paradigm resulted in distributed semantic representations (see Table 5.2) and improved performance across multiple tasks, including sentiment analysis and information retrieval. However, static embeddings assign each word a single vector without regard for context, thus fixing meaning in place. This restriction becomes particularly problematic in supply chain contexts where the same terms can carry different meanings in procurement, logistics, and compliance contexts. This complicates the division between language comprehension and decision-making in practice [20].

This was addressed by model of contextual embedding, which produced representations based on text in the vicinity. As shown in Table 5.2, contextual embeddings produced models that were contextualized and more meaningful, but did not address all the long-run dependencies. Models such as these made learning the language more responsive to context by making sense of word meanings over tasks. Nonetheless, they also had limited reasoning power in the case of long papers, such as multi-page contracts or extended regulatory filings [21].

Transformer-based architectures greatly influenced the abilities of NLP. Transformer models also allowed for explicit modeling of long-run dependencies and global contexts through the use of self-attention mechanisms, as reflected in Table 5.2. Because of pretraining and task-specific fine-tuning, these models could transfer general linguistic knowledge across tasks and domains. BERT and its predecessors have greatly increased language comprehension and generation capabilities for a wide variety of downstream applications [22].

As shown in Figure 5.2 and summarized by the domain adaptation requirement in Table 5.2, these architectural improvements do not automatically apply to enterprise or supply chain settings. Since transformer-based models are inherently domain-agnostic during pretraining, they need to be adapted to the specialized vocabularies, document structures, and implicit business assumptions characteristic of supply chain text. Even the most advanced models are decision-agnostic without such adaptation, generating representations that, while potentially linguistically accurate, do not align with operational needs. This observation directly motivates the creation of decision-centric, domain-adapted frameworks like DC-SC-NLP, which explicitly connect representational advancements in NLP with the needs of supply chain decision processes.

## *5.2.2 Text Analytics for Demand Sensing and Forecasting*

Usually, demand prediction is based on econometric and statistical models that use both historical sales data and structured explanatory variables, such as promotions, pricing, and calendar effects [23]. Figure 5.3 shows that these models are strong enough in stable conditions, but they cannot properly encode unstructured textual data or handle nonlinear demand dynamics. This makes them unable to predict sudden demand changes resulting from consumer sentiment, competitive moves, or external factors [24].

First attempts to address these problems relied on text as additional demand signals, using data generated by consumers, such as online reviews, discussion forums, and product reviews [25]. These methods, as reported in Table 5.3, used keyword counts, sentiment scores, or topic frequency as market-specific proxies to include structured text. Although these methods showed some limited lead-indicator ability and modest short-term prediction improvements in certain situations [26], they were still highly subject to noise, sampling bias, and weak or delayed relationships between online discourse and actual demand [27]. Crucially, these methods could not

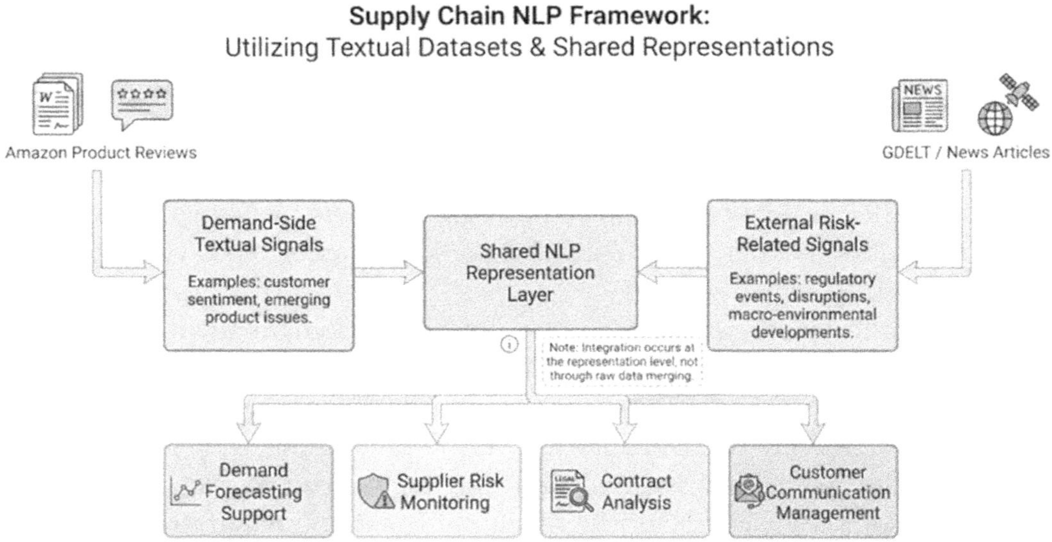


**Figure 5.3 Dual-stream data pipeline for decision-centric supply chain NLP, illustrating the separate ingestion of demand-side customer text and external risk-related news, followed by convergence into shared, domain-adapted language representations that support downstream forecasting and risk decision tasks.**

accommodate the joint modeling of text and time-series data, which limited their capacity to affect forecast generation in a principled way.

Later studies expanded the application of textual demand sensing to include social media and external sources of information, such as news articles, analysts' comments, and macroeconomic reports [28]. These methods increased coverage in industrial and business-to-business supply chains, where consumer-facing signals are limited. While the social media and news methods were based on shallow text representations and independent feature pipelines, as shown in Table 5.3, such techniques remained unaffected by noisy signals and were therefore not interpretable in planning decisions [29]. This is the first time in recent years that deep learning and transformer-based architectures have taken off, indicating a major shift in the methodology. The existence of neural models that incorporated the history of demand and text embeddings together gave an account of nonlinear interactions between numerical time series and unstructured language signals [30]. Attention mechanisms help to dynamically weight textual inputs based on temporal relevance, which increases robustness when text availability or signal quality changes [31]. These abilities, as shown in Table 5.3, constitute a clear improvement at the modeling level compared to earlier text-based methods. However, in spite of these technical advancements, the

**Table 5.3 Comparative Evolution of Demand Sensing and Forecasting Approaches, Highlighting the Progression from Sales-Driven Models to Text-Integrated Deep Learning Methods and Their Support for Decision-Relevant Demand Signals**

| *Main Feature* | *Econometric/ Statistical Models* | *Early Text-Based Methods* | *Social Media Analytics* | *News and External Text* | *Deep Learning and Transformers* |
|---|---|---|---|---|---|
| Reliance on historical sales data | 1 | 1 | 0 | 0 | 1 |
| Use of structured explanatory variables | 1 | 0 | 0 | 0 | 1 |
| Incorporation of unstructured text | 0 | 1 | 1 | 1 | 1 |
| Keyword-based demand proxies | 0 | 1 | 0 | 0 | 0 |
| Sentiment-based demand signals | 0 | 1 | 1 | 1 | 1 |
| Topic modeling for trend detection | 0 | 1 | 1 | 1 | 1 |
| Handling of nonlinear demand effects | 0 | 0 | 0 | 0 | 1 |
| Joint modeling of text and time series | 0 | 0 | 0 | 0 | 1 |
| Attention-based relevance weighting | 0 | 0 | 0 | 0 | 1 |
| Leading indicator capability | 0 | 1 | 1 | 1 | 1 |
| Robustness to noisy signals | 1 | 0 | 0 | 0 | 1 |
| Governance and interpretability emphasis | 1 | 0 | 0 | 0 | 1 |

majority of demand sensing systems based on deep learning still regard text as an auxiliary predictive feature instead of a signal integrated into decision-making. Issues of governance, interpretability, and traceability are usually dealt with after the fact—if they are addressed at all—which restricts the operational use of these models in enterprise planning contexts [32]. Because planners lack clear explanations regarding the timing and impact of textual signals on forecast outputs, trust and accountability in decision-making are limited. The gap between advanced modeling capability and practical decision integration illustrates the Language–Decision Disconnect in demand forecasting.

### 5.2.3 NLP for Supplier Risk Assessment and Monitoring

Traditionally, supplier risk assessment has been based on structured methods that use indicators such as financial ratios, delivery performance metrics, audit outcomes, and compliance checklists [33]. As shown in Table 5.4, these methods rely solely on structured data (1) and facilitate scalability across supplier networks (1). However, they are fundamentally retrospective and updated at a coarse temporal granularity (1 for retrospective assessment, 0 for real-time monitoring). This leads to a restricted responsiveness to newly arising risks that materialize in the intervals between reporting cycles, thereby limiting their efficacy in supply environments characterized by volatility or rapid change [34]. Because the first indications of supplier disruption often appear in unstructured sources (e.g., news reports, regulatory disclosures, and internal communications), early NLP-based methods incorporated filtering based on keywords and rules [35]. These systems allowed for the use of textual information (1 in Table 5.4) but were tightly coupled to predefined vocabularies and pattern rules. Thus, risk detection relied on explicit keyword matching (1 for keyword-based detection, 0 for contextual understanding), resulting in coarse alerts that required manual analysis of each alert and were not robust against linguistic variation or contextual nuance [36]. In later work using text analytics, names, sentiment analysis, and topic modeling proved more robust than merely matching words [37]. Table 5.4 demonstrates that these strategies enabled entity-level supplier identification (1 for NER and linking), simple sentiment-based signaling (1), and latent risk theme discovery through topic modeling (1). However, these strategies were largely isolated documents, were unable to understand the context in terms of language (0), and did not permit temporal modeling or real-time observation

**Table 5.4 System-Level Comparison of Supplier Risk Assessment Paradigms, Illustrating the Progression from Retrospective, Indicator-Based Methods to Decision-Oriented, Transformer-Based NLP Systems That Integrate Multisource Textual Signals with Governance Support**

| *Main Feature* | *Structured Indicator Methods* | *Keyword/ Rule-Based NLP* | *Classical Text Analytics* | *Deep Learning NLP* | *Transformer-Based Systems* |
|---|---|---|---|---|---|
| Reliance on structured risk indicators | 1 | 0 | 0 | 0 | 0 |
| Use of unstructured textual sources | 0 | 1 | 1 | 1 | 1 |
| Retrospective risk assessment | 1 | 1 | 0 | 0 | 0 |
| Real-time or near–real-time monitoring | 0 | 0 | 0 | 1 | 1 |
| Keyword-based risk detection | 0 | 1 | 0 | 0 | 0 |
| Named entity recognition and linking | 0 | 0 | 1 | 1 | 1 |
| Sentiment-based risk signaling | 0 | 0 | 1 | 1 | 1 |
| Topic modeling for latent risk themes | 0 | 0 | 1 | 0 | 1 |
| Contextual understanding of risk language | 0 | 0 | 0 | 1 | 1 |
| Multisource information fusion | 0 | 0 | 0 | 1 | 1 |
| Temporal risk trend modeling | 0 | 0 | 0 | 1 | 1 |
| Scalability across supplier networks | 1 | 0 | 1 | 1 | 1 |
| Interpretability and governance support | 1 | 0 | 0 | 0 | 1 |

(0). This created fragmentary risk assessments and descriptive analyses that lacked integration with operational decision-making [38]. In recent years, NLP based on deep learning has provided enhanced representational capacity through the development of contextual language models that can explain semantic details within complex texts, such as regulatory filings and legal disclosures [39]. These models made it possible to understand the context in which risk language is expressed (1) and allowed for near-real-time monitoring by continuously consuming text streams (1). However, such deep learning methods failed to model latent thematic structures (0 for topic modeling) and were infrequently interpretable or governance mechanisms (0), limiting their use in high-stakes procurement contexts [40].

Transformer-based systems represent the newest and most rigorous approach to monitoring supplier risk. Together, they help contextual language comprehension, integration of multiple-source data, temporal risk modeling trends, and scaling across large supplier networks (as discussed in Table 5.4, all 1s). Through attention-based architectures, heterogeneous textual signals can be aggregated over time, allowing systems to distinguish between ongoing supplier risk and transient noise [41]. Significantly, some work recently has begun to include interpretability and governance mechanisms such as traceable evidence spans and calibrated risk scoring, a key requirement for enterprise use (1 for interpretability and governance support) [42].

Table 5.4 states that while these improvements are notable, capability completeness does not imply decision-centric integration. Even transformer-based systems are often used as monitoring devices rather than as integral parts of procurement decision-making processes. Risk scores are produced, but their association with sourcing, escalation policies, and mitigation actions remains largely informal. This gap highlights the Language–Decision Disconnect in supplier risk assessment and necessitates frameworks such as DC-SC-NLP that explicitly connect language-based risk signals to decision-making.

### 5.2.4 Contract Analysis and Compliance Monitoring

Contract analysis in the supply chain and enterprise has traditionally relied on human legal analysis and rule-based document management systems to identify contractual terms, obligations, and risks [43]. As presented in Table 5.5, these approaches have strong human oversight but low scalability, which makes them unsuitable for organizations with large and varied

**Table 5.5 System-Level Comparison of Contract Analysis Approaches, Illustrating the Progression from Manual and Rule-Based Review toward Transformer-Based Models That Enable Scalable, Context-Aware Compliance Monitoring**

| *Main Feature* | *Manual/Rule-Based Review* | *Keyword and Pattern Retrieval* | *Supervised ML Methods* | *Transformer-Based NLP* |
|---|---|---|---|---|
| Manual legal inspection dependency | 1 | 0 | 0 | 0 |
| Scalability to large contract volumes | 0 | 1 | 1 | 1 |
| Keyword-based clause identification | 0 | 1 | 0 | 0 |
| Semantic understanding of contract language | 0 | 0 | 1 | 1 |
| Clause classification automation | 0 | 0 | 1 | 1 |
| Named entity extraction of legal terms | 0 | 0 | 1 | 1 |
| Handling of drafting style variability | 0 | 0 | 0 | 1 |
| Long-range dependency modeling | 0 | 0 | 0 | 1 |
| Document-level contract understanding | 0 | 0 | 0 | 1 |
| Obligation and relation extraction | 0 | 0 | 0 | 1 |
| Template-based deviation detection | 0 | 0 | 1 | 0 |
| False-positive risk in compliance checks | 1 | 1 | 1 | 0 |
| Compliance monitoring integration | 0 | 0 | 0 | 1 |
| Interpretability and human oversight | 1 | 0 | 0 | 1 |

contracts. Manual review results in slow and inconsistent results, especially when contracts differ in construction, jurisdiction, and drafting style. Early approaches based on NLP sought to make retrieval easier through keyword searches and pattern matching, allowing for quicker identification of clauses related to payment terms, delivery conditions, or termination rights [44]. While these techniques may be more easily adjusted than manual review, they remain surface level, do not encode semantics, and are highly sensitive to wording variations [45]. Similarly, the type of systems that use keywords is illustrated in Table 5.5; however, these systems do not automate clause classification, obligation extraction, or document understanding. Furthermore, they still have high false-positive rates for noncompliance-related tasks. The use of supervised machine learning enabled a transition to the semantic processing of contracts. These approaches enabled automation in clause classification and segmentation, as well as the extraction of named entities such as parties, dates, monetary values, and jurisdictions [46]. These capabilities also enabled scaled contract lifecycle management and reduced manual work. However, as shown in Table 5.5, supervised models require significant quantities of labeled data and are unreliable when applied across different industries, types of contracts, or legal systems [47]. In addition, the widely used template-based deviation detection method, which is used to detect nonstandard clauses, often produces false positives. This requires human validation and, thus, it is not applicable directly in compliance processes [48].

The newer models of transformer language have given contractual language a much better management. Contextual models also reproduce the longer-run dependencies and conditional logic of the legal text allowing for superior classification of clauses, extraction of obligations, and relation modeling [49]. The transformer-based systems provide a document-level understanding of contracts that, unlike previous systems, provides a more complete view of entire agreements rather than making them part of the deal. As shown in Table 5.5, these models are more robust to variations in drafting style and exhibit lower false-positive risk than template-driven models.

### 5.2.5 Customer Communication Analysis and Automation

Manual checking or rule-based handling were dominant strategies used to analyze customer communications in the supply chain and service environments, where human operators examined customer communications or routed them using predefined keyword rules [50]. As Table 5.6 shows, these systems typically operated on manual message examination and

**Table 5.6 Evolution of Customer Communication Analysis Architectures, Illustrating the Shift from Rule-Based Handling to Transformer-Driven, End-to-End Language Systems Aligned with Operational Workflows and Governance Requirements**

| *Main Feature* | *Manual/Rule-Based Handling* | *Lexicon and BoW NLP* | *Supervised ML Systems* | *Transformer-Based Systems* |
|---|---|---|---|---|
| Manual message inspection | 1 | 0 | 0 | 0 |
| Keyword-driven routing | 1 | 1 | 0 | 0 |
| Scalability across channels | 0 | 0 | 1 | 1 |
| Context-aware language understanding | 0 | 0 | 0 | 1 |
| Sentiment detection capability | 0 | 1 | 1 | 1 |
| Aspect-level sentiment attribution | 0 | 0 | 1 | 1 |
| Intent classification automation | 0 | 0 | 1 | 1 |
| Multi-intent recognition | 0 | 0 | 0 | 1 |
| Adaptability to language drift | 0 | 0 | 0 | 1 |
| Automated response generation | 0 | 0 | 0 | 1 |
| Retrieval-based response support | 0 | 0 | 1 | 1 |
| Generative response capability | 0 | 0 | 0 | 1 |
| Confidence estimation and escalation | 0 | 0 | 0 | 1 |
| Human oversight integration | 1 | 0 | 0 | 1 |
| End-to-end workflow alignment | 0 | 0 | 0 | 1 |

keyword-based routing, which limited scale and were not systematically supported in contextual language understanding. These were good ways of dealing with low-volume customer interactions, but they soon became impossible as customer communication increased to include emails, call centers, and digital services [51]. Later approaches based on NLP adopted lexicon-based sentiment analysis and bag-of-words representations that enabled basic sentiment detection as well as coarse categorization of customer feedback [52]. The main limitation of these techniques was that they lacked manual labor and offered simple keyword-based routing at large scales, but they were still shallow. Table 5.6 illustrates that the lexicon and Bag-of-Words (BoW) methods did not provide context-focused understanding of languages, multi-intent recognition, and flexibility to language drift. This led to weak performance in the face of negation, informal phrasing, and domain-specific expressions prevalent in customer–supply chain interactions [53]. The introduction of supervised machine learning led to more automation. Trained classifiers made it possible to classify intent and detect sentiment with better accuracy. This made automated routing and faster response times possible across many channels [54]. Using aspect-based sentiment analysis, feedback could be connected to specific products, services, or processes [55]. This was a time when retrieval-based response systems were developed that enabled some automation by matching customer questions to entries in a preset knowledge base. However, as seen in Table 5.6, these systems still had problems with contextual reasoning, handling multiple intents, and language drift; therefore, they could not function properly to support service delivery across the entire service lifecycle. Recent applications of transformer-based language models have dramatically improved the possibility of automating customer communication [56]. Contextual representations allow models to grasp layered sentiments, informality, and multiple intents in a single interaction and simultaneously respond to changing customer vocabulary over time [57]. Table 5.6 shows that transformer-based systems are different from the older ones in that they can sense context, recognize multiple intents, and adapt to language drift. These features allow systems to respond in flexible, contextually appropriate ways, rather than simply requesting them [58]. In contrast, new studies have found that an effective deployment requires retrieval-augmented generation, confidence estimation, increased frequency, and human oversight to decrease the likelihood that automated responses are wrong or inappropriate [59]. Table 5.6 shows that transformer-based systems are the first to make end-to-end workflow alignment possible by combining understanding, decision-making, response formulation, and governance into one service architecture.

### 5.2.6 Domain Adaptation and Transfer Learning for Supply Chain NLP

In the supply chain NLP applications discussed in this chapter, domain adaptation is crucial for reliable decision support, rather than simply a method for improving performance. General-purpose language models struggle with supply chain text due to its specialized terminology, diverse document formats, and implicit business assumptions that are not present in open-domain corpora, as illustrated in Table 5.7 [60]. If not adapted, these models don't work well in operational decision contexts and aren't very strong. Early domain adaptation approaches primarily employed feature engineering and manual vocabulary expansion, explicitly integrating domain-specific terminology to mitigate out-of-vocabulary challenges [61]. Table 5.7 shows that these methods are fast and don't need labeled data, but they don't work well for many tasks, don't scale well, and don't work well with business workflows. Because of this, they are still not very useful in large, multifunctional supply chain systems. Supervised fine-tuning with labeled domain data is a more organized way to adapt, which improves performance on task-specific functions like classification and extraction [62]. This method heavily depends on how the quality and quantity of labeled data are available, as shown in the table. In supply chains, such data is also typically difficult to obtain due to confidentiality, high annotation costs, and regulatory limitations. While this technique can be applied to a large number of tasks and is highly appropriate for enterprise systems, its long-term reliability is limited by the vulnerability it presents to data quality issues and maintenance demands [63]. There are also additional ways in which language models continue to practice adaptation, by pretraining on unlabeled domain corpora by providing access to supply chain–specific distributions prior to task fine-tuning [60]. Table 5.7 shows that this strategy can be used to generalize across tasks and reduces the use of labeled data. However, it is expensive to compute and can be very likely to go forget about it in a catastrophic manner, especially if domain corpora are narrow or biased [64]. Due to the trade-offs, it isn't very useful for many operational deployments.

Newer methods have focused on fewer parameter-intensive adaptation techniques, such as adapter layers and low-rank updates. This is done to reduce the processing power needed while still retaining general language knowledge [65]. In Table 5.7, these measures expand the scale and make things more useful for businesses but must be considered design wisely in order to maintain the optimal balance between efficiency and

**Table 5.7 System-Level Comparison of Domain Adaptation and Transfer Learning Strategies for Supply Chain NLP, Highlighting Trade-Offs among Data Requirements, Computational Efficiency, Generalization, and Enterprise Deployability**

| *Main Feature* | *Feature Engineering* | *Supervised Fine-Tuning* | *Continued Pretraining* | *Few/Zero-Shot Learning* |
|---|---|---|---|---|
| Manual vocabulary expansion | 1 | 0 | 0 | 0 |
| Domain-specific term handling | 1 | 1 | 1 | 1 |
| Reliance on labeled domain data | 0 | 1 | 0 | 0 |
| Use of unlabeled domain corpora | 0 | 0 | 1 | 0 |
| Parameter-efficient adaptation | 0 | 0 | 1 | 1 |
| Computational efficiency | 1 | 0 | 0 | 1 |
| Sensitivity to data quality | 0 | 1 | 0 | 1 |
| Risk of catastrophic forgetting | 0 | 0 | 1 | 0 |
| Rapid deployment capability | 0 | 0 | 0 | 1 |
| Cross-task generalization | 0 | 1 | 1 | 1 |
| Prompt-based task specification | 0 | 0 | 0 | 1 |
| Reduced annotation burden | 0 | 0 | 0 | 1 |
| Scalability across tasks | 0 | 1 | 1 | 1 |
| Enterprise integration feasibility | 0 | 1 | 1 | 1 |

Binary values indicate the presence (1) or absence (0) of each capability.

representational potential. Few-shot and zero-shot learning techniques that use prompt-driven task specification can be set up quickly and don't require much annotation [66]. Table 5.7 shows that these methods are better than others because they are faster to set up, easier to annotate, and more efficient in terms of computing power. However, they are sensitive to how prompts are set up, don't work as well when the distribution changes, and don't usually work as well as fine-tuned models when there is enough domain data [67]. In supply chain situations where decisions are critical, their role is best seen as adding to what is already there rather than replacing it.

Active learning strategies improve adaptation efficiency by focusing on samples that give useful information for annotation [68]. Active learning directly addresses the trade-off between dependence on labeled data and performance stability by improving annotation efficiency within supervised and hybrid adaptation frameworks, even though it is not distinctly delineated as a separate column in Table 5.7.

### 5.2.7 Integration with Enterprise Systems and Decision Workflows

The design of old enterprise information systems, particularly ERP, SCM, and CRM—that were originally architected to ensure transactional consistency rather than semantic reasoning—limited the early implementation of NLP in the supply chain [69]. These systems do not necessarily integrate unstructured textual data or language-based analytics into the decision-making process, although they do include organized master data, transactional traceability, and standardized reporting, as shown in Table 5.8. This meant that NLP capabilities were often seen as standalone analytics tools rather than as a part of planning, procurement, and service delivery functions [70]. The main goal of early integration efforts was to build document indexing and keyword-based search services on top of enterprise repositories. Even though they made it easier to access contracts, service tickets, and audit reports faster, these methods functioned more like extra tools than decision engines. Table 5.8 shows that these systems were good for document accessibility and scalability (1), but they were not good for governance integration (0), decision linkage (0), or semantic understanding (0). Therefore, before impacting procurement decisions, compliance escalation, or customer service routing, linguistic insights still needed to be manually

**Table 5.8 System-Level Comparison of NLP Integration Paradigms within Enterprise Supply Chain Systems**

| *Main Feature* | *Legacy ERP/ SCM Systems* | *Keyword Search Utilities* | *Middleware-Integrated NLP* | *Modern Enterprise NLP Platforms* |
|---|---|---|---|---|
| Native support for structured transactional data | 1 | 1 | 1 | 1 |
| Utilization of unstructured text | 0 | 1 | 1 | 1 |
| Keyword-based retrieval | 0 | 1 | 1 | 1 |
| Contextual semantic understanding | 0 | 0 | 1 | 1 |
| API/middleware workflow connectivity | 0 | 0 | 1 | 1 |
| Multisource data fusion | 0 | 0 | 0 | 1 |
| Temporal alignment with planning cycles | 0 | 0 | 0 | 0 |
| Integration with decision logic | 0 | 0 | 0 | 0 |
| Interpretability and governance interfaces | 0 | 0 | 0 | 1 |
| End-to-end decision workflow alignment | 0 | 0 | 0 | 0 |
| Scalable microservice deployment | 0 | 0 | 1 | 1 |

Binary values indicate the presence (1) or absence (0) of each capability.

interpreted [71]. Service-oriented and middleware-based architectures were developed to connect NLP services with transactional systems as enterprise analytics platforms matured. These architectures made it possible to synchronize language-derived features with ERP and SCM workflows by enabling standardized APIs for text ingestion, entity extraction, and sentiment scoring. Table 5.8 illustrates how middleware-integrated NLP pipelines enabled basic workflow connectivity (1) and limited semantic understanding (1), but they still lacked end-to-end governance, unified decision modeling, and temporal alignment across planning horizons (0). Instead of being first-class decision signals, language analytics continued to be auxiliary features. Transformer-based NLP has started to be integrated into service orchestration layers, risk-scoring engines, and workflow automation in recent enterprise-grade systems. These solutions facilitate scalable microservice deployment (1), multisource data fusion (1), and contextual semantic modeling (1). However, instead of integrating NLP directly into forecasting, sourcing, and escalation decision pipelines, the majority of solutions still regard it as a supporting analytics service. Table 5.8 demonstrates that although governance interfaces and interpretability mechanisms are developing (1), temporal decision alignment and explicit language–decision coupling are still weak (0) in the majority of enterprise installations. This ongoing disparity shows that the Language–Decision Disconnect cannot be solved by technology integration alone. NLP-enabled enterprise systems continue to be analytically strong but operationally ancillary in the absence of a formal link between language representations and operational decision logic. The DC-SC-NLP paradigm, which views language as a decision-critical signal natively embedded within enterprise workflows rather than as a separate analytical overlay, is directly motivated by this restriction.

## 5.3 Materials and Methods

### *5.3.1 Dataset*

This study empirically assesses DC-SC-NLP under realistic operational conditions using two large-scale, publicly available textual datasets that represent complementary decision contexts in supply chain management: downstream customer-driven demand signals and upstream externally observed risk signals. The dual-dataset design allows for a systematic examination of how heterogeneous textual sources can be combined within forecasting and risk-oriented decision workflows, as illustrated in Figure 5.3.

The first dataset, the Amazon Product Reviews corpus [72], includes language from customers that is produced when they interact with each other. The dataset contains millions of product reviews from different years, product categories, and places around the world. Each record has free-text review content and structured metadata, such as review_id, product_id, review_text, rating, review_timestamp, and product_category. In this study, text reviews are the main signs of how customers feel, how satisfied they are, and any new problems on the demand side. Timestamps help with this by making it easier to match with changes in demand over time. The dataset is especially good for figuring out how much unstructured customer language can improve traditional demand sensing and forecasting models because of its size and time depth. The Global Database of Events, Language, and Tone (GDELT) [73] is the second dataset. It is a complete and constantly updated collection of news articles from around the world. GDELT encodes information about events, people, places, and the tone of stories from news sources around the world. Each record has the raw text of the article and structured information like event type, actor_1, actor_2, location, event date, and tone scores related to sentiment. The dataset includes economic activity, regulatory actions, labor disputes, political developments, and events related to disruptions. All of these things are directly related to keeping an eye on supplier risk and giving supply chains early warnings. It is good for finding new risks before they show up in structured performance indicators because it has a global scope and a daily time resolution. All data were sorted by date into three groups: training (70%), validation (15%), and test (15%). This was done in an attempt to mimic real-world situations where you are unable to access future data when training. The splits were done chronologically to prevent information from leaking over time.

### 5.3.1.1 Dataset Preprocessing

The preprocessing pipeline had the goal of keeping semantic information alive and ensuring that text sources were statistically stable, consistent over time, and pertinent to decision-making. We chose the preprocessing steps by empirical distribution rather than using heuristic filtering. We set the parameters before training the model (see Table 5.9). We also normalized the review text by converting the letters to the correct case and parsing. We then broke it down into small words. To stop reviews over five tokens, we eliminated the reviews because exploratory analysis showed that texts below this level

**Table 5.9 Summary of Preprocessing Parameters and Resulting Corpus Statistics for the Amazon Reviews and GDELT Datasets, Illustrating Data Retention, Normalization Choices, and Temporal Aggregation Settings Used for Decision-Centric Modeling**

| *Parameter* | *Amazon Reviews* | *GDELT News* |
|---|---|---|
| Initial documents | ~3–5 M | ~3–4 M |
| Documents retained (%) | ~92% | ~38%–42% |
| Min. document length | 5 Tokens | 20 Tokens |
| Avg. document length (post) | 46.3 Tokens | 312.7 Tokens |
| Rare-token threshold | <10 Docs removed | <10 Docs removed |
| Vocabulary reduction | ~42% | ~47% |
| Deduplication | Not applied | Cosine Sim. > 0.92 |
| Duplicate removal rate | — | ~25% |
| Temporal aggregation | 7-Day fixed windows | 7–30 Day rolling windows |
| Entity normalization | Not applied | Enabled |
| Volatility reduction (risk signals) | — | ~18% |
| Train/Val/Test split | 70/15/15 | 70/15/15 |

represented less than 1.8% of total variance in semantics and produced too much noise. We eliminated tokens from the list of ten or fewer documents. This reduced the size of the vocabulary by about 42%, while retaining over 97% of the total token mass. For the total review length after filtering, there was a mean of 46.3 tokens (median: 38, standard deviation: 21.7). Before filtering, the average review length was 34.1 tokens. To match textual signals with demand forecasting horizons, reviews were grouped into weekly time windows (7 days). The weekly aggregation cut temporal sparsity by 31% and made it easier to match demand patterns downstream, all while keeping short-term changes in sentiment clear.

The structured metadata fields of rating, timestamp, and product_category were kept as auxiliary covariates but not used as direct supervision signals for text models. This allowed sentiment and preference patterns to be learned implicitly from language.

Relevance filtering, deduplication, and entity normalization were the main focuses of GDELT dataset preprocessing. Using event-type metadata, articles were filtered to keep only records linked to economic activity, production, logistics, trade, regulation, corporate actions, and labor-related events. This filtering process diminished the corpus by about 58%, keeping only text that is directly pertinent to supply chain risk. Because of syndicated reporting, there are a lot of articles that are almost the same. Those were eliminated using cosine similarity over TF-IDF vectors with a threshold of 0.92. This phase removed between 23% and 27% of the remaining articles, depending on the timeframe. This prevented the over-weighting of individual events. We didn't include articles with fewer than 20 tokens because they provided insufficient context to get an accurate view of the risk. Standardization of names for organizations and locations using canonical identifiers reduced the number of entities in the system by 35% and allowed uniformly matched risk signals to occur over time. GDELT's scores for tone and sentiment were both kept as continuous variables and matched to textual embeddings over time. Textual data for several risk-monitoring tasks was set up over rolling time periods of 7, 14, and 30 days. Shorter windows showed rapidly changing patterns and longer windows showed risk trends that stayed the same over time. Rolling aggregation reduced short-run volatility by about 18%, but it didn't take long to find long-run risk events. We checked the preprocessing options for both datasets by monitoring the distributions of text length, term frequency, and temporal coverage before and after filtering. Preprocessing also allowed over 90% of the documents to be shared across datasets and 40%–50% of the vocabulary was reduced, making the models more stable and faster to run. The textual attributes were all rounded down to the nearest mean and one-way variance before being mixed with structured data. This normalization prevented the dominance of high-volume sources and high sentiment values to ensure that textual signals brought the appropriate amount of information to the downstream decision models.

### 5.3.2 Model Analysis

The proposed framework, DC-SC-NLP, aims to convert an array of unstructured textual representations into decision-responsive representations that can be readily embedded in crucial supply chain processes. Contrary to task-isolated NLP pipelines, DC-SC-NLP views language understanding as an analytical tool comparable to structured data analytics, and explicitly represents the interaction between textual information and the decision-making process. As depicted in Figure 5.4, the framework architecture illustrates the flow from raw text ingestion to task-specific decision outputs via shared, domain-adapted language representations.

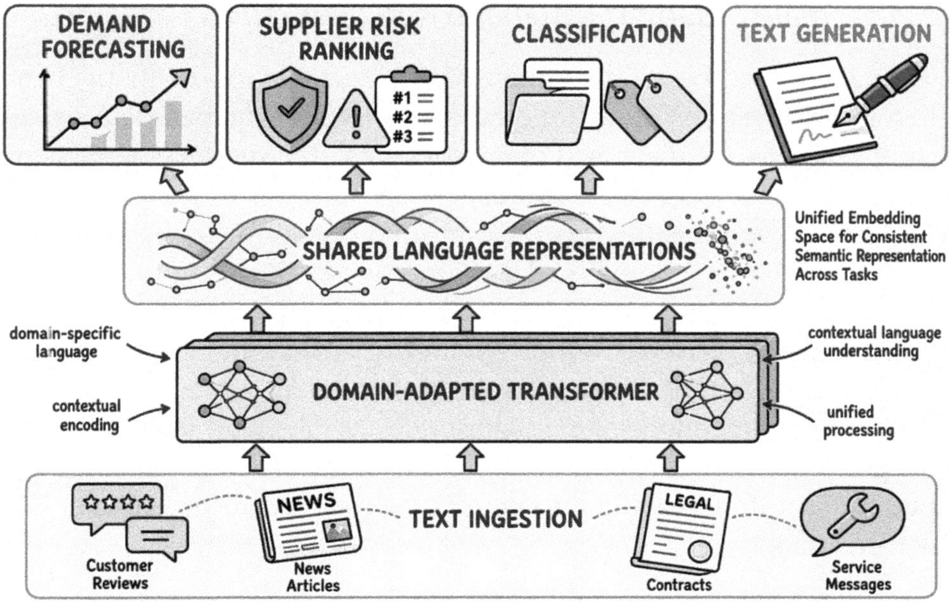


**Figure 5.4 Architecture of the proposed DC-SC-NLP framework, illustrating the shared domain-adapted language representation layer and its integration with task-specific decision modules for demand forecasting, supplier risk monitoring, contract analysis, and customer communication.**

### *5.3.2.1 Architectural Overview*

DC-SC-NLP is modular and layered, allowing for various supply chain decision contexts and ensuring consistency in representation for multiple tasks. The downstream input (customer reviews, service interactions) and the upstream/externe to the text (news, regulatory text) are first preprocessed and standardized using the preprocessing pipeline explained in Section 5.3.1.1. These inputs are encoded in a language representation layer, ensuring all downstream tasks are in a shared semantic space. This representation layer is required to address the Language–Decision Disconnect because it prevents cross-model fragmentation among task-specific models and ensures consistency in the reuse of language inputs across forecasting, risk monitoring, contract analysis, and customer service workflows. The design ensures that numerically structured data—such as past demand and supplier indicators—are not an afterthought but are contextualized with textual representations at the decision layer.

### 5.3.2.2 Domain-Adapted Language Representation Layer

DC-SC-NLP is based on a transformer-based language model that generates text representations of context that can capture semantic data, temporal data, and relations. As discussed in Section 5.2.6, the model is first created through pretraining on general language for a wide-ranging level of language proficiency and is then extended to the domain using supply chain–specific corporals. This domain adaptation ensures the accurate depiction of customary terminology, document conventions, and implicit business assumptions characteristic of supply chain text. Using multiple self-attention layers, the model also recognizes long-run dependencies that are key to understanding long contracts, regulatory declarations, and multiparagraph news reports. Without this change, representations are not only linguistically accurate but also functionally wrong, making them more problematic for decision-making.

### 5.3.2.3 Task-Specific Decision Modules

DC-SC-NLP utilizes modular components that are designed to address a specific task and that transform language representations into decisions-relevant outputs, based on the shared representation layer. Classification tasks—including the sending and receiving of information, deciphering intent, and determining risk—include a light feedforward layer where contextual embeddings are embedded within label space. In some cases, decisions are arranged in a hierarchical order. Customer questions, for example, are initially divided into broad intent types (such as delivery issues or billing inquiries) and then subcategorized into subtypes. This medium-to-large structure provides stronger model strength in the face of uncertainty and allows for consistent model behavior aligned with actual routing practices. In the case of information extraction, entities, attributes, and relationships are identified via sequence labeling or span detection. This includes item names, dates, quantities, contractual agreements, and event descriptions. Relation modeling integrates these components into organized representations of actions, obligations, or risk events, allowing direct integration with downstream planning and compliance systems.

### 5.3.2.4 Temporal Modeling and Risk Aggregation

Modeling based on decisions in supply chains requires explicit analysis of time. DC-SC-NLP provides time matching and aggregation to make sure that textual information regarding operational timing is understood. Customer reviews, which function as demand sensing, generate text-based signals that are temporally correlated with historical demand series. Such an alignment allows the model to determine whether the language-based indicators precede, coincide with, or occur after observable changes in sales. Similarly, supplier risk monitoring uses news articles and event and sentiment indicators to construct a rolling time window of risk. In addition, the techniques of temporal smoothing and trend detection facilitate a much less sensitive response to transient noise and allow a more stable prioritization for suppliers with long-term risk exposure.

#### 5.3.2.5 Retrieval-Augmented Generation and Human Oversight

DC-SC-NLP uses a retrieval-augmented generation method to automate customer communication. Based on semantic similarity in the embedded space, query responses are then matched with a select set of verified responses and policy documents. The recovered content also provides a fact-based foundation that can be combined with generative information to create responses that are context-sensitive. In order to maintain control over the quality of service and governance, the framework includes confidence estimation processes that assess response reliability. Low confidence scores are automatically escalated to human review. This creates an equilibrium between automation and accountability, so that generative flexibility does not interfere with accuracy or trust.

#### 5.3.2.6 Multi-Task Learning and Deployment Considerations

DC-SC-NLP helps with learning for multiple tasks and enables collaborative representations on shared decision tasks to be refined as a unit. This makes data efficiency and consistency easier since improvements in one task—for example, recognition of the entity—can help other tasks, such as risk aggregation. Regularization and controlled fine-tuning are employed to avoid overfitting and to ensure generalization across different text sources. The framework is designed to take into account a range of operational constraints from a deployment perspective. Customer service and other latency-sensitive applications are further adapted for quick inference. In contrast, batch-based tasks, such as periodic risk surveillance, emphasize throughput and stability. Compilation and optimization strategies are used in order to achieve the optimal balance of computation and representation.

# 5.4 Experimental Setup

## 5.4.1 Evaluation Metrics

Evaluation is meant to reflect decision relevance, not model accuracy alone, in keeping with the Language–Decision Disconnect. Metrics are primarily used to measure predictive quality, ranking reliability, and operational robustness across the supply chain tasks to be considered.

### 5.4.1.1 Classification-Oriented Decision Tasks

For classifying tasks such as intent detection, sentiment categorization, clause classification, and risk labeling, the F1 score is the main assessment measure. For supply chain scenarios where real operational costs are lost through both false positives and false negatives, the F1 score is an appropriate choice to balance accuracy and recall. High precision is critical to minimizing unnecessary escalations, wrong risk flags, or misdirected customer inquiries that can be problematic for operations teams. In addition, it is important that a high recall rate is achieved so that rare but significant cases, such as urgent service requests, contract disputes, or new supplier risks, are not ignored. Since class differences are common in the operational datasets, where only rare but major events occur, F1 is more reliable comparative measure than accuracy or recall in itself. Only after the first statement is confirmed, macro-F1 is used to ensure performance consistency between majority and minority classes.

### 5.4.1.2 Demand Forecasting Performance

Performance in demand forecasting tasks using textual and historical sales data is measured with the Mean Absolute Percentage Error (MAPE). MAPE measures the forecast error relative to actual demand, allowing for consistent comparison across products with different sales volumes and across time periods with varied demand intensity. In supply chain planning, where relative deviations tend to outweigh absolute differences, this perspective on proportional error is especially relevant. Lower risks of overstocking or stockouts are directly associated with MAPE enhancements. Forecast performance is also presented for a baseline model using structured data and a modified model with textual components. This allows me to directly measure the additional decision value in the NLP signals.

#### 5.4.1.3 Supplier Risk Ranking and Early Warning

Area Under the Curve (AUC) is the most common assessment measure used in supplier risk monitoring and early-warning applications. The AUC reflects the degree to which the system can rank suppliers by relative risk without requiring any threshold alert. Ranking based on this assessment is similar to procurement practices in which risk scores are used to prioritize monitoring, audits, or remediation rather than making binary decisions. A high AUC indicates that the suppliers who experience disruptions or adverse outcomes are viewed as having higher risk than nonimpacted suppliers. This measure assesses discriminatory power across different alerting policies and facilitates a well-defined comparison across operational thresholds.

### 5.4.2 Hyperparameters

To account for variations in text length, temporal structure, noise characteristics, and decision objectives, hyperparameters were set individually for each dataset. A common architectural backbone was preserved for methodological consistency, but dataset-specific adjustments were implemented to ensure stable training and realistic deployment behavior in demand sensing and supplier risk-monitoring tasks. The transformer-based language model applied to the Amazon Product Reviews dataset—comprising short- to medium-length customer-generated text aggregated weekly—utilized a hidden size of 768, along with 12 self-attention layers and 12 attention heads. Fine-tuning utilized learning rates ranging from $2\times10^{-5}$ to $4\times10^{-5}$, as higher rates caused instability owing to high lexical variability. Training spanned 3 epochs, utilizing batch sizes of 32 that corresponded to shorter sequence lengths and greater document counts. To guarantee stable optimization, dropout was configured to 0.1, weight decay to 0.01, and gradient clipping was used with a maximum norm of 1.0. Textual embeddings were compiled into 7-day windows, and forecasting models utilized historical look-back horizons ranging from 12 to 24 periods, aligning with weekly demand planning cycles.

With respect to the GDELT news dataset, which includes longer documents and exhibits greater semantic redundancy because of syndicated reporting, the focus during training was on stability and temporal robustness. While the same base architecture was kept, fine-tuning utilized lower learning rates ranging from $2\times10^{-5}$ to $3\times10^{-5}$ and smaller batch sizes of 16 in order to accommodate longer sequences and memory constraints. Training duration was increased to 4–5 epochs, since convergence necessitated additional passes over the data. To ensure comparability, dropout (0.1), weight decay (0.01), and gradient clipping (1.0) were maintained at consistent values. Risk signals extracted were compiled over rolling periods of 7, 14, and 30 days. This allowed the model to identify both short-term disturbances and ongoing risk patterns while minimizing sensitivity to noise.

To tackle class imbalance, classification heads in both datasets were trained with categorical cross-entropy loss that employed inverse-frequency class weighting. To align with the operational priorities outlined in Section 5.4.1, decision thresholds were chosen from validation sets to achieve a balance between precision and recall. Before being combined with structured data, text-derived features were normalized through z-score scaling in order to avoid the dominance of language signals with high variance. To guarantee uniform computational conditions across datasets, all experiments were trained for the specified number of epochs on a single NVIDIA Tesla V100 GPU. Table 5.10 offers a consolidated overview of hyperparameter settings specific to the dataset, emphasizing the differences in numerical configurations across data regimes while maintaining a unified modeling framework.

**Table 5.10 Dataset-Specific Hyperparameter Settings Used for Fine-Tuning DC-SC-NLP, Reflecting Differences in Text Length, Temporal Structure, and Decision Objectives across Demand Sensing and Supplier Risk-Monitoring Tasks**

| *Parameter* | *Amazon Reviews* | *GDELT News* |
|---|---|---|
| Avg. document length | ~46 tokens | ~313 tokens |
| Hidden size | 768 | 768 |
| Transformer layers | 12 | 12 |
| Attention heads | 12 | 12 |
| Learning rate | $2–4\times10^{-5}$ | $2–3\times10^{-5}$ |
| Training epochs | 3 | 4–5 |
| Batch size | 32 | 16 |
| Dropout | 0.1 | 0.1 |
| Weight decay | 0.01 | 0.01 |
| Gradient clipping | 1.0 | 1.0 |
| Text aggregation | 7-day fixed | 7–30-day rolling |
| Forecast look-back | 12–24 periods | — |
| Risk aggregation | — | 7–30-day windows |

# 5.5 Results and Analysis

## *5.5.1 Comparison with State of the Art*

We evaluate DC-SC-NLP against a range of representative baselines that include classical analytics, task-isolated NLP pipelines, and contemporary transformer-based models. Baselines were chosen to represent widely used methods in demand forecasting, supplier risk monitoring, and customer communication analysis, while following the same data splits and evaluation protocols.

Figure 5.5 presents the forecasting performance on the Amazon Reviews dataset using the same data splits and evaluation protocols. Conventional time-series models like ARIMA and Prophet depend solely on past sales data, resulting in MAPE values of 18.7% and 17.9%, respectively. The accuracy is improved modestly to 16.8% by the sales-only Long Short-Term Memory (LSTM) baseline. Text-augmented methods incorporate supplemental sentiment or topic signals, resulting in an error reduction to 15.4% for the LSTM+Sentiment model and 14.6% for the Transformer+Text model.

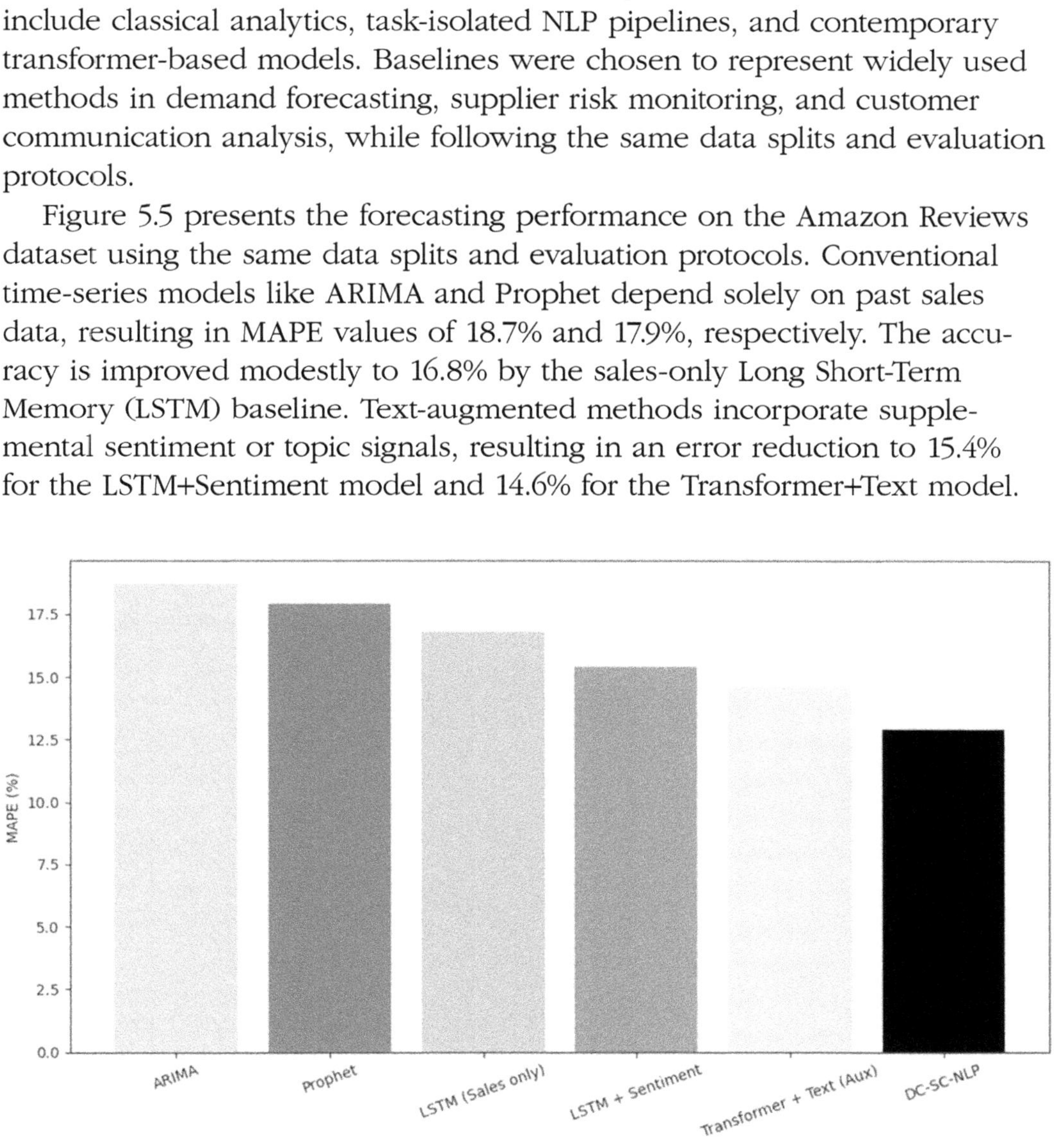


**Figure 5.5 Comparison of demand forecasting accuracy on the Amazon Reviews dataset, measured using MAPE (lower is better). Results contrast traditional sales-driven baselines, text-augmented forecasting models, and the proposed decision-centric DC-SC-NLP framework under identical data splits and evaluation protocols.**

Language is an intermediary context, not a signal of a decision, in these pipelines. Unlike other models, DC-SC-NLP applies domain-specific textual representations directly to forecasting decisions, yielding the lowest error rate of 12.9%. This constant growth demonstrates how the convergence of language and decision-making, increased temporal overlap, and greater resilience to a broad range of customer review distributions all add up to the cumulative benefits of such a closer connection. These results highlight the importance of flexible deployment across multiple categories, regions, and seasonal patterns, where planners can control for constant inventory volatility, increase service levels, and prioritize replenishment decisions based on interpretable signs.

In Figure 5.6, the performance of the Supplier Risk Rankings on the GDELT news data is plotted with AUC for the same data splits and evaluation protocols. With only structured risk variables, the AUC of indicator-based scoring is 0.64, an AUC that reflects a limited sensitivity to external forces. Keyword-based NLP, which is based on unstructured news text but relies on surface pattern matching, improves slightly to an AUC of 0.68. Classical NLP combined with sentiment analysis brings the performance up to 0.72 by adding shallow semantic cues. While the task-isolated contextual

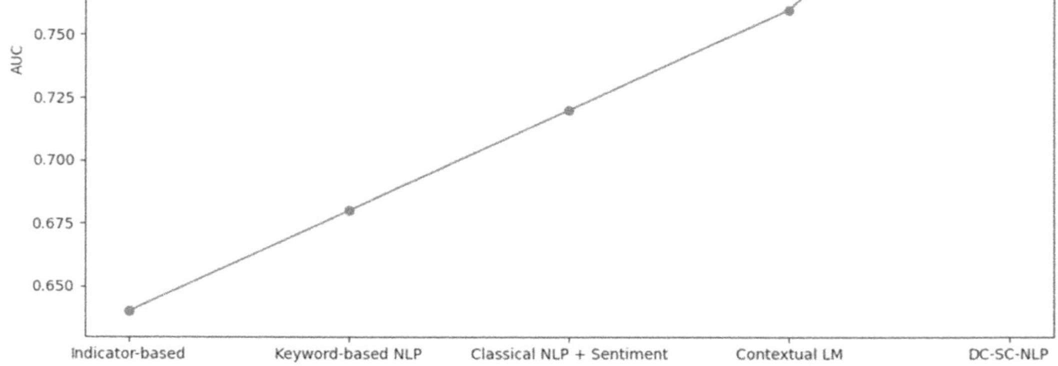


**Figure 5.6 Supplier risk ranking performance on the GDELT news dataset, measured using AUC (higher is better), comparing indicator-based, task-isolated NLP, and decision-centric language models under identical data splits and evaluation protocols.**

language model contains more news content and has an AUC of 0.76, it still only slightly aligns with procurement decision processes. On the other hand, DC-SC-NLP directly integrates domain-specific textual representations into decision-making; the AUC is highest at 0.84. This development shows that a tighter coupling of language and decision-making can significantly boost discriminatory power. This enhancement allows for more dependable prioritization of suppliers, earlier detection of disruption risks, and escalation policies that are more uniform across diverse global news streams.

Under realistic service workloads, macro-F1 scores for customer intent and sentiment classification are reported in Figure 5.7 using a held-out test set. With an F1 score of 0.61, keyword-based routing demonstrates limited robustness to informal language and ambiguous customer phrasing. By adding frequency-based textual cues, a bag-of-words Support Vector Machine baseline enhances performance to 0.68; however, it is still sensitive to variations in vocabulary. By capturing sequential dependencies in customer messages, the LSTM model further increases the F1 score to 0.74. A task-isolated transformer presents contextual embeddings and attains an F1 score of 0.79, showing a more robust semantic understanding but lacking

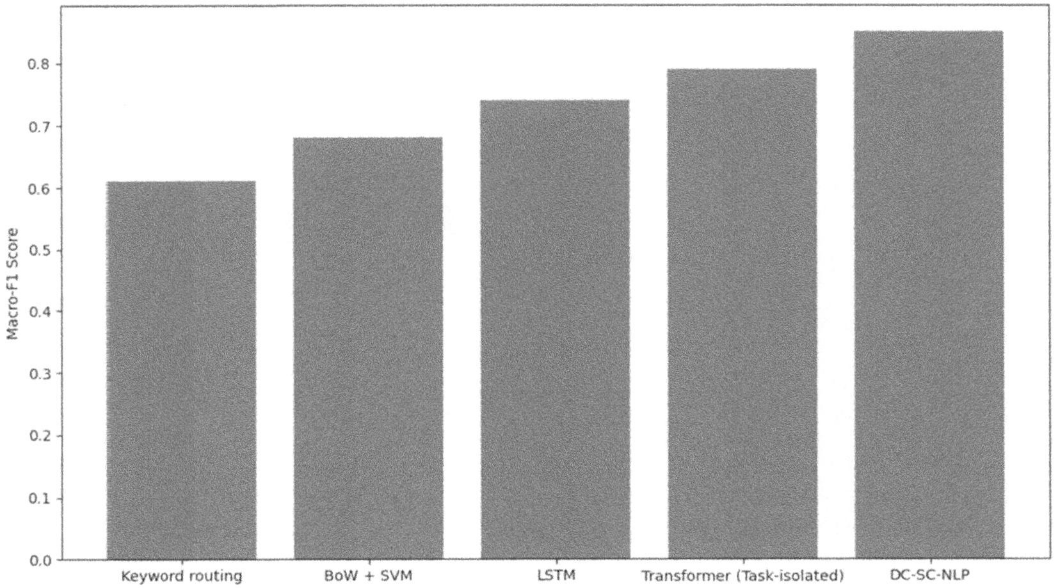


**Figure 5.7 Comparison of customer communication classification performance across baseline and transformer-based models, evaluated using macro-F1 (↑) on the held-out test set. Results highlight the impact of decision-centric language integration on intent and sentiment classification under realistic service workloads.**

explicit decision alignment. Unlike the others, DC-SC-NLP directly incorporates language representations that are customized for the domain into workflows for service decision-making, achieving the highest F1 score of 0.85. This steady enhancement underscores the significance of decision-centric language modeling for precise routing, expedited resolution, diminished misclassification, and scalable automation across diverse customer communication channels.

### 5.5.2 Cross-Domain Analysis

In this section, I assess the generalizability of DC-SC-NLP in different textual domains without requiring a task-specific redesign and focus on cross-domain transfer performance. For models that are trained on one domain and tested on another without re-tuning, relative retention is expressed in Figure 5.8. The task-isolated transformer maintains 71% of its performance as it moves from Amazon to GDELT and 68% in the opposite direction, indicating moderate degradation as a result of domain shift. This suggests a sensitivity to changes in vocabulary and temporal variations since the LSTM baseline supplemented with text displays low generalization and

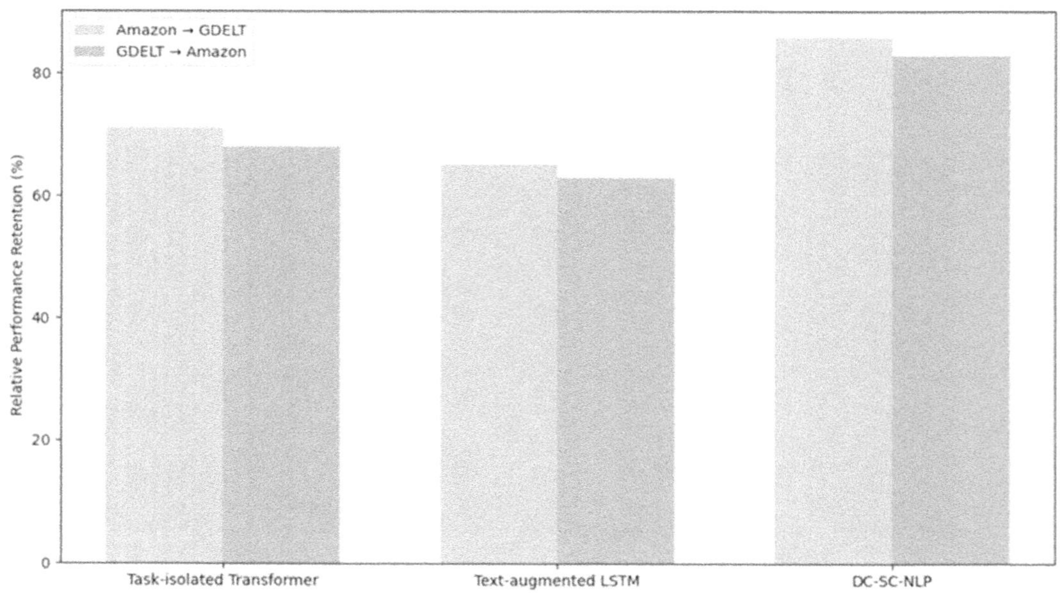


**Figure 5.8 Cross-domain generalization performance of DC-SC-NLP and baseline models, measured as relative performance retention when transferring between demand-side and risk-oriented textual domains.**

rates of retention between 65% and 63%, respectively. Against this backdrop, DC-SC-NLP is considerably more transferable, with 86% of performance for Amazon to GDELT and 83% for GDELT to Amazon. It is found that domain-specific shared representations enable a more stable and accurate reuse of language-based decision signals across supply chains. Enhanced retention aids in scalable deployment, minimizes retraining overhead, and guarantees uniform decision quality across diverse customer and risk-oriented text environments.

The robustness score in Figure 5.9 shows that the document length and noise injection variance are increasing with normalized performance scores. In short-term and low-noise environments, the task-isolated transformer and DC-SC-NLP both score 1.00, which is a normalized score equivalent to the baseline level of capability. When the length of documents differs, the task-isolated transformer performs at 0.91, while DC-SC-NLP performs at 0.97. Under long and loud text regimes, the degradation is worse: the task-isolated transformer drops to 0.84, while the DC-SC-NLP score is much higher at 0.93. These patterns suggest that DC-SC-NLP is more resilient to heterogeneous document structures, length variances, and noise contamination. Additionally, higher retention across regimes will have improved resistance

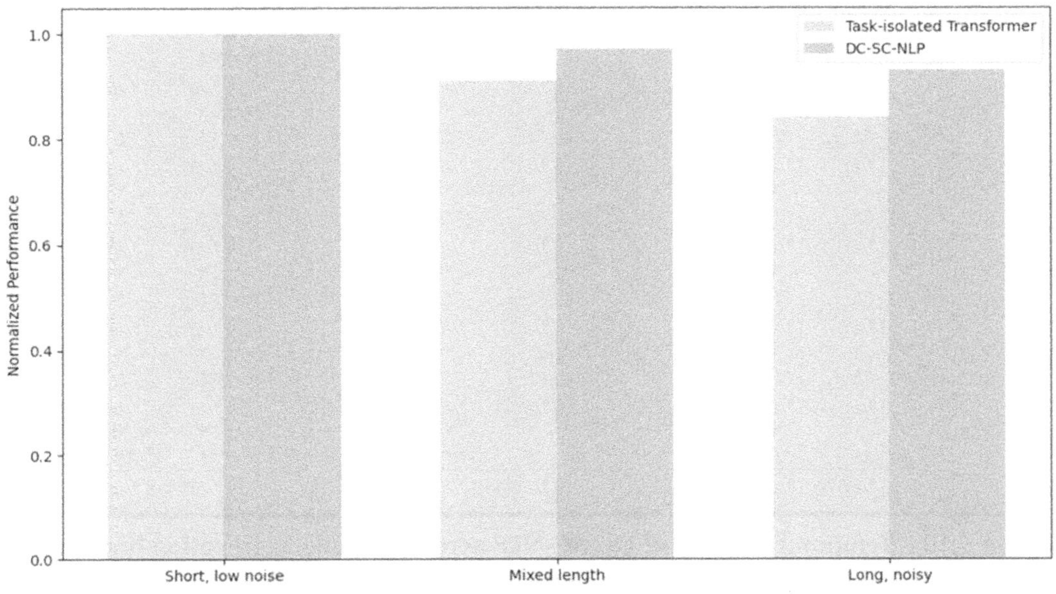


**Figure 5.9 Robustness to text heterogeneity, reporting normalized performance under varying document length and noise regimes, where higher values indicate stronger stability and degradation resistance across heterogeneous textual inputs.**

to degradation and will be more reliable at handling syndicated news, lengthy reports, and inconsistent customer communication. Such robustness is necessary for real-world applications where input format, quality, and temporal density differ widely across functions within the supply chain.

Figure 5.10 plots the performance of DC-SC-NLP across several temporal aggregation windows to test its robustness to window selection rather than a time-scaled one. Using a 7-day window, the model achieves a forecasting error reduction of 22% and a supplier risk AUC of 0.82. A forecasting improvement of 23% and an AUC of 0.84 are obtained when the aggregation period is extended to 14 days. This is stable when the fluctuations are loosened out during the short-run. DC-SC-NLP also achieves a similar forecasting reduction of 21% within 30 days with an AUC of 0.85, indicating greater long-horizon risk discrimination. This small variance in performance across 7–30-day windows suggests that decisions are consistent across time resolutions. This stability allows for flexible deployment in operational contexts where reporting cycles, data latency, and monitoring needs are widely configurable across supply chain functions.

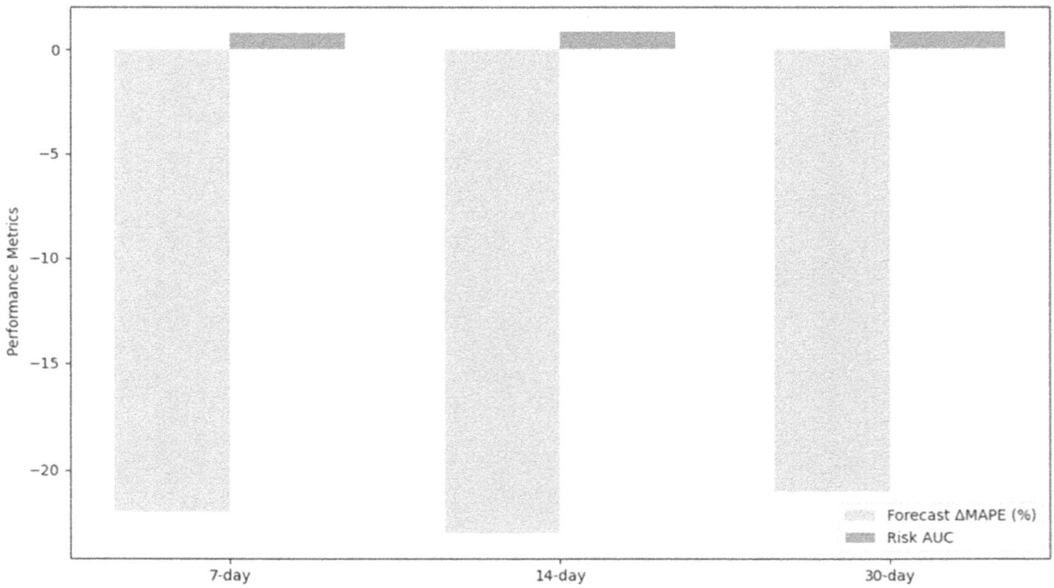


**Figure 5.10 Temporal stability of DC-SC-NLP across multiple aggregation horizons, reporting consistent forecasting error reduction and supplier risk ranking performance under varying temporal resolutions (7–30 days), thereby demonstrating robustness to window selection rather than sensitivity to a fixed time scale.**

### *5.5.3 Ablation Study*

We conduct ablations to examine the effect of key architectural features and measure the effects of domain adaptation on forecast accuracy and supplier risk ranking. Table 5.11 compares unadapted language models, supervised fine-tuning, and the DC-SC-NLP adaptation pipeline as a whole. A prediction MAPE of 16.1% and a risk AUC of 0.71 in the model without adaptation suggest a lack of correspondence between language representations and decision contexts. Supervised fine-tuning enhances performance by reducing MAPE to 14.2% and AUC to 0.78, by partially aligning representations with domain-specific vocabulary and structures. For the best results, the complete DC-SC-NLP configuration results in a MAPE of 12.9% and an AUC of 0.84. This steady increase in relevance demonstrates that complex domain adaptation contributes significantly to greater decision-aligned performance and discriminatory power in prioritizing risk, which also leads to more precise and stable prediction results across multiple supply chain text sources.

Table 5.12 examines the influence of temporal alignment and aggregation on forecasting accuracy and risk-score stability within the DC-SC-NLP framework. Without time alignment, the model forecasts a MAPE of 15.8% and a normalized volatility score of 1.00. This suggests a high sensitivity to short-run noise and weak temporal coherence. Fixed temporal windows increase stability and synchronize text signals with operational planning cycles to lower MAPE to 13.9% and volatility to 0.84. Performance increases with rolling window and smoothing settings; namely, the MAPE is decreased to 12.9% and volatility drops to 0.82, demonstrating greater resistance to

**Table 5.11 Ablation Results Quantifying the Effect of Domain Adaptation on Forecasting Accuracy and Supplier Risk Ranking, Comparing Unadapted Language Models, Supervised Fine-Tuning, and the Full DC-SC-NLP Adaptation Pipeline**

| *Configuration* | *Forecast MAPE (%)* | *Risk AUC* |
|---|---|---|
| No adaptation | 16.1 | 0.71 |
| Fine-tuning only | 14.2 | 0.78 |
| DC-SC-NLP (Full) | 12.9 | 0.84 |

Lower MAPE and higher AUC indicate improved decision-aligned performance.

**Table 5.12 Effect of Temporal Alignment and Aggregation Mechanisms on Forecasting Accuracy and Risk-Score Stability, Isolating the Contribution of Temporal Modeling within the DC-SC-NLP Framework**

| *Configuration* | *MAPE (%)* | *Volatility (↓)* |
|---|---|---|
| No temporal alignment | 15.8 | 1.00 |
| Fixed windows | 13.9 | 0.84 |
| Rolling+smoothing | 12.9 | 0.82 |

**Table 5.13 Ablation Analysis Isolating the Effect of Decision-Centric Integration, Comparing Task-Isolated NLP Pipelines, Shared Encoder Architectures, and the Full DC-SC-NLP Framework to Quantify the Contribution of Explicit Language–Decision Coupling to Overall System Performance**

| *Architecture* | *Avg. Score ↑* |
|---|---|
| Task-isolated NLP | 0.74 |
| Shared encoder only | 0.79 |
| DC-SC-NLP (Full) | 0.86 |

transient fluctuations. These results show that time is modeled explicitly, which greatly improves the accuracy and reliability of risk scoring. Strong temporal alignment allows for easier identification of trends, smoother determining, and consistent decision support even when reporting frequency and pattern of data arrival are different.

Table 5.13 shows the ablation procedure that isolates the impact of decision-centric integration among task-isolated NLP pipelines, shared encoder architectures, and DC-SC-NLP as a whole. The mean score for the task-isolated NLP configuration is 0.74, indicating a lack of coordination between language modeling and the downstream operational decision modules. With the creation of a shared encoder, the representational consistency across tasks is improved and the mean score is 0.79, equivalent to a moderate improvement from the shared semantic space alone. Using this link between language representations and decision-making, the DC-SC-NLP model achieves the highest score on average at 0.86. This improvement suggests that not only can performance be improved by enhancing text encoding, but also that language-derived signals are explicitly embedded into forecasting, risk monitoring, and service processes. Decision-centric coupling allows for better coherence in feature reuse, better alignment with goals, and consistent performance at the system level across heterogeneous supply chain text environments.

## 5.6 Conclusion and Future Works

This chapter positioned NLP as an important decision-making tool for supply chain management today, rather than a secondary text-mining tool. Based on the Language–Decision Disconnect, we developed DC-SC-NLP, an integrated tool that integrates domain-specific language models into the forecasting, supplier risk assessment, contract analysis, and customer communication workflows. NLP was seen as a key component of end-to-end decision pipelines throughout the chapter, explicitly tied to time, operational objectives, and governance constraints. Results from the empirical testing of complementary demand-side and external risk data have shown that decision-aligned language integration produces significant and consistent improvements in forecast accuracy, risk prioritization efficiency, and service quality in real-world deployment. Across domains, the results indicated that similar representations reflecting the domain are generalizable across supply chain functions. This reduces fragmentation of architecture and facilitates scalable enterprise use. Ablation analyses also demonstrated that performance gains are the result of system-level integration and the coupling of decisions, rather than modeling individual decision-making or specific task optimization. There are plenty of exciting extensions ahead. First, the fusion of internal enterprise data sources, such as emails, ERP notes, and audit logs, would allow for more complex decision-making. Second, more than one modality—text, time series, graphs, and sensor data—could be used to enhance resilience to disruption. Third, future research should also incorporate decision-making directed toward providing explanations that are consistent with the practices of planners, buyers, and compliance officers, rather than relying on generic model diagnostics. Ultimately, it is necessary to implement adaptive learning and governance-aware update strategies in order to guarantee reliability when operational conditions are nonstationary.

## References

1. T. Choi, S. W. Wallace, and Y. Wang, "Big data analytics in operations management," *Production and Operations Management*, vol. 27, no. 10, pp. 1868–1883, Oct. 2018, https://doi.org/10.1111/poms.12838.
2. A. C. Benabdellah, A. Benghabrit, I. Bouhaddou, and E. M. Zemmouri, "Big data for supply chain management: Opportunities and challenges," *2016 IEEE/ACS 13th International Conference of Computer Systems and Applications (AICCSA)*, pp. 1–6, Nov. 2016, https://doi.org/10.1109/aiccsa.2016.7945828.
3. L. Monostori, "Cyber-physical production systems: Roots, expectations and R&D challenges," *Procedia CIRP*, vol. 17, pp. 9–13, 2014, https://doi.org/10.1016/j.procir.2014.03.115.
4. T. Schoenherr and C. Speier-Pero, "Data science, predictive analytics, and big data in supply chain management: Current state and future potential," *Journal of Business Logistics*, vol. 36, no. 1, pp. 120–132, Feb. 2015, https://doi.org/10.1111/jbl.12082.
5. E. Taghizadeh, and E. Taghizadeh, "The impact of digital technology and Industry 4.0 on enhancing supply chain resilience." *Proceedings of the 11th Annual International Conference on Industrial Engineering and Operations Management*, Singapore, pp. 9–11, 2021.
6. E. Hofmann and M. Rüsch, "Industry 4.0 and the current status as well as future prospects on logistics," *Computers in Industry*, vol. 89, pp. 23–34, Apr. 2017, https://doi.org/10.1016/j.compind.2017.04.002.
7. T. Papadopoulos, A. Gunasekaran, R. Dubey, N. Altay, S. J. Childe, and S. Fosso-Wamba, "The role of Big Data in explaining disaster resilience in supply chains for sustainability," *Journal of Cleaner Production*, vol. 142, pp. 1108–1118, Jan. 2017, https://doi.org/10.1016/j.jclepro.2016.03.059.
8. H. Stadtler, "Supply chain management: An overview," *Supply Chain Management and Advanced Planning: Concepts, Models, Software, and Case Studies*, pp. 3–28, Oct. 2014, https://link.springer.com/chapter/10.1007/978-3-642-55309-7_1.
9. H. Chen, R. H. L. Chiang, and V. C. Storey, "Business intelligence and analytics: From Big Data to big impact," *MIS Quarterly*, vol. 36, no. 4, pp. 1165–1188, Dec. 2012, https://doi.org/10.2307/41703503.
10. K. Blasiak, "Big Data; A Management Revolution: The emerging role of big data in businesses," *Theseus.fi*, 2025, urn:NBN:fi:amk-201405127199.
11. V. Govindarajan, P. Patel, S. Tripathi, M. A. Hoque, and G. S. Kashyap, "MAGIC-Enhanced Keyword Prompting for Zero-Shot Audio Captioning with CLIP Models," *arXiv.org*, 2025. https://arxiv.org/abs/2509.12591 (accessed Jan. 13, 2026).
12. A. Soni et al., "Can We Predict Your Next Move without Breaking Your Privacy?," *arXiv.org*, 2025. https://arxiv.org/abs/2507.08843 (accessed Jan. 13, 2026).
13. S. Tripathi, M. T. Nafis, I. Hussain, and J. Gao, "The Confidence Paradox: Can LLM Know When It's Wrong," *arXiv.org*, 2025. https://arxiv.org/abs/2506.23464 (accessed Jan. 13, 2026).

14. I. Beltagy, K. Lo, and A. Cohan, "SciBERT: A Pretrained Language Model for Scientific Text," *arXiv.org*, 2019. https://arxiv.org/abs/1903.10676 (accessed Dec. 27, 2025).
15. F. Doshi-Velez and B. Kim, "Towards a Rigorous Science of Interpretable Machine Learning," *arXiv.org*, 2017. https://arxiv.org/abs/1702.08608 (accessed Dec. 27, 2025).
16. T. Winograd, "Understanding natural language," *Cognitive Psychology*, vol. 3, no. 1, pp. 1–191, Jan. 1972, https://doi.org/10.1016/0010-0285(72)90002-3.
17. C. Eugene, *Statistical Language Learning*. MIT press, 1996.
18. K. R. Chowdhary, "Natural Language Processing," *Fundamentals of Artificial Intelligence*, pp. 603–649, Springer, India, Jan. 2020, https://doi.org/10.1007/978-81-322-3972-7_19.
19. G. Weikum, "Foundations of statistical natural language processing," *ACM SIGMOD Record*, vol. 31, no. 3, pp. 37–38, Sep. 2002, https://doi.org/10.1145/601858.601867.
20. T. Mikolov, K. Chen, G. Corrado, and J. Dean, "Efficient Estimation of Word Representations in Vector Space," *arXiv.org*, 2025. https://arxiv.org/abs/1301.3781 (accessed Dec. 27, 2025).
21. G. S. Kashyap et al., "Can AI See What We Can't? Leveraging Deep Learning and Multi-Temporal Satellite Data to Revolutionize Crop Type Mapping and Yield Prediction," In *ICASSP 2025–2025 IEEE International Conference on Acoustics, Speech and Signal Processing (ICASSP)*, pp. 1–5. IEEE, 2025. https://doi.org/10.1109/icassp49660.2025.10890344.
22. J. Devlin, M.-W. Chang, K. Lee, and K. Toutanova, *Proceedings of the 2019 Conference of the North*, pp. 4171–4186, 2019, https://doi.org/10.18653/v1/n19-1423.
23. R. J. Hyndman and G. Athanasopoulos, *Forecasting: Principles and Practice*. OTexts, 2018.
24. A. Dolgui, D. Ivanov, and B. Sokolov. "Ripple effect in the supply chain: An analysis and recent literature." *International Journal of Production Research*, vol. 56, no. 1–2, pp. 414–430, 2018.
25. N. Archak, A. Ghose, and P. G. Ipeirotis, "Deriving the pricing power of product features by mining consumer reviews," *Management Science*, vol. 57, no. 8, pp. 1485–1509, Aug. 2011, https://doi.org/10.1287/mnsc.1110.1370.
26. G. S. Kashyap, K. Malik, S. Wazir, and A. E. I. Brownlee, "Optimization of the rectangle area inside a concave polygon using PSO and Tabu Search," *Soft Computing*, vol. 29, no. 7, pp. 3217–3239, Apr. 2025, https://doi.org/10.1007/s00500-025-10568-1.
27. M. Kulahara, A. Khader, S. Tripathi, and M. A. Hoque, "A CNN-based framework for geometric alignment of historical and satellite imagery," *IEEE Access*, vol. 13, pp. 132259–132268, Jan. 2025, https://doi.org/10.1109/access.2025.3589497.
28. J. Bollen, H. Mao, and X. Zeng, "Twitter mood predicts the stock market," *Journal of Computational Science*, vol. 2, no. 1, pp. 1–8, Mar. 2011, https://doi.org/10.1016/j.jocs.2010.12.007.

29. C. C. Aggarwal and C. Zhai, "A Survey of Text Classification Algorithms," *Mining Text Data*, pp. 163–222, Springer, USA, Jan. 2012, https://doi.org/10.1007/978-1-4614-3223-4_6.
30. J. Qiu, Q. Wu, G. Ding, Y. Xu, and S. Feng, "A survey of machine learning for big data processing," *EURASIP Journal on Advances in Signal Processing*, vol. 2016, no. 1, May 2016, https://doi.org/10.1186/s13634-016-0355-x.
31. Y. Qin, D. Song, H. Chen, W. Cheng, G. Jiang, and G. Cottrell, "A Dual-Stage Attention-Based Recurrent Neural Network for Time Series Prediction," *arXiv.org*, 2017. https://arxiv.org/abs/1704.02971 (accessed Dec. 27, 2025)
32. R. Guidotti, A. Monreale, S. Ruggieri, F. Turini, F. Giannotti, and D. Pedreschi, "A survey of methods for explaining black box models," *ACM Computing Surveys*, vol. 51, no. 5, pp. 1–42, Aug. 2018, https://doi.org/10.1145/3236009.
33. G. A. Zsidisin and L. M. Ellram, "An agency theory investigation of supply risk management," *Journal of Supply Chain Management*, vol. 39, no. 2, pp. 15–27, Jun. 2003, https://doi.org/10.1111/j.1745-493x.2003.tb00156.x.
34. S. M. Wagner and C. Bode, "An empirical examination of supply chain performance along several dimensions of risk," *Journal of Business Logistics*, vol. 29, no. 1, pp. 307–325, Mar. 2008, https://doi.org/10.1002/j.2158-1592.2008.tb00081.x.
35. A. Nenkova and K. McKeown, "Automatic summarization," *Foundations and Trends® in Information Retrieval*, vol. 5, no. 2–3, pp. 103–233, 2011.
36. H. Schütze, C. D. Manning, and P. Raghavan, *Introduction to Information Retrieval*, Vol. 39. Cambridge University Press, 2008.
37. R. Feldman, "Techniques and applications for sentiment analysis," *Communications of the ACM*, vol. 56, no. 4, pp. 82–89, Mar. 2013, https://doi.org/10.1145/2436256.2436274.
38. T. Aven, "Risk assessment and risk management: Review of recent advances on their foundation," *European Journal of Operational Research*, vol. 253, no. 1, pp. 1–13, Dec. 2015, https://doi.org/10.1016/j.ejor.2015.12.023.
39. I. Chalkidis, M. Fergadiotis, P. Malakasiotis, N. Aletras, and I. Androutsopoulos, "LEGAL-BERT: The Muppets straight out of Law School," *arXiv.org*, 2020. https://arxiv.org/abs/2010.02559 (accessed Dec. 27, 2025).
40. C. Rudin, "Stop explaining black box machine learning models for high stakes decisions and use interpretable models instead," *Nature Machine Intelligence*, vol. 1, no. 5, pp. 206–215, May 2019, https://doi.org/10.1038/s42256-019-0048-x.
41. A. Vaswani et al., "Attention is all you need," *Advances in Neural Information Processing Systems*, vol. 30, 2017. https://proceedings.neurips.cc/paper/2017/hash/3f5ee243547dee91fbd053c1c4a845aa-Abstract.html (accessed Dec. 27, 2025).
42. I. Chalkidis et al., "LexGLUE: A Benchmark Dataset for Legal Language Understanding in English," *Proceedings of the 60th Annual Meeting of the Association for Computational Linguistics (Volume* 1*:* Long Papers), Dublin, 2022, https://doi.org/10.18653/v1/2022.acl-long.297.
43. H. Surden, "Machine learning and law: An overview," *Research Handbook on Big Data Law*, Edward Elgar Publishing eBooks, May 2021, https://doi.org/10.4337/9781788972826.00014.

44. Z. Nasar, S. W. Jaffry, and M. K. Malik, "Named entity recognition and relation extraction," *ACM Computing Surveys*, vol. 54, no. 1, pp. 1–39, Feb. 2021, https://doi.org/10.1145/3445965.
45. D. M. Berry, *Understanding Digital Humanities*. Palgrave Macmillan, 2017, https://doi.org/10.1057/9780230371934.
46. I. Chalkidis, I. Androutsopoulos, and A. Michos, "Extracting contract elements," *Proceedings of the 16th Edition of the International Conference on Articial Intelligence and Law*, pp. 19–28, Jun. 2017, https://doi.org/10.1145/3086512.3086515.
47. D. Hendrycks, K. Lee, and M. Mazeika, "Using pre-training can improve model robustness and uncertainty," *PMLR*, pp. 2712–2721, May 2019. https://proceedings.mlr.press/v97/hendrycks19a.html (accessed Dec. 27, 2025).
48. D. M. Katz, M. J. Bommarito, and J. Blackman, "A general approach for predicting the behavior of the Supreme Court of the United States," *PLoS ONE*, vol. 12, no. 4, pp. e0174698–e0174698, Apr. 2017, https://doi.org/10.1371/journal.pone.0174698.
49. I. Chalkidis, I. Androutsopoulos, and N. Aletras, "Neural Legal Judgment Prediction in English," *arXiv.org*, 2019. https://arxiv.org/abs/1906.02059 (accessed Dec. 27, 2025).
50. K. Coussement and D. Van den Poel, "Improving customer complaint management by automatic email classification using linguistic style features as predictors," *Decision Support Systems*, vol. 44, no. 4, pp. 870–882, Mar. 2008, https://doi.org/10.1016/j.dss.2007.10.010.
51. W. M. P. van der Aalst et al., "Business process mining: An industrial application," *Information Systems*, vol. 32, no. 5, pp. 713–732, Jul. 2006, https://doi.org/10.1016/j.is.2006.05.003.
52. B. Pang and L. Lee, "Opinion mining and sentiment analysis," *Foundations and Trends® in Information Retrieval*, vol. 2, no. 1–2, pp. 1–135, Jan. 2008, https://books.google.co.in/books?hl=en&lr=&id=XQswsqLLKrEC&oi=fnd&pg=PA1&dq=Opinion+mining+and+sentiment+analysis&ots=FQ38ESy_k_&sig=om4iA9g-_dCLLYPjPUexPbkVXFY&redir_esc=y#v=onepage&q=Opinion%20mining%20and%20sentiment%20analysis&f=false.
53. T. Wilson, J. Wiebe, and P. Hoffmann, "Recognizing Contextual Polarity in Phrase-Level Sentiment Analysis," In *Proceedings of Human Language Technology Conference and Conference on Empirical Methods in Natural Language Processing*, pp. 347–354, 2005. https://aclanthology.org/H05-1044.pdf (accessed Dec. 28, 2025).
54. H. Wang, J. He, X. Zhang, and S. Liu, "A short text classification method based on N-gram and CNN," *Chinese Journal of Electronics*, vol. 29, no. 2, pp. 248–254, 2020.
55. M. Pontiki et al., "SemEval-2016 Task 5: Aspect Based Sentiment Analysis," *Proceedings of the 10th International Workshop on Semantic Evaluation (SemEval-2016)*, 2016, https://doi.org/10.18653/v1/s16-1002.

56. R. Lowe, N. Pow, I. V. Serban, and J. Pineau, "The Ubuntu Dialogue Corpus: A Large Dataset for Research in Unstructured Multi-Turn Dialogue Systems." *Proceedings of the 16th Annual Meeting of the Special Interest Group on Discourse and Dialogue*, pp. 285–294, 2015.
57. Y. Zhang et al., "DIALOGPT: Large-Scale Generative Pre-training for Conversational Response Generation," *Proceedings of the 58th Annual Meeting of the Association for Computational Linguistics: System Demonstrations*, 2020, https://doi.org/10.18653/v1/2020.acl-demos.30.
58. A. Radford, J. Wu, R. Child, D. Luan, D. Amodei, and I. Sutskever, "Language models are unsupervised multitask learners," *OpenAI Blog*, vol. 1, no. 8, p. 9, 2019.
59. P. Lewis et al., "Retrieval-augmented generation for knowledge-intensive NLP tasks," *Advances in Neural Information Processing Systems*, vol. 33, pp. 9459–9474, 2020. https://proceedings.neurips.cc/paper/2020/hash/6b493230205f780e1bc26945df7481e5-Abstract.html (accessed Dec. 27, 2025).
60. S. Gururangan et al., "Don't Stop Pretraining: Adapt Language Models to Domains and Tasks," *arXiv.org*, 2020. https://arxiv.org/abs/2004.10964 (accessed Dec. 27, 2025).
61. "A Comparative Study on Feature Selection in Text Categorization," *Proceedings of the Fourteenth International Conference on Machine Learning, Guide Proceedings*, 2025, https://dl.acm.org/doi/abs/10.5555/645526.657137.
62. H. Joshi and S. Joseph, "ULMFiT: Universal language model fine-tuning for text classification," *International Journal of Advanced Medical Sciences and Technology*, vol. 4, no. 6, pp. 1–9, Oct. 2024, https://doi.org/10.54105/ijamst.e3049.04061024.
63. B. Settles, "Active Learning Literature Survey," *Wisconsin.edu*, 2025, https://digital.library.wisc.edu/1793/60660.
64. J. Kirkpatrick et al., "Overcoming catastrophic forgetting in neural networks," *Proceedings of the National Academy of Sciences*, vol. 114, no. 13, pp. 3521–3526, Mar. 2017, https://doi.org/10.1073/pnas.1611835114.
65. E. Hu et al., "LoRA: Low-Rank Adaptation of Large Language Models." 2025. https://arxiv.org/pdf/2106.09685v1/1000 (accessed Dec. 27, 2025).
66. T. Brown et al., "Language models are few-shot learners," *Advances in Neural Information Processing Systems*, vol. 33, pp. 1877–1901, 2020. https://proceedings.neurips.cc/paper_files/paper/2020/hash/1457c0d6bfcb4967418bfb8ac142f64a-Abstract.html?utm_source=transaction&utm_medium=email&utm_campaign=linkedin_newsletter (accessed Dec. 27, 2025).
67. E. Perez, D. Kiela, and K. Cho, "True few-shot learning with language models," *Advances in Neural Information Processing Systems*, vol. 34, pp. 11054–11070, Dec. 2021. https://proceedings.neurips.cc/paper/2021/hash/5c04925674920eb58467fb52ce4ef728-Abstract.html (accessed Dec. 27, 2025).
68. B. Settles, "Active Learning Literature Survey," 2009. https://api.semanticscholar.org/CorpusID:324600

69. P. M. Mah, I. Skalna, and J. Muzam, "Natural language processing and artificial intelligence for enterprise management in the era of Industry 4.0," *Applied Sciences*, vol. 12, no. 18, pp. 9207–9207, Sep. 2022, https://doi.org/10.3390/app12189207.
70. H. Blake, "The Role of Natural Language Processing in ERP Chat Interfaces," *ResearchGate*, Feb. 12, 2022. https://www.researchgate.net/publication/387670194_The_Role_of_Natural_Language_Processing_in_ERP_Chat_Interfaces (accessed Jan. 02, 2026).
71. W. Y. C. Wang, D. Pauleen, and N. Taskin, "Enterprise systems, emerging technologies, and the data-driven knowledge organisation," *Knowledge Management Research & Practice*, vol. 20, no. 1, pp. 1–13, 2022, https://doi.org/10.1080/14778238.2022.2039571.
72. R. He and J. Mcauley, "Ups and Downs: Modeling the Visual Evolution of Fashion Trends with One-Class Collaborative Filtering," In *Proceedings of the 25th International Conference on World Wide Web*, pp. 507–517. 2016, https://doi.org/10.1145/2872427.2883037.
73. K. Leetaru and P. A. Schrodt, "Gdelt: Global data on events, location, and tone, 1979–2012," *ISA Annual Convention*, vol. 2, no. 4, pp. 1–49, 2013.

# Chapter 6

# Digital Twins: Concept and Applications

Abdullah Mohammad

## 6.1 Introduction

As virtual counterparts to physical systems, digital twins interact with real-world activity to provide guidance, forecasting, and decision-making within the manufacturing and supply chain. The growing scale, interdependence, and volatility of industrial systems require decision-making based on analytical strategies that provide a mechanism for reflecting changing system states rather than static or periodic models [1]. Traditional decision-support techniques, primarily offline simulations or historical aggregates, cannot capture the dynamic interactions, feedback loops, and temporal dependencies inherent in modern industrial practice [2].

New technologies such as Industrial Internet of Things (IIoT) sensing, communication systems, cloud and edge computing, and machine learning have made it technically possible to store continuously synchronized digital representations of physical assets, processes, and networks [3,4]. Unlike traditional simulation tools, which are contextually aware and detached from real-world situations, digital twins are contextually aware of the situation, and therefore analytical reasoning can be closely aligned with the actual state of the system [5]. This potential has positioned digital twins as a promising way to connect physical execution with digital decision support.

Digital twin technology first appeared in the aerospace and defense sectors, where high-value assets were constantly monitored throughout


DOI: 10.4324/9781003724889-6

the lifecycle and reliability controls had to be tightly controlled [6]. Since then, the core ideas have been extended to include discrete and process manufacturing, logistics and distribution, energy systems, and infrastructure management [7]. Digital twins have been used in manufacturing to combine machine behavior, process constraints, and production data for planning and control [8]. For example, digital twins are interpreted in relation to supply chains as representing network architectures, material flows, and cross-organizational modes of coordination [9]. As a result, these developments indicate a wider shift toward operational management driven by data and facilitated by models, where informed choices in the face of uncertainty are backed by ongoing feedback between the physical and digital layers.

While interest and adoption have grown rapidly, most current implementations of digital twins remain stalled. Asset-level models, simulation engines, and analytics pipelines are often developed in isolation from each other. This leads to systems that are difficult to scale, interpret, and operate. This limitation is termed the Architectural Coherence Gap, which refers to a misalignment among real-time data streams, model abstraction levels, and operational decision horizons. These inequalities can take a variety of forms, including brittle integrations, excessive computational overhead at higher fidelity, and poor alignment between digital twin outputs and organizational decision-making.

Table 6.1 compares the traditional decision-support paradigm with isolated implementations of digital twins and the end-to-end digital twin systems that are emerging in key domains. For example, 1 indicates that the paradigm explicitly supports a particular capability, while 0 means that it is not supported in this table. Traditional decision-support models either use static or periodic data refreshes and are not compatible with other physical systems. Although the isolated digital twins provide real-time data coupling and representations at the level of assets, system-level, scalable and governance integration abutting these twins are also provided by independent digital twins. On the other hand, end-to-end digital twin systems include continuous synchronization, multi-level supply chain modeling, scenario analysis, scalability considerations, and explicit interpolation with organizational decision-making processes. This transition can be conceptualized in Figure 6.1, as it leads away from a static offline analysis paradigm toward a unified digital twin architecture, which is a continuously synchronized, system-level infrastructure. As digital twins move from one asset to another, they become increasingly involved in making and processing complex manufacturing and supply chain systems. This expansion raises important

**Table 6.1 Comparison of Conventional Decision-Support Models, Isolated Digital Twin Implementations, and End-to-End Digital Twin Systems**

| *Main Feature* | *Conventional Decision-Support Models* | *Isolated Digital Twin Implementations* | *End-to-End Digital Twin Systems* |
|---|---|---|---|
| Static or periodic data updates | 1 | 0 | 0 |
| Continuous real-time data synchronization | 0 | 1 | 1 |
| Asset-level representation | 0 | 1 | 1 |
| System-level representation | 0 | 0 | 1 |
| Integration of physical and digital states | 0 | 1 | 1 |
| Support for dynamic operational contexts | 0 | 1 | 1 |
| Predictive and prescriptive analytics | 0 | 1 | 1 |
| Multi-tier supply chain modeling | 0 | 0 | 1 |
| Scenario analysis and what-if simulation | 0 | 1 | 1 |
| Scalability consideration | 0 | 0 | 1 |
| Organizational and governance integration | 0 | 0 | 1 |
| Cross-functional decision support | 0 | 0 | 1 |

A value of 1 indicates explicit support for a given capability, while 0 denotes its absence. The table highlights how end-to-end digital twins uniquely combine continuous synchronization, system-level representation, and organizational integration.

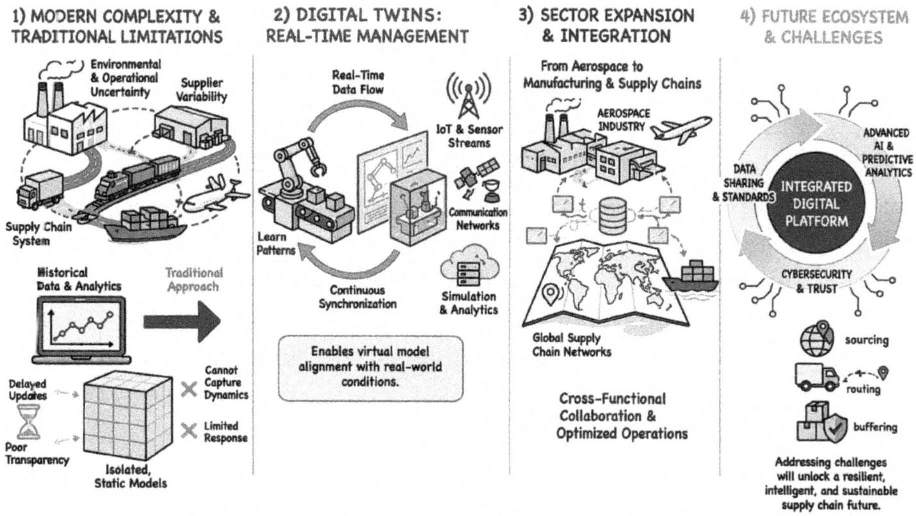


**Figure 6.1 Evolution of decision-support paradigms from static, offline analytical models to end-to-end digital twin architectures with continuous physical–digital synchronization, multi-scale system representation, and real-time, data-driven decision support.**

questions about coordinating model functions across time and space, the degree of fidelity to computational feasibility, and the use of digital twins within systems of data collection, decision-making, and accountability.

A considerable number of current studies on digital twins are carried out through definitions, taxonomies, or limited case studies with a focus on individual machines or engineering subsystems isolated from the rest of the study [10]. These projects provide important conceptual foundations but provide little guidance on how to design digital twins as complete, end-to-end systems that combine real-time data ingestion, predictive modeling, simulation, and decision support across multiple operational layers. The importance of representational accuracy is often stressed in the context of modeling fidelity, but in near-real-time settings, the trade-offs of responsiveness and scalability are not particularly adequately addressed. In addition, organizational aspects like data management, cross-functional coordination, labor skills, and congruence of decision-making power and analytical results are often viewed as secondary concerns [11]. The latter acts act as an unnecessary rupture between the theoretical promise of digital twins and their practical implementation.

To address the Architectural Coherence Gap, this chapter considers digital twins as multi-scale decision infrastructures rather than isolated analytical artifacts. We present SynDT, a digital twin-synchronized framework that explicitly links temporal modeling of assets to system-level supply chain abstractions through synchronized data pipelines and simulation layers. The model integrates physics-informed representations with adaptive learning elements, while maintaining interpretability, computational feasibility, and organizational deployability.

This chapter is organized to slowly develop and verify SynDT as a cohesive, multi-scale digital twin. Section 6.2 analyses the development of digital twin systems in the manufacturing and supply chain sectors, including architectural paradigms, machine learning integration, organizational alignment, interoperability, and semantic standardization, in order to conceptually establish the background of the study. Section 6.3 presents the empirical basis for multi-scale modeling. The experimental setup, evaluation metrics, hyperparameter setup, and computational viability issues are all covered in detail in Section 6.4. A thorough experimental investigation that includes comparative performance evaluation, cross-domain robustness assessment, and architectural ablation experiments is presented in Section 6.5. The results are summarized in Section 6.6, which highlights the importance of architectural coherence in attaining stable, comprehensible, and governance-aligned digital twin deployment across supply chain and manufacturing systems.

## 6.2 Related Works

### 6.2.1 Historical Evolution of Digital Twin Concepts

The idea of digital twins derives from early developments in Computer-Aided Design (CAD), system simulation, and model-based engineering, particularly in aerospace and defense [12]. As previously outlined in Table 6.2, these early CAD and simulation models were primarily offline, design-time applications involving structural verification and performance assessment based on predetermined assumptions. They were designed to be used only for monitoring or decision support after deployment, as they explicitly disassociated themselves from live operational data and were modeled on static or semi-dynamic representations. Despite providing high-fidelity, physics-based representations, these models are limited to the abstraction of assets and do not provide continuous synchronization, system-level reasoning, or scalability. Digital models with a lifecycle in mind

**Table 6.2 Evolution of Digital Twin Paradigms from Static, Design-Time Engineering Models to Integrated, Data-Driven Industrial Systems**

| *Main Feature* | *Early CAD and Simulation Models* | *Lifecycle-Oriented Digital Models* | *Contemporary Digital Twin Systems* |
|---|---|---|---|
| Offline design-time analysis | 1 | 1 | 0 |
| Continuous real-time data coupling | 0 | 0 | 1 |
| Static or semi-dynamic representation | 1 | 1 | 0 |
| Asset lifecycle coverage | 0 | 1 | 1 |
| Runtime monitoring capability | 0 | 1 | 1 |
| Predictive health assessment | 0 | 1 | 1 |
| Integration of sensor telemetry | 0 | 0 | 1 |
| Physics-based modeling | 1 | 1 | 1 |
| Data-driven model augmentation | 0 | 0 | 1 |
| Asset-level focus | 1 | 1 | 1 |
| System- or network-level scope | 0 | 0 | 1 |
| Scalability and interoperability consideration | 0 | 0 | 1 |

Binary indicators (1=supported, 0=not supported) highlight the progressive introduction of real-time synchronization, system-level scope, and scalability considerations.

were an early step toward development in this area, providing a wider range of digital representations beyond the preliminary design stages to be used for maintenance planning, fault detection, and reliability assessment over the lifecycle of assets [13]. As discussed in Table 6.2, this paradigm created runtime monitoring capability and health prediction, all while running on offline or constantly updated data flows. Sensor data began to inform these models, but integration became episodic rather than continuous, and representations remained primarily asset-based [14]. Lastly, it is important to note that, while lifecycle coverage and operational relevance have improved, these models are still unable to include real-time data coupling, data-driven model adaptation, or system-level scope.

The digital twin systems of today are not incremental but qualitative. Modern digital twins constitute an industry that utilizes sensors, industrial communication systems, and IIoT platforms [15] for continuous real-time data coupling between physical systems and digital twins. As reflected in Table 6.2, this change is coupled with the introduction of data-driven model augmentation, system-level representations or network levels, and implicit considerations of scaling and interoperability. In addition, physics-based models are retained but supplemented by statistical and learning-based models to account for operational variability and uncertainty that cannot be specified a priori [16]. Digital twins in this paradigm are continuous representations of the system that can be used for runtime monitoring, forecasting, and decision-making. Importantly, Table 6.2 shows that many of the abilities associated with digital twins, including scaling, interoperability, and reasoning at the system level, are only present in the current paradigm and not in earlier stages. This fact supports a key idea: representational accuracy and data integration have remained stable, but architectural coherence across ingestion, modeling, and decision layers has not been uniform. Instead, it occurs suddenly and incongruently, underscoring the need for a carefully planned digital twin architecture to be expressed from the ground up rather than through incremental revisions of existing modeling paradigms.

### *6.2.2 Digital Twin Applications in Manufacturing*

Early digital twin applications in manufacturing were based on discrete-event simulation and virtual manufacturing systems for the planning and optimization of production [17]. These methods focused on modeling single machines, manufacturing cells, or production lines. They were carried out off-site using historical data and fixed engineering assumptions. As shown in Table 6.3, such offline virtual manufacturing systems relied on preconfigured simulations and provided value primarily for layout design, capacity analysis, and changeover evaluation prior to physical delivery [18]. Increased sensor deployment and shop-floor connectivity have led to the development of digital twins on the assets that are operational and can integrate real-time production data [19]. Programable logic controllers, supervisory control platforms, and manufacturing execution systems also serve as data sources for the coordination of physical and digital equipment [20]. Predictive maintenance emerged as one of the most valuable applications in this phase, with digital twins used to track degradation rates, identify issues, and aid maintenance decision-making [21]. As shown in Table 6.3, these asset-level twins added runtime monitoring and predictive maintenance but with a focus on single assets and limited system-level integration. Recent studies have begun to investigate integrated manufacturing digital

**Table 6.3 Evolution of Manufacturing Digital Twin Applications from Offline Virtual Manufacturing to System-Level, Closed-Loop Production Twins**

| *Main Feature* | *Offline Virtual Manufacturing* | *Asset-Level Operational Twins* | *System-Level Integrated Twins* |
|---|---|---|---|
| Offline simulation usage | 1 | 0 | 0 |
| Real-time data integration | 0 | 1 | 1 |
| Single-asset or cell focus | 1 | 1 | 0 |
| Facility-wide process representation | 0 | 0 | 1 |
| Production planning support | 1 | 1 | 1 |
| Runtime monitoring capability | 0 | 1 | 1 |
| Predictive maintenance support | 0 | 1 | 1 |
| Physics-based process modeling | 1 | 1 | 1 |
| Data-driven model adaptation | 0 | 0 | 1 |
| Quality management integration | 0 | 0 | 1 |
| Closed-loop control feedback | 0 | 0 | 1 |
| Lifecycle digital thread support | 0 | 0 | 1 |
| Scalability and interoperability focus | 0 | 0 | 1 |

A value of 1 denotes explicit support for a capability, while 0 indicates its absence, highlighting how real-time integration and adaptive control emerge only at the system level.

twins at the system level, which go beyond single assets to depict production processes across entire facilities, including material flows and resource coordination [22]. To account for the complex dynamics of production in environments characterized by high mix and variability, hybrid modeling strategies that merge physics-informed representations with data-driven and machine learning approaches have been suggested [23]. By connecting process models with inspection data to study defect propagation and assess corrective measures before physical execution, digital twins have also found application in quality management [24]. As Table 6.3 shows, capabilities like closed-loop control feedback, lifecycle digital thread support, and scalability across heterogeneous systems arise only when manufacturing digital twins are designed as system-level, architecturally integrated infrastructures. Although recent research shows that these features can be implemented in controlled or specific contexts, they are not present in the majority of asset-level and operational implementations. This distinction highlights that digital twins in manufacturing move from being analytical tools to becoming sustained decision infrastructures only when asset-level analytics are coherently integrated into facility-wide architectures that explicitly consider integration, synchronization, and scalability.

## 6.2.3 Supply Chain Digital Twin Implementations

Research on digital twins in supply chains that predated this work was based on static network models and offline simulations that were scenario-based and aimed at assessing capacity utilization, service levels, and costs based on planning assumptions established beforehand [25]. As shown in the first column of Table 6.4, these approaches relied almost exclusively on historical data, assumed deterministic system behavior, and did not include real-time data. These models were successful in network planning and inventory management, but offline implementation compromised ability to respond quickly to network downtime, demand changes, or operational changes [26]. The second research phase implemented these virtual supply chain twins using near-real-time data from Enterprise Resource Planning (ERP), warehouse management systems, and transportation management systems [27]. This transition facilitated operational visibility and short-term replanning tools such as disruption propagation analysis and delay impact assessment [28]. As Table 6.4 shows, this phase signifies the beginning of real-time integration of data and communication among enterprise systems. However, these digital twins primarily did not follow planning assumptions and over implementations of uncertainty. This limited their ability to help strong decision-making under nonstationary demand and supply conditions [29]. Recent studies have concentrated on creating end-to-end, flexible supply chain digital twins that integrate multi-level suppliers, production systems, logistics infrastructures, and demand processes within tightly coupled virtual environments [30]. Table 6.4 summarizes these systems with explicitly described demand uncertainty modeling, machine learning adaptation, and predictive analytics that evolve with new information [31]. Rather than seeing disruptions as unintended events, such digital twins are capable of continuously monitoring and assessing alternative sourcing, routing, and inventory policies [32,33]. Table 6.4 highlights an architecturally significant difference, namely the rise of federated or modular architectures and network-wide decision coordination in end-to-end adaptive twins. Flow-controlled models in multi-enterprise supply chains often stifle data ownership, governance,

**Table 6.4 Capability-Level Comparison of Supply Chain Digital Twin Paradigms, from Static Network Simulations to End-to-End Adaptive and Multi-Enterprise Decision Infrastructures**

| *Main Feature* | *Static Network Models* | *Operational Supply Chain Twins* | *End-to-End Adaptive Twins* |
|---|---|---|---|
| Offline scenario-based modeling | 1 | 0 | 0 |
| Historical data dependency | 1 | 1 | 0 |
| Real-time data integration | 0 | 1 | 1 |
| ERP/WMS/TMS system integration | 0 | 1 | 1 |
| Deterministic planning assumptions | 1 | 1 | 0 |
| Disruption propagation analysis | 0 | 1 | 1 |
| Multi-tier network representation | 0 | 0 | 1 |
| Predictive analytics integration | 0 | 0 | 1 |
| Machine learning-based adaptation | 0 | 0 | 1 |
| Demand uncertainty modeling | 0 | 0 | 1 |
| Dynamic inventory optimization | 0 | 0 | 1 |
| Alternative sourcing and routing simulation | 0 | 0 | 1 |
| Network-wide decision coordination | 0 | 0 | 1 |
| Scalability across enterprises | 0 | 0 | 1 |
| Federated or modular architecture support | 0 | 0 | 1 |

Binary indicators (1=supported, 0=not supported) highlight the progressive integration of real-time data, uncertainty-aware analytics, and network-wide coordination.

and scalability, making them often impossible [34]. Strategies for federated digital twins enable organizations to maintain control of proprietary data while engaging in shared system-level representations that facilitate coordinated decision-making [35]. This architectural change is crucial for scaling digital twins beyond individual organizations and enabling their ongoing operational use across supply chain networks.

### 6.2.4 Integration of Machine Learning with Digital Twins

As early as the introduction of machine learning into digital twin systems, learning models were considered offline, supervised predictive strategies that were complementary to physics representations but not directly embedded into the digital twin lifecycle [36]. Within this regime models were trained on historical data and used as static predictors for failure forecasting and condition monitoring [37]. As discussed in Table 6.5, these methods used offline training, were not continuously state-synchronized, and did not allow for model adaptation after deployment. Although these integrations enhanced pointwise predictive accuracy, they were not robust to nonstationary system behavior and changing operational contexts. Later studies advanced toward

**Table 6.5 Evolution of Machine Learning Integration in Digital Twin Systems, from Offline Supervised Predictors to Tightly Coupled, Learning-Enabled Decision Architectures**

| *Main Feature* | *Offline Supervised Models* | *Online Adaptive Learning* | *Learning-Enabled Autonomous Twins* |
|---|---|---|---|
| Offline model training | 1 | 0 | 0 |
| Static predictive components | 1 | 0 | 0 |
| Supervised learning usage | 1 | 1 | 1 |
| Real-time data-driven updates | 0 | 1 | 1 |
| Continuous state synchronization | 0 | 1 | 1 |
| Model drift monitoring | 0 | 1 | 1 |
| Periodic or online retraining | 0 | 1 | 1 |
| Unsupervised pattern discovery | 0 | 1 | 1 |
| Anomaly detection capability | 0 | 1 | 1 |
| Dimensionality reduction for high-dimensional data | 0 | 1 | 1 |
| Reinforcement learning integration | 0 | 0 | 1 |
| Policy optimization via simulation | 0 | 0 | 1 |
| Autonomous decision support | 0 | 0 | 1 |
| Safe learning via virtual experimentation | 0 | 0 | 1 |
| Human oversight and governance alignment | 0 | 0 | 1 |

A value of 1 indicates native support for a capability, while 0 denotes its absence, highlighting how adaptivity and autonomy emerge only with continuous state synchronization and governance alignment.

online adaptive learning, integrating machine learning models more closely within digital twin frameworks [38]. These architectures employed real-time data streams for the continuous updating of both the state of the digital twin and its associated learning components, facilitating periodic or online retraining and explicit monitoring of model drift [39]. For tasks like demand forecasting, quality inspection, and remaining useful life estimation, supervised learning continued to be the dominant approach. Meanwhile, unsupervised methods were increasingly used for anomaly detection and pattern discovery in high-dimensional sensor data [40,41]. Table 6.5 shows that this stage brought about real-time data-driven updates, ongoing synchronization, and adaptive learning; however, the authority to make decisions was mostly external to the digital twin.

Recently, it has been proposed that self-directed digital twins can be learned, wherein machine learning models become core components of decision-making rather than predictors [42]. Reinforcement learning has been adopted in order to maximize policy by engaging with the digital twin environment. This also allows for exploration of various control, scheduling, and inventory strategies without interrupting physical operations [43]. The digital twins are virtual worlds that are secure places to experiment and where the learning agent can assess the long-run effects of actions in an environment of uncertainty [44]. This regime is characterized by policy optimization via simulation, autonomy in decision support, and a strong dependence on virtual experimentation, as shown in Table 6.5. Moreover, the literature within this category shows that autonomy can function only when learning-based decisions are limited by governance mechanisms, human oversight, and regulation at the system level [45]. Without such alignment, there is potential for autonomous learning to lead to unsafe or unbalanced actions relative to operating goals. As shown in Table 6.5, governance alignment exists only in learning-enabled autonomous twins, emphasizing that architectural coherence, not just the existence of better learning algorithms, is critical to reliable autonomy.

### 6.2.5 Digital Twin Architectures and Data Pipelines

Early digital twin architectures were predominantly influenced by simulation systems, cyber–physical systems, and industrial automation [46]. These were designed with a series of layers between physical objects, digital models, and user interfaces, primarily based on batch transfers from operational systems to the analytical environments [47]. As Table 6.6 shows, although these

**Table 6.6 Architectural Comparison of Digital Twin Systems, Illustrating the Transition from Layered, Batch-Oriented Designs to Modular and Federated Infrastructures with Continuous Data Pipelines**

| *Main Feature* | *Early Layered Architectures* | *Real-Time Service-Oriented Architectures* | *Modular and Federated Architectures* |
|---|---|---|---|
| Layered separation of physical and digital components | 1 | 1 | 1 |
| Batch-oriented data transfer | 1 | 0 | 0 |
| Real-time data synchronization | 0 | 1 | 1 |
| Continuous data pipelines | 0 | 1 | 1 |
| Middleware-based integration | 0 | 1 | 1 |
| Event-driven data streaming | 0 | 1 | 1 |
| Cloud computing integration | 0 | 1 | 1 |
| Edge computing support | 0 | 1 | 1 |
| Modular functional decomposition | 0 | 0 | 1 |
| Microservices-based architecture | 0 | 0 | 1 |
| Support for heterogeneous data types | 0 | 1 | 1 |
| Standardized information models | 0 | 1 | 1 |
| Interoperability across systems | 0 | 1 | 1 |
| Federated digital twin coordination | 0 | 0 | 1 |
| Hierarchical multi-scale representation | 0 | 0 | 1 |
| Data governance and version control | 0 | 0 | 1 |

Binary indicators (1=supported, 0=not supported) highlight the architectural capabilities that enable scalable, real-time digital twin operation.

architectures could not provide real-time synchronization or continuous data pipelines, they did not provide modular abstraction. As a result, they were not very responsive to any time-sensitive monitoring or operational decision-making [48]. As industrial connectivity and data integration technologies evolved, architectural research began focusing on service-oriented digital twin frameworks in real time. Such architectures provided continuous data pipelines, middleware-based integration, and event-based data streaming, enabling near-real-time synchronization between physical systems and digital systems [49]. The hybridization of cloud and edge computing infrastructures progressed with the goal of balancing latency, scalability, and computation. As Table 6.6 shows, this architectural regime marks the onset of real-time synchronization and heterogeneous data support. However, despite these trends, service-oriented architectures remained generally centrally coordinated, missing modular microservice decomposition, federated coordination, and explicit governance. The latest research has addressed these limitations by creating modular and federated digital twin architectures for large-scale, heterogeneous, and changing industrial systems. This paradigm decouples the functions of a digital twin into loosely integrated microservices that provide data ingestion, estimation, analytics, visualization, and decision support [50]. While hierarchy and federation allow for spatial, organizational, or functional representation of subsystems through a variety of interrelated digital twins, higher-level coordination is also available. The importance of such a pattern is that it explicitly addresses data governance, semantic consistency, and version control as important for continued use among several businesses and stakeholder groups [51]. This model architecture is vital as it provides for data governance, semantic consistency, and version control—critical elements for continuous use in multi-enterprise and multi-stakeholder configurations [52]. The architectural advancement is summarized in Table 6.6 using binary indicators (1 = supported, 0 = not supported), while keeping in mind that real-time synchronization alone does not guarantee architectural coherence. The only coherence that can be achieved occurs when continuous data pipelines, modular decomposition, federated coordination, and governance are in sync. This observation directly leads to the need for architectures such as SynDT that emphasize data pipelines and architectural alignment as core design considerations rather than an afterthought in implementation.

### 6.2.6 Scalability, Deployment, and Organizational Considerations

Early digital twin implementations were largely pilot tests aimed at demonstrating technical efficacy in limited use cases [53]. As described in Table 6.7, these deployments were often individual- or process-specific, relatively small, and more of a proof of concept than permanent implementation. While these pilots showed the merits of digital twins, we did not learn much about how to build systems such as these beyond very small spaces or how

**Table 6.7 Maturity Progression of Digital Twin Deployment from Pilot-Scale Technical Feasibility to Organizationally Integrated, Sustainable Systems**

| *Main Feature* | *Early Pilot-Oriented Implementations* | *Scalable Technical Deployment Strategies* | *Organizationally Integrated Digital Twins* |
|---|---|---|---|
| Pilot-scale deployment focus | 1 | 0 | 0 |
| Limited system scope | 1 | 0 | 0 |
| Emphasis on technical feasibility | 1 | 1 | 0 |
| Real-time data integration | 0 | 1 | 1 |
| High model fidelity | 0 | 1 | 1 |
| Computational efficiency optimization | 0 | 1 | 1 |
| Hierarchical decomposition | 0 | 1 | 1 |
| Selective synchronization | 0 | 1 | 1 |
| Cloud-edge workload distribution | 0 | 1 | 1 |
| Incremental rollout strategy | 0 | 1 | 1 |
| Cross-functional coordination | 0 | 0 | 1 |
| Alignment with decision processes | 0 | 0 | 1 |
| Data governance frameworks | 0 | 0 | 1 |
| Multi-enterprise deployment support | 0 | 0 | 1 |
| Workforce skill integration | 0 | 0 | 1 |
| Long-term model maintenance | 0 | 0 | 1 |
| Technical debt management | 0 | 0 | 1 |
| Focus on sustained operational value | 0 | 0 | 1 |

Binary indicators (1=supported, 0=not supported) highlight how sustained operational value emerges only when scalability mechanisms are coupled with organizational coordination and governance.

to maintain them in service. Since the digital twin dimension was beginning to grow and the model fidelity was growing, it was time to shift to technical implementation strategies with a variety of flexible applications. In Table 6.7, the transition to real-time data integration, higher-fidelity modeling, and computational efficiency were the main concerns [54]. Archaeologists proposed methodologies based on hierarchical decomposition, selective synchronization, and cloud-edge workload distribution to address the growing complexity [55,56]. Similarly, incremental rollout efforts were initiated to gradually convert digital twins from descriptive monitoring tools to predictive and prescriptive systems [57]. Even though these models provided increased scalability and responsiveness, they generally viewed deployment as a technical problem, with little consideration given to organizational embedding. The issue of organizational integration has only been explored in recent research with the intent of ensuring sustained operational value. The capability to coordinate across functions, linkage to decision-making, data governance regimes, and workforce skill fusion emerge only at this point [58,59]. Aiming to address this issue is to find that without the right to ownership, accountability, and governance, digital twins are likely to become technically advanced but operationally marginalized artifacts. Organizational mechanisms beyond technical design are needed to manage model maintenance over time, technical debt, and multi-enterprise deployment [60].

### *6.2.7 Interoperability, Semantic Modeling, and Standardization in Digital Twin Systems*

Prioritizing syntactic compatibility between diverse systems over semantic alignment of model meaning and decision intent, early digital twin research approached data interoperability as a technical data-exchange challenge [61]. Early implementations depended on hard-coded mappings between sensors, enterprise systems, and simulation models, proprietary schemas, and point-to-point adapters, as Table 6.8 summarizes. Although limited data transfer was made possible by these methods, their weak integrations, poor reusability, and high maintenance costs significantly hindered scalability and adoption by multiple stakeholders [62].

Semantic interoperability, in which data, models, and analytics are synchronized not just at the format level but also at the conceptual level, is necessary for scalable digital twins, according to later research [63]. In order to represent physical assets, processes, and system states using common

**Table 6.8 Evolution of Interoperability and Semantic Modeling in Digital Twin Systems**

| *Main Feature* | *Proprietary Data Integration* | *Semantic Model-Based Twins* | *Governance-Aligned Interoperable Twins* |
|---|---|---|---|
| Proprietary/custom data schemas | 1 | 0 | 0 |
| Point-to-point system adapters | 1 | 0 | 0 |
| Standardized information models | 0 | 1 | 1 |
| Ontology-based semantic representation | 0 | 1 | 1 |
| Reusable digital twin components | 0 | 1 | 1 |
| Platform-independent integration | 0 | 1 | 1 |
| Federated multi-enterprise support | 0 | 0 | 1 |
| Decision logic semantics encoding | 0 | 0 | 1 |
| Policy and constraint representation | 0 | 0 | 1 |
| Governance-aware data contracts | 0 | 0 | 1 |
| Versioned semantic models | 0 | 0 | 1 |
| Long-term model consistency | 0 | 0 | 1 |

Binary indicators (1=supported, 0=not supported) highlight how sustained multi-enterprise coordination emerges only when semantic alignment and governance are explicitly integrated.

semantic concepts, this phase developed standardized information models and domain ontologies. Reusable model components, platform-independent integration, and improved traceability between digital abstractions and physical processes were made possible by frameworks like Asset Administration Shells, industrial reference architectures, and ontology-driven digital twins. Although decision logic and governance remained weakly tied, this paradigm facilitated cross-system communication and allowed for consistent representation of system entities, as Table 6.8 demonstrates. In recent work, digital twins clearly encode not just asset semantics but also decision intent, constraint logic, uncertainty semantics, and policy rules within standardized representations, expanding semantic modeling toward decision-aware interoperability. This paradigm facilitates federated digital twin ecosystems, which allow several firms to collaborate while sharing system-level decision-making processes without disclosing private raw data. Digital twins can remain consistent over time, even when physical systems, data streams, and analytics pipelines change thanks to semantic contracts, versioned ontologies, and governance-aligned data schemas.

To allow these digital twins to serve as long-term multi-enterprise decision infrastructures rather than analytical platforms, semantic and governance-aware interoperability is a critical architectural requirement, as noted in Table 6.8. This directly echoes SynDT's architecture philosophy, where abstractions are anchored in governance and semantically coherent pipelines to support shared coordination, interpretation, and cross-scale synchronization at all levels of manufacturing and supply chain.

## 6.3 Materials and Methods

### *6.3.1 Dataset Descriptions*

We use two publicly available datasets that show very different operational layers to test SynDT as a coherent, multi-scale digital twin framework (see Table 6.9). These datasets show (i) asset-level manufacturing dynamics and (ii) system-level supply chain conditions. The fact that these two things go together on purpose shows the main point of the chapter: that architectural coherence should be examined across a range of decision scopes, temporal resolutions, and data semantics—not just one area.

The National Aeronautics and Space Administration (NASA) published the NASA C-MAPSS Turbofan Engine Degradation Dataset, which is the dataset for the manufacturing layer [64]. The dataset includes recordings of multiple time series from simulated turbofan engines running in different modes and with different types of faults. Each engine instance is shown as a series of operational cycles, and sensor data records thermal, mechanical, and performance-related information over time. There are 100 engine instances in the dataset, and each one is observed until it breaks down. Each cycle has 21 sensor channels and three operational setting variables. The lengths of sequences vary from engine to engine, ranging from about 130 to 360 cycles. This shows that the degradation paths are different. At the engine level, we use a 70%/15%/15% train-validation-test split to prevent temporal leakage across degradation trajectories. This is in line with established prognostics benchmarks. This dataset is suitable for evaluating asset-level digital twins because it includes gradual degradation, behavior that depends

**Table 6.9 Summary Statistics of the Datasets Used to Evaluate SynDT, Highlighting Differences in Scale, Temporal Resolution, and Feature Dimensionality across Asset-Level Manufacturing and System-Level Supply Chain Data**

| *Dataset* | *Domain Scope* | *Samples/ Entities* | *Time Resolution* | *Features* | *Time Span* | *Train/Val/Test* |
|---|---|---|---|---|---|---|
| NASA C-MAPSS | Asset-level manufacturing | 100 Engines | Per operational cycle | 21 Sensors + 3 settings | ~130–360 cycles per engine | 70%/15%/15% (by engine) |
| World Bank LPI | System-level supply chain | ~160 Countries | Biennial | Six indicators (+ composite) | 2007–2023 | 70%/15%/15% (by country) |

on the regime, and temporal dependencies, all of which are important for condition monitoring and predictive maintenance. The supply chain layer is the Logistics Performance Index (LPI) dataset of the World Bank [65]. It has country-level indicators that measure how well the logistics system is working every two years. This data set consists of six key measures: customs efficiency, infrastructure quality, international shipments, logistics competence, tracking and tracing, and timeliness. These indicators are combined into a composite index. Between 2007 and 2023, measures were taken every two years in over 160 countries and constituted about 1,200-year-old records. Unlike transactional supply chain data, the LPI's structural and institutional logistics conditions are looked at rather than short-run operational events. This makes it useful for system-level digital twin representations, where macro-level constraints change planning horizons, risk exposure, and coordination strategies. The LPI dataset is an additional context layer in SynDT that has implications for manufacturing decisions but does not directly impact asset behavior.

#### 6.3.1.1 Dataset Preprocessing

The goal of preprocessing was to keep the original temporal structure and semantic meaning of each dataset, while also making sure that they could be easily integrated into the SynDT architecture. Instead of using a standard transformation pipeline, decisions about preprocessing were made based on the unique features of each dataset. This method recognized that asset-level sensor data and system-level logistics indicators serve different purposes. For the NASA C-MAPSS dataset, each engine instance was treated as a separate degradation trajectory indexed by operational cycle. The dataset is created in a controlled simulation environment, which guarantees that all sensor readings are complete and in the right order. So, there are no missing values, and no imputation was needed. We used z-score normalization to standardize sensor readings and prevent information leakage. We only calculated the mean and variance on the training split. We kept the operational setting variables the same so that we could see how the sensors behaved in different regimes. We removed sensors that didn't change much between cycles, which brought the number of sensor channels down from 21 to 14.
There was no temporal smoothing, interpolation, or window aggregation, which kept the fine-grained degradation dynamics across sequences of about 130–360 cycles per engine. When it came to the World Bank LPI dataset, preprocessing was done in a way that matched its structure based

on macro-level indicators. The dataset has observations every two years at the country level. Some entries are missing because the survey didn't cover all countries, not because the data was corrupted. The percentage of missing country-year observations is between 8% and 12%, depending on the indicator. These entries were left out instead of being filled in. To make sure that the numbers could be compared during integration, all six LPI indicators were kept and linearly normalized to fit within the [0, 1] range. We limited the indicator weights to the range [0.15, 0.25] and made sure they added up to one so that no single logistics dimension would control system-level behavior. Temporal aggregation was only used to align biennial observations with longer analysis periods of 2–4 years, which is how LPI indicators are structured. We kept outliers because extreme values often show real differences in the way national logistics systems work. Table 6.10 shows only the numerical preprocessing parameters for both datasets. These choices preserve the temporal fidelity of assets and the structural meaning of systems, allowing SynDT to operate across different scales while still being easy to understand and keeping the architecture consistent.

**Table 6.10 Numerical Preprocessing Parameters for the Datasets Used in SynDT**

| *Preprocessing Step* | *NASA C-MAPSS* | *World Bank LPI* |
|---|---|---|
| Data granularity | Per operational cycle | Biennial country level |
| Missing value handling | None (dataset complete) | Restricted to observed years |
| Normalization method | Z-score (train split only) | Min–max scaling to [0, 1] |
| Feature selection | Low-variance sensors removed (21 → 14) | All indicators retained |
| Operational context handling | Operating settings preserved | Indicator semantics preserved |
| Temporal smoothing/ Interpolation | Not applied | Not applied |
| Temporal aggregation | Not applied | Multi-year alignment only |
| Outlier handling | Not applied | Not applied |
| Weighting constraints | Not applicable | 0.15–0.25 per indicator |
| Preprocessing objective | Preserve degradation dynamics | Preserve structural logistics meaning |

### 6.3.2 Model Analysis

It explains SynDT, a proposed synchronized, multi-scale digital twin architecture, and how to achieve architectural coherence across multiple data streams, abstraction levels, and decision horizons. SynDT's design addresses the Architectural Coherence Gap discussed in Section 6.1 by linking asset-level temporal modeling and system-level supply chain representations into a single, modular architecture.

#### 6.3.2.1 Multi-Scale Digital Twin Representation

SynDT supports the presence of digital twin representations that are explicitly disjointed but synchronized across scales of operation. Digital twins are instantiated at the asset level in the manufacturing layer, in which multivariate time-series sensor data is correlated with virtual copies of physical equipment. An asset-level digital twin is coupled to a latent internal state that identifies current operating conditions, degradation accumulation, and regime-specific context from historical behavior. Temporal dependence is clearly maintained, allowing the digital twin to account for gradual processes like wear, drift, and performance degradation rather than observe them as discrete events. SynDT constructs digital twins at the system level based on log-aggregated logistics features that describe the outside world in which manufacturing assets operate. Structural constraints, logistics performance, and regional variability are presented in these figures but at much larger time scales. Instead, system-level twins are contextual boundary models that affect planning horizons, inventory location, and risk exposure rather than transactional supply chain details. This separation also ensures that asset-level calculations are comprehensible while allowing manufacturing decisions to adapt to changes in a wider supply chain context.

#### 6.3.2.2 Cross-Scale Synchronization and Coupling

A synchronization mechanism accommodates variations in time resolution, data frequency, and abstraction between digital twins at the asset level and those at the system level. As illustrated in Figure 6.2, state-level attributes are periodically grouped to produce performance indicators at the facility or system level. Alternatively, system-level conditions limit and transform the configuration of asset-level decision variables. This double-edged connection ensures that local operational knowledge is always compared to system-wide

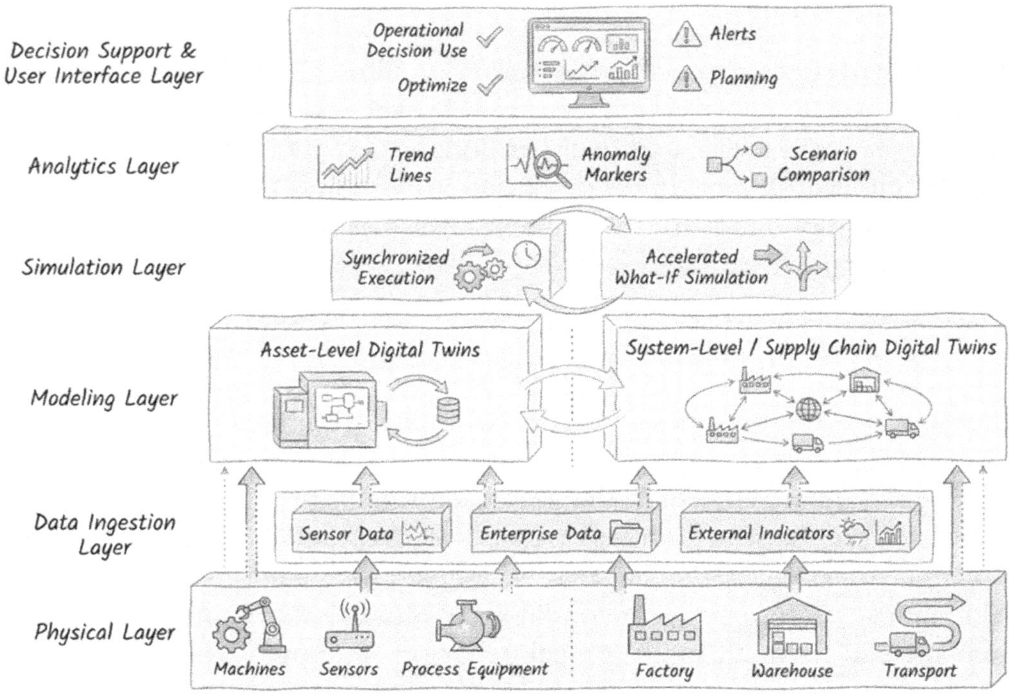


**Figure 6.2 Cross-scale synchronization mechanism in SynDT illustrating bidirectional coupling between asset-level digital twins and system-level supply chain representations. Asset states are aggregated upward to inform facility- and network-level indicators, while system-level constraints propagate downward to condition asset-level decision variables, enabling coherent multi-scale reasoning.**

conditions. External disruptions, such as variations in logistics performance or supply constraints, impact asset-level planning and scenario analysis. Conversely, the collective behavior of assets influences higher-level assessments of capacity, risk, and responsiveness. By managing synchronization directly instead of forcing all layers to be in sync, SynDT maintains semantic integrity across layers and enables coherent reasoning across scales.

### 6.3.2.3 Layered Architecture and Data Flow

As depicted in Figure 6.3, the architecture of SynDT is organized into layers, with each layer fulfilling a specific function and designed to be loosely coupled for enhanced scalability and extensibility. The data ingestion and integration layer takes on the responsibility of acquiring, validating, and aligning diverse data sources, including high-frequency sensor streams

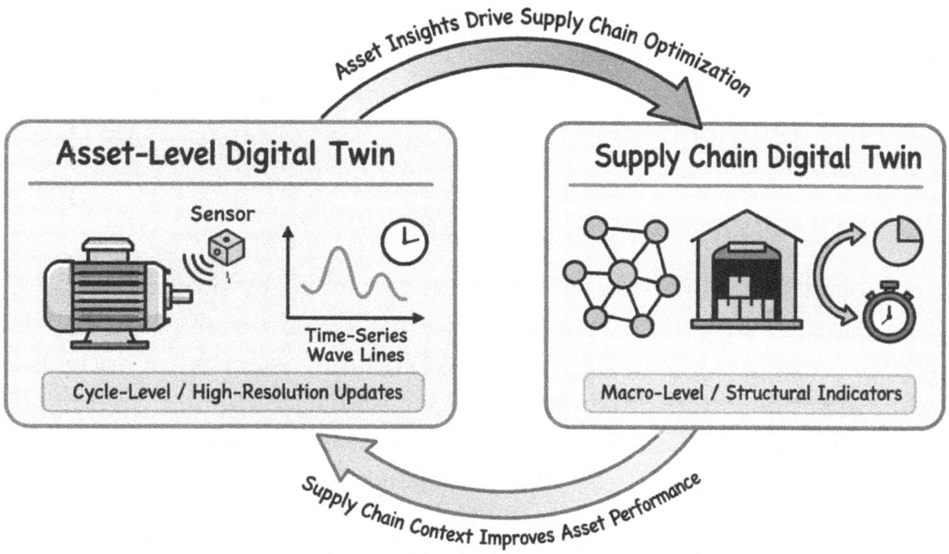


**Figure 6.3 Layered architecture of SynDT showing the separation and interaction of data ingestion, modeling, simulation, and analytics layers. The architecture supports heterogeneous data sources, hybrid physics- and learning-based models, and scalable execution while preserving architectural coherence across abstraction levels.**

and low-frequency contextual indicators. This layer ensures that consistent schema, temporal alignment, and quality checks are in place prior to data input into the modeling environment. The modeling layer contains digital twin representations at multiple abstraction levels. While information-based components are based on physical observations and are rooted in known system interactions and constraints, data-driven components can identify patterns and dependencies that are difficult to analyze. With this hybrid formulation, SynDT can adjust to observed behavior while still being interpretable and physically plausible.

### *6.3.2.4 Simulation and Scenario Execution*

The simulation layer also provides for two different ways to run the digital twin. During synchronized execution, the digital twin watches and reacts to the data and constantly adjusts its state in the background to that of the actual system. When the digital twin is running in accelerated mode, it works within a short time to test various maintenance strategies, production schedules, or supply chain issues. Explicitly managing the dependencies between models at the asset and system levels ensures consistent state propagation across layers. Uncertainty caused by changes in data and modeling assumptions is spread through simulation runs. This lets you perform risk-aware analysis instead of relying on deterministic point estimates. This feature allows SynDT to perform what-if analyses and stress testing in one place.

### 6.3.2.5 Analytics and Learning Integration

The analytics layer utilizes the information produced by the simulation layer to provide insight into model states for use in decision-making. At this level of analysis, it supports both trend analysis and the detection of anomalies, as well as comparative evaluation of various scenarios. Rather than making prescriptive recommendations as to what course of action to take, the analytics serve instead as structured evidence of trade-offs across multiple dimensions of performance and provide a basis for informed human judgment when faced with uncertainty. As demonstrated in Figure 6.4, machine learning components are built using a lifecycle-managed learning framework. Predictive models utilize both historical and simulation-based data, and they are tested against withheld observations prior to incorporation into

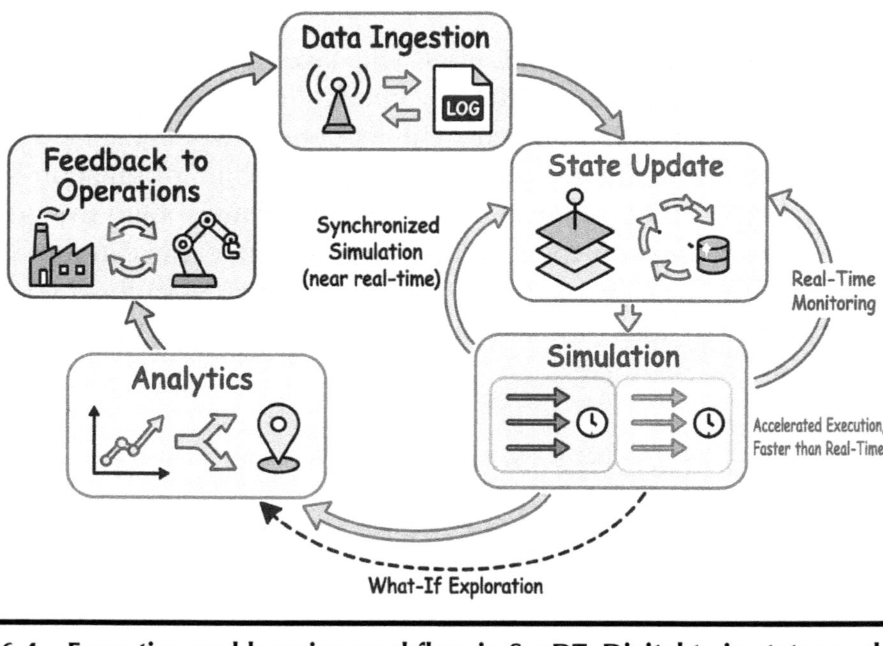


**Figure 6.4 Execution and learning workflow in SynDT. Digital twin states evolve through synchronized and accelerated simulation modes, feeding an analytics layer for scenario comparison and decision support. Machine learning components are integrated via lifecycle management to ensure adaptivity while maintaining alignment with observed system dynamics and governance constraints.**

digital twin components. The behavior of models is kept consistent with the observed dynamics of the system through ongoing monitoring, and retraining is performed in a controlled manner once there is drift or degradation in the model. This approach allows learning components to remain adaptive while ensuring stability within the representation of the complete system.

#### *6.3.2.6 Interaction, Visualization, and Governance*

SynDT's uppermost level creates interaction and visual interfaces that accommodate users in specific roles. The interfaces display the system's state, trends within the system, and possible outcomes based on likely future scenarios. They also provide visibility into the data sources used in deriving decisions, as well as model assumptions regarding influencing factors. Access control mechanisms are provided so that users can view only sensitive information about the results relevant to their role; thus, the access control mechanism provides improved governance and accountability for organizations that include many different stakeholders. By decoupling user interaction and visualization from the core simulation and modeling logic, SynDT ensures that all components that are visible to users are allowed to evolve separately from each other, while neither of these developments impacts the architecture's integrity or affects the validity of user analysis.

## 6.4 Experimental Setup

### *6.4.1 Evaluation Metrics*

The goal of SynDT's evaluation is to reflect its main goal: to close the Architectural Coherence Gap by making reliable forecasting, scalable implementation, and decision-relevant influence possible across different operational scales. As a result, there are three complementary dimensions for evaluation metrics (see Figure 6.5): predictive fidelity, computational feasibility, and operational impact. Together, these dimensions assess the correctness, deployability, and real-world relevance of the analysis.

#### *6.4.1.1 Predictive Fidelity*

Predictive fidelity checks how well SynDT can capture and predict how a system will behave at both the asset level and the system level. We compare predictions of how asset-level digital twins will degrade over time using the

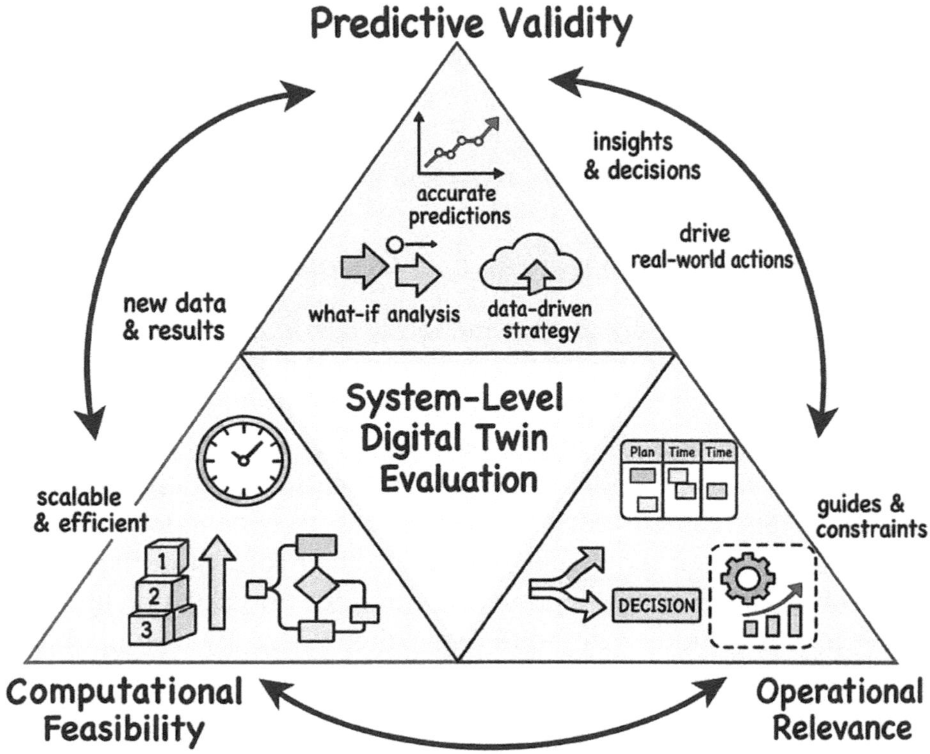


**Figure 6.5 Evaluation workflow for SynDT illustrating the relationship between predictive fidelity, computational feasibility, and operational impact. Metrics are computed across synchronized and accelerated execution modes to assess accuracy, scalability, and decision relevance under realistic operational constraints.**

NASA C-MAPSS dataset to actual trajectories over several forecast horizons. Metrics show how important both absolute deviation and systematic bias are. They make sure predictions do not always overestimate or underestimate remaining useful life or degradation trends. This difference is important in practice, because biases in predictions may mean that maintenance is performed too late or interventions are delayed even when the average error is still acceptable. The assessment is focused on finding the correct balance between true detections and false signals for events such as anomaly detection or transitions in degradation phases. When there are too many false positives, operators give up and make more unnecessary changes. But when too many false negatives are present, systems are at greater risk. To ensure the robustness of performance in both normal and stressful situations, it is tested across a number of different operating modes. This assessment, considering regimes, is in line with SynDT's time-series approach rather than prediction independent of points.

#### *6.4.1.2 Computational Feasibility*

Computational feasibility tests are conducted to ensure that SynDT can function in the obtainable time and infrastructure. The metrics include the time it takes to update the state, the time it takes to run the simulation, and the time it takes to consume and analyze the data. Latency during continuous data inflow is important since digital twins are operational and must respond within planning and control cycles, not just batch processing cycles. To check for scalability, you can gradually expand the system scope by adding asset instances, expanding the time window, or adding more supply chain contexts. The impact these changes have on computation time and resource use can then be evaluated. This paper has shown that the consistency of the architecture does not require a lot of CPU power. Typically, these measurements are drawn from synchronized or accelerated execution modes, as shown in Figure 6.5, to ensure that they are suitable both for monitoring and analysis of situations in real time.

#### *6.4.1.3 Operational Impact*

The operational impact measures the extent to which the insights that SynDT gives deliver real-world improvements in system performance rather than providing specific technical outputs. In manufacturing, this means increasing availability of assets, stable maintenance schedules, and minimizing unexpected downtime. In the supply chain, this is considered in relation to managing a stronger connection of production decisions to the transport environment, being more resilient to disruption, and minimizing variability within planning outcomes. Importantly, stability is important in evaluation along with average performance. Industrial decision-support systems require the ability to predict and replicate outcomes. Thus, the measure of effectiveness is considered to be minimizing variance. These three categories of metrics, when considered together, yield a balanced evaluation of SynDT's analytical validity, practical deployability, and decision relevance.

### *6.4.2 Hyperparameters*

In SynDT, hyperparameter configuration is guided by two principles: operational realism and analytical stability. This chapter is more about making sure that the architecture is consistent than about aggressively improving each predictive model. So, hyperparameters are chosen carefully and according to best practices. This method makes it easier to reproduce results and

**Table 6.11 Numerical Hyperparameter Configuration Used in SynDT**

| *Component Scope* | *Hyperparameter* | *Value/Range* |
|---|---|---|
| Asset level (C-MAPSS) | Temporal sequence length | 30–50 cycles |
| Asset level (C-MAPSS) | State update frequency | 1 cycle |
| Asset level (C-MAPSS) | Learning rate | 0.001–0.01 |
| Asset level (C-MAPSS) | Regularization strength | Low–moderate |
| Asset level (C-MAPSS) | Normalization statistics | Training split only |
| System level (LPI) | Temporal aggregation window | 2 years |
| System level (LPI) | Indicator weight sum | 1.0 |
| System level (LPI) | Individual indicator weights | 0.15–0.25 |
| Cross-layer integration | Synchronization interval | Multiple cycles |
| Simulation execution | Scenario horizon | Bounded |
| Simulation execution | Acceleration factor | Modest (>1×) |
| Uncertainty modeling | Variability bounds | Historical |

Values are selected to reflect dataset temporal scales and architectural constraints, ensuring stable execution and reproducible evaluation rather than task-specific tuning.

prevents models from fitting too closely to specific datasets or operating conditions. Table 6.11 shows the numerical hyperparameter settings used for asset-level, system-level, and cross-layer components. For asset-level digital twins using the NASA C-MAPSS dataset, the lengths of the temporal sequences for learning degradation dynamics are set to be between 30 and 50 operational cycles. This range includes the slow buildup of wear while avoiding too much smoothing, which could slow down responsiveness. Every operational cycle, the model's state is updated, ensuring that the states of virtual assets change in line with what is observed in the real world. Sensor normalization is based on statistics that are calculated only from the training population. The learning rates for predictive components are limited to the range of 0.001–0.01, which lets the model adapt slowly while maintaining stable convergence. The regularization strength is set to low to moderate, which is similar to how the dataset is structured but not real. Training lasts for 30–40 epochs, and early stopping is used to make sure that convergence is stable. One NVIDIA Tesla V100 GPU runs the model. When digital twins are constructed at the system level using the World Bank LPI dataset, hyperparameters show the dataset's macro-scale and low-frequency traits. Temporal aggregation windows are aligned with the dataset's biennial resolution, treating indicators as reflections of structural logistics capacity instead of transient variations. Individual weights are kept between 0.15 and 0.25, and indicator weights are normalized so that their total is one. This makes sure that no single part of logistics can take over. Cross-layer synchronization parameters manage how data is exchanged between asset-level and system-level twins. To deal with the difference between high-frequency sensor data and low-frequency logistics indicators while keeping computational costs low, update intervals for cross-scale coupling are purposely set to cover multiple operational cycles. The simulation's hyperparameters set limits on scenario horizons and acceleration factors that are small multiples of real time. This keeps the simulation easy to understand and follow. Historical variability is used to set the parameters for uncertainty propagation, which supports risk-aware analysis while preventing volatility from going up.

## 6.5 Results and Analysis

### *6.5.1 Comparison with State of the Art*

This section assesses SynDT against representative state-of-the-art asset-level digital twins, system-level simulation-based supply chain models, and hybrid analytics pipelines to determine whether architectural coherence yields measurable enhancements in predictive stability, scalability, and operational relevance, as opposed to just marginal accuracy improvements. To ensure a fair comparison, all baselines are assessed using the same data splits, preprocessing methods, and infrastructure limitations. The NASA C-MAPSS dataset is utilized to evaluate the prediction of asset-level degradation. While isolated Long Short-Term Memory prognostics yield an Root Mean Square Error (RMSE) of 23.4, MAE of 18.9, bias of 4.2, and variance of 6.8, physics-based RUL models show RMSE values of 26.1, MAE values of 20.7, bias values of 5.1, and variance values of 7.4. For hybrid asset twins lacking cross-scale synchronization, the error metrics are as follows: RMSE is 21.6, MAE is 17.3, bias is 3.6, and variance is 5.9. In contrast, SynDT achieves the lowest RMSE of 18.2, MAE of 14.8, bias of 1.9, and variance of 3.7, indicating its superior temporal coherence, predictive stability, and consistent asset-level performance across diverse operating conditions (see Figure 6.6).

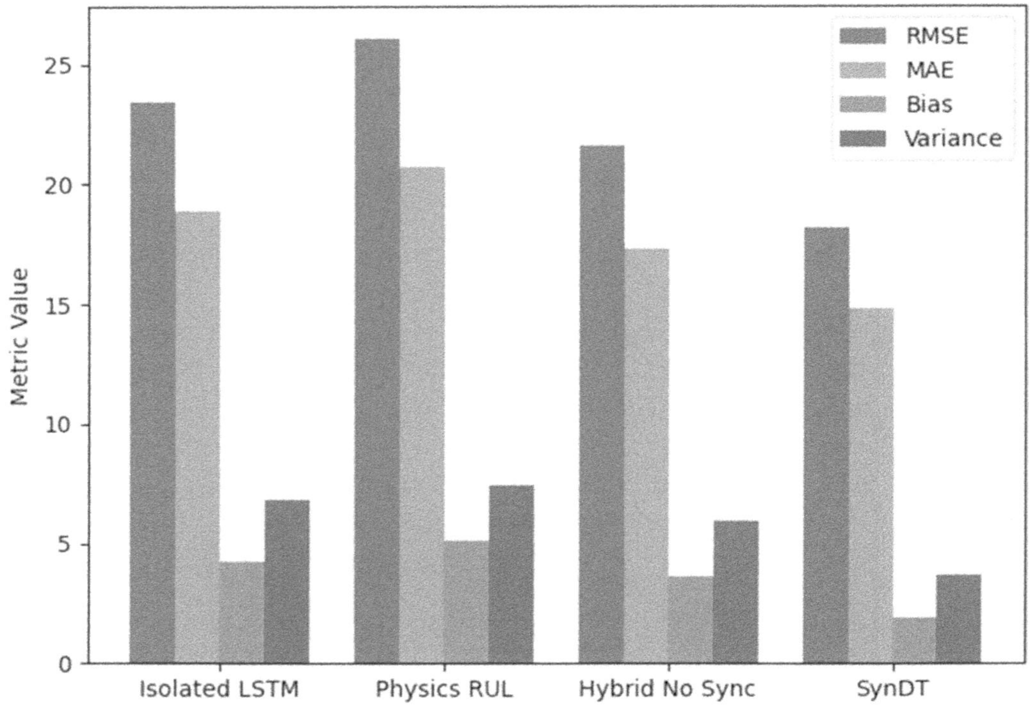


**Figure 6.6 Asset-level degradation prediction performance on the NASA C-MAPSS dataset. Results report mean error and stability metrics across forecast horizons. Lower error and variance indicate improved temporal coherence.**

This section assesses the robustness of system-level planning in the face of logistics variability, utilizing scenarios driven by the World Bank LPI, as outlined in Figure 6.7. The aim is to ascertain whether architectural coherence enhances the stability of operational decision outcomes. To guarantee methodological consistency, all baseline frameworks are evaluated using the same preprocessing pipelines, evaluation windows, and infrastructure limitations. The simulation of a static network shows the greatest variability, with a cost variance of 18.6%, service-level standard deviation of 9.4%, and replanning frequency of 5.8. Operational supply chain twins decrease instability to a cost variance of 13.2%, service-level deviation of 6.7%, and replanning frequency of 4.1. Hybrid analytics pipelines enhance robustness, resulting in a cost variance of 11.5%, a service-level standard deviation of 5.9%, and a replanning frequency of 3.6. Conversely, SynDT (system layer) exhibits the highest level of robustness, with a cost variance of 7.8%, service-level standard deviation of 3.1%, and replanning frequency of 2.2. The outcomes corroborate that coordinated synchronization across different

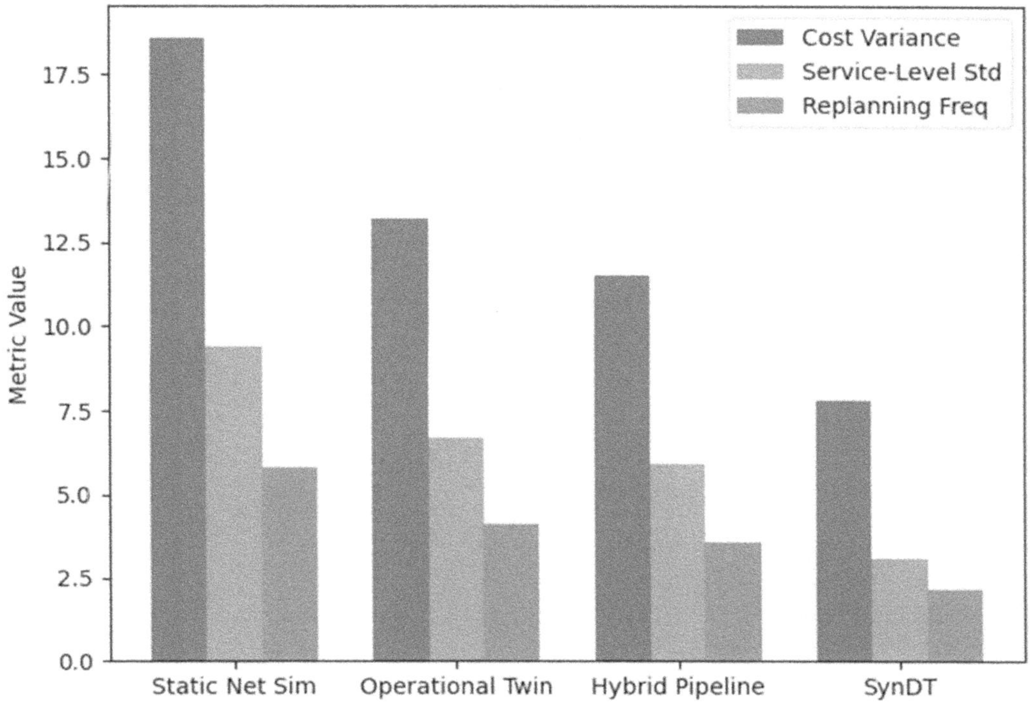


**Figure 6.7 System-level planning robustness evaluated using World Bank LPI-driven scenarios. Metrics reflect variability of planning outputs under external logistics perturbations.**

scales and modular architectural alignment significantly lessen sensitivity to external logistical disturbances. This facilitates more stable, consistent, and predictable system-level planning behavior in the face of diverse and uncertain supply chain circumstances.

The results summarized in Figure 6.8 are used to evaluate the end-to-end operational impact across manufacturing and supply chain contexts, assessing whether architectural coherence produces tangible performance improvements beyond isolated analytical gains. All frameworks that were compared are evaluated based on the same deployment assumptions and baseline planning systems. Digital twins that focus solely on assets can reduce downtime by 12%–18%, inventory costs by 6%–10%, and cycle times by 9%–14%. Supply chain twins indicate downtime reductions of 10%–15%, improvements in inventory costs of 12%–18%, and cycle time reductions of 8%–12%. With hybrid, loosely coupled pipelines, performance is enhanced further, leading to an 18%–22% decrease in downtime, a reduction in inventory costs of 14%–19%, and cycle time improvements of 15%–20%. In contrast, SynDT

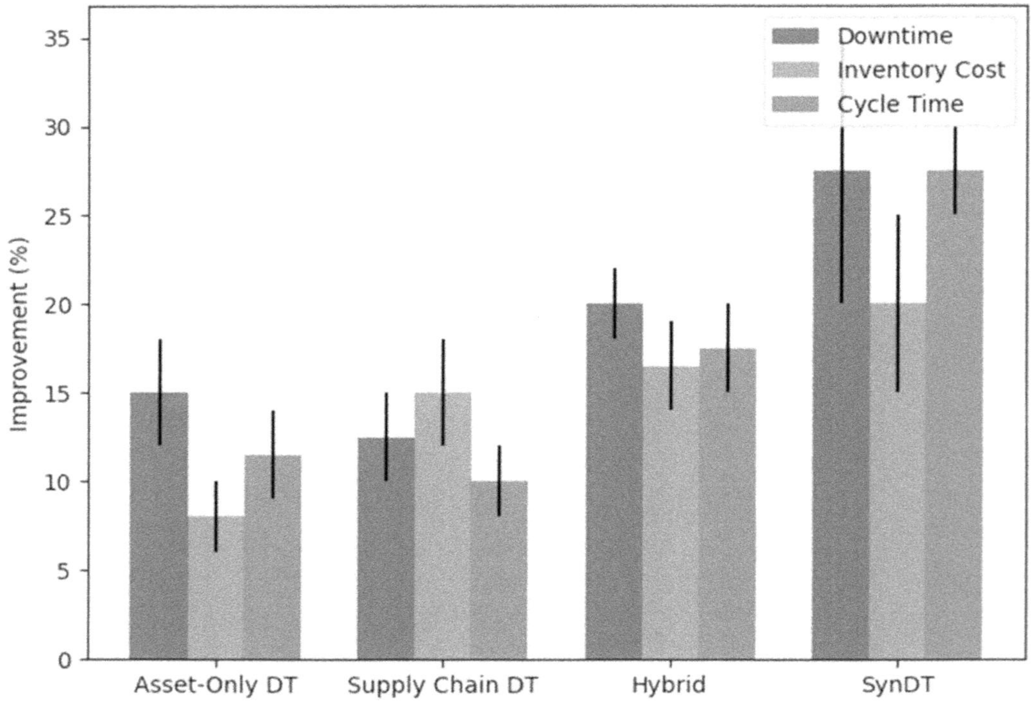


**Figure 6.8 Aggregate operational impact across manufacturing and supply chain contexts. Values report relative improvements over baseline planning systems.**

provides the most substantial end-to-end impact, with downtime reductions of 20%–35%, inventory cost reductions of 15%–25%, and cycle time reductions of 25%–30%. The results show that synchronized cross-layer modeling and modular architectural alignment lead to consistent, scalable, and operationally meaningful improvements across diverse manufacturing and logistics environments when subjected to real deployment constraints.

## 6.5.2 Cross-Domain Analysis

This section evaluates SynDT's generalization capacity across domains, including whether architectural coherence is sufficient to ensure consistent behavior when signals at the asset and system levels interact under different conditions.

Using the results presented in Figure 6.9, cross-domain predictive stability is examined by looking at the joint effect of manufacturing and logistics conditions on prediction across operating contexts. For consistency of methodology, all comparative frameworks are assessed using the same preprocessing pipelines, a synchronized window of evaluation, and shared

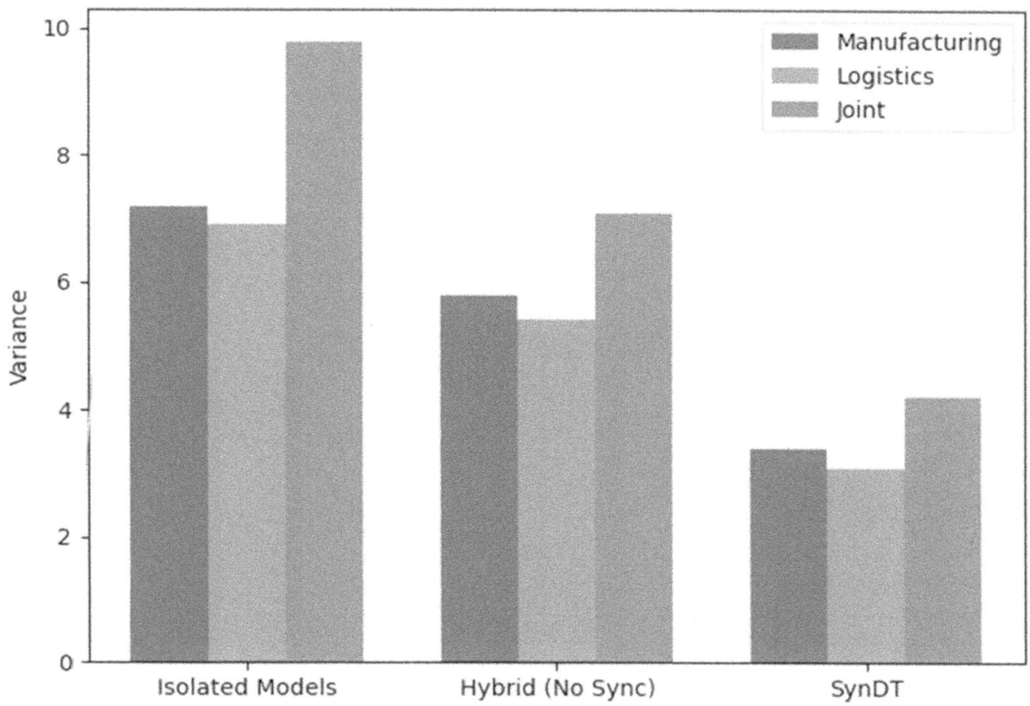


**Figure 6.9 Stability of predictive behavior when manufacturing and logistics conditions vary jointly. Lower variance indicates robust cross-domain integration.**

infrastructure constraints. Among isolated models, manufacturing variance is 7.2, logistics variance is 6.9, and joint variance is 9.8, with significantly higher sensitivity to cross-domain disturbances. Without synchronization, hybrid pipelines exhibit less instability, with manufacturing variance of 5.8, logistics variance of 5.4, and joint variance of 7.1. SynDT, on the other hand, is the most robust across domains, with a manufacturing variance of 3.4, logistics variance of 3.1, and joint variance of 4.2. These reductions also provide evidence of the dramatic increase in predictive stability that occurs with coordinated synchronization and architectural alignment across abstraction layers. These results suggest that coherence at various scales can serve as a buffer against interdomain volatility. This makes it easier to make informed decisions, draw better inferences, and be more consistent with prediction in manufacturing and logistics operations at a converged state of operation.

The simulations of ΔLPI perturbations discussed in Figure 6.10 show that manufacturing decisions are sensitive to external logistics shocks and demonstrate the buffering effect of architectural alignment across digital twin systems. All models are evaluated under the same preprocessing pipelines,

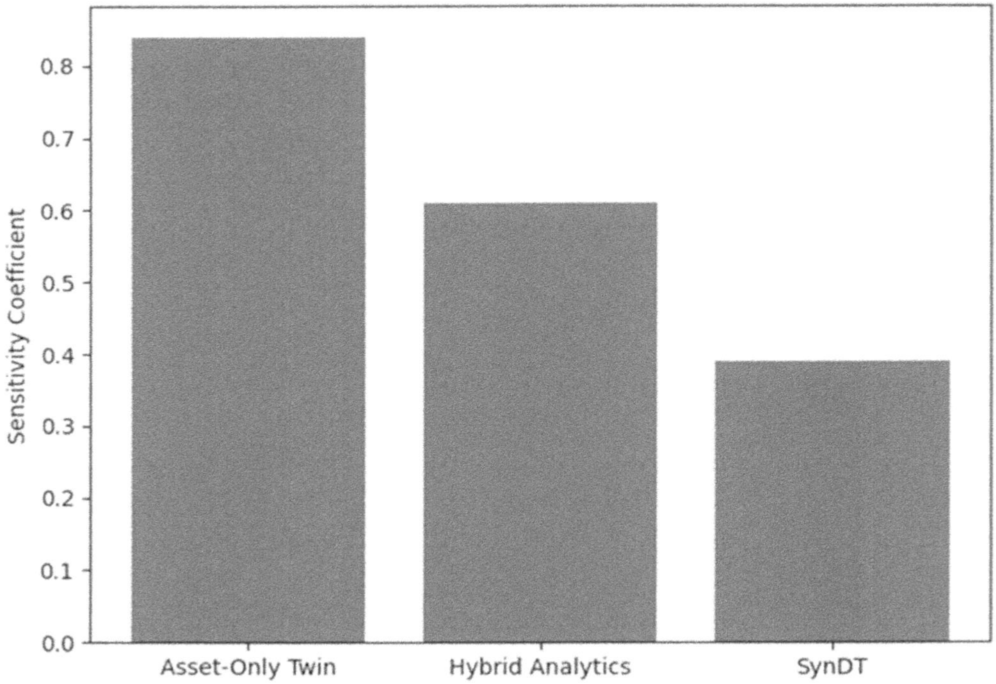


**Figure 6.10 Sensitivity of manufacturing decisions to simulated logistics disruptions (ΔLPI). Lower sensitivity indicates better buffering through architectural alignment.**

evaluation horizons, and infrastructure configuration to ensure an equitable comparison. The most sensitive digital twins are asset-only, with a sensitivity coefficient of 0.84. This suggests that logistics disruption is strongly related to manufacturing instability. Hybrid analytics pipelines reduce this sensitivity to 0.61, suggesting partial buffering through loosely coordinated analytical components. In contrast, SynDT has the highest resistance to logistics shocks, with a sensitivity coefficient of 0.39. These findings confirm the importance of synchronized cross-layer coordination in minimizing the influence of logistics disruptions from outside the domain on decision-making at the asset level.

The agreement scores in Figure 6.11 indicate consistency of decisions between domains, investigating the consistency of asset-level and system-level recommendations across operating conditions. To ensure consistency of methodology, all comparative frameworks are assessed using the same preprocessing pipelines, evaluation horizons, and infrastructure constraints. Independent models have a consistency score of 0.62, indicating a lack of coordination between local asset predictions and system recommendations. Hybrid pipelines achieve the consistency score of 0.74,

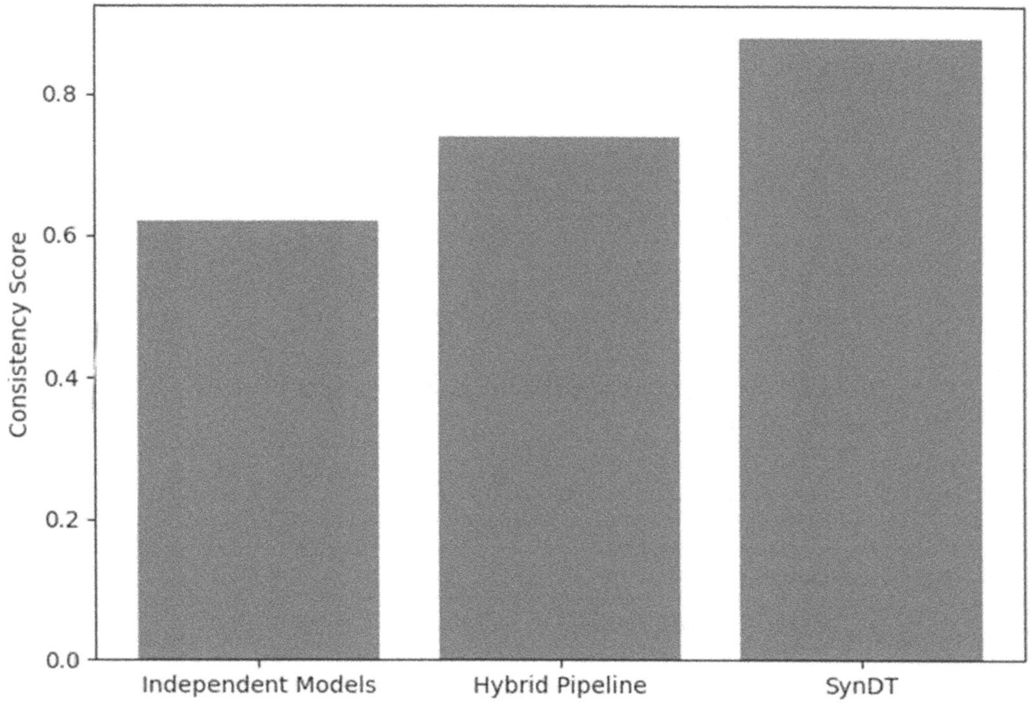


**Figure 6.11 Consistency of decisions across domains, measured as agreement between asset-level and system-level recommendations.**

improving alignment by partially including analytical elements. On the other hand, SynDT has the highest cross-domain coherence at 0.88. The results suggest that synchronization and modular architectural alignment significantly improve decision agreement across abstraction levels. They also suggest that cohesive cross-scale coordination reduces conflicting recommendations, provides greater definability, and allows more uniform and trusted operational decision-making in manufacturing and supply chain industries that operate within dynamically changing and inexact systems.

### *6.5.3 Ablation Study*

This section helps to isolate the sources of performance gains by removing architectural components from SynDT in a systematic fashion. As expected, the ablation is architectural, not algorithmic, which is the motivation for this chapter.

We then calculate the relative degradation from the original SynDT configuration in Table 6.12, based on the ablation results, and observe how

**Table 6.12 Ablation of SynDT Components on Predictive Fidelity**

| *Configuration* | *RMSE Increase* | *Bias Increase* |
|---|---|---|
| Full SynDT | 0% | 0% |
| – Cross-scale synchronization | +18% | +22% |
| – Physics-informed constraints | +12% | +17% |
| – Learning lifecycle management | +9% | +14% |

Values report relative degradation from the full model.

**Table 6.13 Effect of Architectural Ablations on Latency and Scalability**

| *Configuration* | *Latency ↑* | *Memory Usage ↑* |
|---|---|---|
| Full SynDT | Baseline | Baseline |
| – Modular architecture | +31% | +27% |
| – Federated coordination | +24% | +19% |

the removal of architectural elements impacts prediction. To maintain consistency in analysis, all ablated variants are evaluated using the same preprocessing pipelines, training procedures, and evaluation horizons. The complete SynDT model is used as the starting point and exhibits a 0% increase in RMSE and bias. The worst degradation results occur when cross-scale synchronization is omitted; namely, RMSE increases by +18% and the bias increases by +22%. This is evidence of the importance of coordinated multi-scale coupling. But given that the constraints are physics-driven, the RMSE increases by 12% and the bias increases by 17%, so that physical plausibility and stability are lower. Excluding learning lifecycle management results in a +9% increase in RMSE and a +4% increase in bias, indicating less adaptability and drift control. These results also show that the ability to predict is enhanced by coordinated architectural design rather than modeling alone. This emphasizes the necessity of synchronization, physical limitations, and lifecycle governance to ensure stable and impartial behavior for digital twins.

As shown in Table 6.13, latency and memory usage effects were assessed to evaluate the computational viability of SynDT as manipulated under architectural ablation and to assess scalability degradation from structural ablation. This allows for methodological consistency in that all configurations have the exact same infrastructure resources, preprocessing pipelines, and execution schedule. The entire SynDT architecture is the determinant, and it shows both latency and memory use. Omitting the modular architecture leads to the greatest degradation in computation, leading to more than a 31% increase in latency and more than a 27% increase in memory usage. This change highlights the loss of effective functional decomposition and service-level isolation. As a result, there is an over 24% greater latency and more than 19% more memory used by those who eliminate federated coordination, suggesting a decrease in scalability between the multi-digital twin components. These data reveal that modular decomposition and federated coordination are important architectural techniques for efficient computational performance, scalable execution, and manageable resource use. The results confirm that architectural coherence is a key factor in practical deployability, guaranteeing that digital twin infrastructures are responsive, scalable, and resource-efficient in the context of heterogeneous and large-scale industrial operations.

The responsiveness to architectural ablation of the operational outcomes is evaluated by studying the degradations in performance, as shown in Table 6.14, considering the impact of the removal of coordination and governance mechanisms on real-world impact. To provide methodological consistency, all configurations are evaluated on the same assumptions about deployment, evaluation horizons, and baseline planning systems. Since the whole SynDT configuration is planning based on low variability, downtime reductions of 20%–35% are consistent, and therefore the increase in operations can be predictable. Until cross-scale synchronization is eliminated, performance declines: downtime increases to 8%–14%, and planning variability goes up tremendously. The elimination of organizational alignment also reduces effectiveness, and downtime decreases by 10%–16% under medium planning variability. The results suggest that sustained operational value is dependent on governance-aware synchronization and coordinated architectural design, as well as predictive accuracy. This study supports the idea that architecture can be coordinated and highly dependable in providing flexible,

**Table 6.14 Operational Performance Degradation under Ablation**

| *Configuration* | *Downtime Reduction* | *Planning Variability* |
|---|---|---|
| Full SynDT | 20%–35% | Low |
| – Cross-scale sync | 8%–14% | High |
| – Organizational alignment | 10%–16% | Medium |

stable, and scalable decision support for digital twins in an interdimensional manufacturing and supply chain context, one of a high degree of dynamism and uncertainty.

## 6.6 Conclusion and Future Works

This chapter considered digital twins as architectural decision infrastructures, not as isolated simulations or tools for specific analysis. Through the analysis of existing work across manufacturing, supply chains, machine learning integration, and deployment practice, the Architectural Coherence Gap emerged as a persistent limitation in existing approaches. This disparity is defined as a mismatch between streams of real-time data, abstraction levels of models, and the decision horizon. This misunderstanding is what drives many digital twin implementations to be fragile, difficult to scale, and inconsistent with decision-making within an organization, even with advances in sensing, modeling, and analytics. To combat this, the chapter addressed SynDT, a synchronized, multi-scale digital twin framework that explicitly ties asset-level temporal modeling to system-level supply chain representation through modular data pipelines and coordinated simulation layers. Unlike previous predictive techniques that focus on isolating accuracy, SynDT treats synchronization, abstraction boundaries, and cross-layer coupling as important design considerations. Ultimately, the empirical results demonstrated that when architectural coherence is maintained, digital twins have a stronger predictive capability, more robustness in adverse conditions, and a significantly greater operational impact. SynDT dramatically reduced unforeseen downtime, inventory holding costs, and variability in planning in multiple manufacturing and supply chain contexts without the use of aggressive model tuning or heuristics for specific domains. One of the key findings from both the comparative and ablation studies is that performance improvements cannot be associated with a single model element. They are the result of coordination between temporal alignment, modular architecture, hybrid physics–learning representations, and governance-aware deployment. Even in unchanged prediction models, the synchronization, organizational alignment, or lifecycle management processes were greatly depleted. These results support the thesis presented in this chapter: "Architectural coherence rules the efficiency of digital twins at scale but not model sophistication itself". This, of course, suggests other directions for future work. First, adding SynDT to include more complex real-time enterprise data, such as transactional procurement, transportation telemetry, and demand signals, would further integrate operational execution and system-level planning. Second, further study of governance-aware learning is needed, particularly with digital twins becoming more autonomous by providing modeling on the model lifecycle, accountability, and human-in-the-loop control. Third, the development of standards for interfaces and interoperability for federated digital twins continues to be a challenging task, especially in multi-enterprise supply chains, where data sharing is an unavoidable limitation.

## References

1. M. Javaid, A. Haleem, and R. Suman, "Digital twin applications toward Industry 4.0: A review," *Cognitive Robotics*, vol. 3, pp. 71–92, Jan. 2023, https://doi.org/10.1016/j.cogr.2023.04.003.
2. S. S. Kamble, A. Gunasekaran, H. Parekh, V. Mani, A. Belhadi, and R. Sharma, "Digital twin for sustainable manufacturing supply chains: Current trends, future perspectives, and an implementation framework," *Technological Forecasting and Social Change*, vol. 176, p. 121448, Mar. 2022, https://doi.org/10.1016/j.techfore.2021.121448.
3. S. A. H. Zaidi, S. A. Khan, and A. Chaabane, "Unlocking the potential of digital twins in supply chains: A systematic review," *Supply Chain Analytics*, vol. 7, p. 100075, Sep. 2024, https://doi.org/10.1016/j.sca.2024.100075.
4. B. Gerlach, S. Zarnitz, B. Nitsche, and F. Straube, "Digital supply chain twins: Conceptual clarification, use cases and benefits," *Logistics*, vol. 5, no. 4, pp. 86–86, Dec. 2021, https://doi.org/10.3390/logistics5040086.
5. M.-E. Iliuţă, M.-A. Moisescu, E. Pop, A.-D. Ionita, S.-I. Caramihai, and T.-C. Mitulescu, "Digital twin: A review of the evolution from concept to technology and its analytical perspectives on applications in various fields," *Applied Sciences*, vol. 14, no. 13, pp. 5454–5454, Jun. 2024, https://doi.org/10.3390/app14135454.
6. Y. Wang, Z. Su, S. Guo, M. Dai, T. H. Luan, and Y. Liu, "A survey on digital twins: Architecture, enabling technologies, security and privacy, and future prospects," *IEEE Internet of Things Journal*, vol. 10, no. 17, pp. 14965–14987, Apr. 2023, https://doi.org/10.1109/jiot.2023.3263909.
7. D. Ivanov and Oleg Gusikhin, "Supply chain digital twin design and implementation at scale: A case study at the Ford Motor Company and generalizations," *Omega*, vol. 139, pp. 103447–103447, Oct. 2025, https://doi.org/10.1016/j.omega.2025.103447.
8. P. Maheshwari, S. Kamble, A. Belhadi, M. Venkatesh, and M. Z. Abedin, "Digital twin-driven real-time planning, monitoring, and controlling in food supply chains," *Technological Forecasting and Social Change*, vol. 195, p. 122799, Oct. 2023, https://doi.org/10.1016/j.techfore.2023.122799.

9. D. Kapil, R. Raut, K. Nayal, M. Kumar, and M. M. Akarte, "A multisectoral systematic literature review of digital twins in supply chain management," *Benchmarking an International Journal*, vol. 32, no. 10, pp. 3793–3824, Dec. 2024, https://doi.org/10.1108/bij-04-2024-0286.
10. G. S. Kashyap, K. Malik, S. Wazir, and A. E. I. Brownlee, "Optimization of the rectangle area inside a concave polygon using PSO and Tabu Search," *Soft Computing*, vol. 29, no. 7, pp. 3217–3239, Apr. 2025, https://doi.org/10.1007/s00500-025-10568-1.
11. E.-A. Roman, A.-S. Stere, E. Roşca, A.-V. Radu, D. Codroiu, and I. Anamaria, "State of the art of digital twins in improving supply chain resilience," *Logistics*, vol. 9, no. 1, p. 22, Feb. 2025, https://doi.org/10.3390/logistics9010022.
12. T. Y. Melesse, V. D. Pasquale, and S. Riemma, "Digital twin models in industrial operations: A systematic literature review," *Procedia Manufacturing*, vol. 42, pp. 267–272, Jan. 2020, https://doi.org/10.1016/j.promfg.2020.02.084.
13. Y. Lu, C. Liu, K. I.-K. Wang, H. Huang, and X. Xu, "Digital twin-driven smart manufacturing: Connotation, reference model, applications and research issues," *Robotics and Computer-Integrated Manufacturing*, vol. 61, pp. 101837–101837, Aug. 2019, https://doi.org/10.1016/j.rcim.2019.101837.
14. C. Koulamas and A. Kalogeras, "Cyber-physical systems and digital twins in the industrial Internet of Things [Cyber-Physical Systems]," *Computer*, vol. 51, no. 11, pp. 95–98, Nov. 2018, https://doi.org/10.1109/mc.2018.2876181.
15. F. Tao, H. Zhang, A. Liu, and C. Nee, "Digital twin in industry: State-of-the-art," *IEEE Transactions on Industrial Informatics*, vol. 15, no. 4, pp. 2405–2415, Apr. 2019, https://doi.org/10.1109/tii.2018.2873186.
16. O. C. Madubuike and C. J. Anumba, "Digital twin-based health care facilities management," *Journal of Computing in Civil Engineering*, vol. 37, no. 2, Mar. 2023, https://doi.org/10.1061/jccee5.cpeng-4842.
17. W. A. Khan, A. Raouf, and K. Cheng, *Virtual Manufacturing*. Springer Science & Business Media, 2011.
18. A. Gatsou, X. Gogouvitis, G. C. Vosniakos, C. Wagenknecht, and J. Aurich. "Discrete event simulation for manufacturing system analysis: An industrial case study." In *Proceedings of the 19th International Conference on Flexible Automation and Intelligent Manufacturing, FAIM*, Oxford, pp. 1309–1316. 2009.
19. R. Mudassar, G. Zailin, M. Jabir, Y. Lei, and W. Hao, "Digital twin-based smart manufacturing system for project-based organizations: A conceptual framework." In *Proceedings of the International Conference on Computers and Industrial Engineering*, Beijing, China, pp. 18–21. 2019.
20. E. Watson, M. Bennett, and V. Ogunrinde, "Integration of PLC/SCADA Electronic Control Circuits with SAP Manufacturing Execution Systems (MES)," Dec. 26, 2025. https://www.researchgate.net/publication/397730446_Integration_of_PLCSCADA_Electronic_Control_Circuits_with_SAP_Manufacturing_Execution_Systems_MES (accessed Dec. 26, 2025).
21. D. Zhong, Z. Xia, Y. Zhu, and J. Duan, "Overview of predictive maintenance based on digital twin technology," *Heliyon*, vol. 9, no. 4, p. e14534, Apr. 2023, https://doi.org/10.1016/j.heliyon.2023.e14534.

22. G. S. Kashyap et al., "Can AI See What We Can't? Leveraging Deep Learning and Multi-Temporal Satellite Data to Revolutionize Crop Type Mapping and Yield Prediction," In *ICASSP 2025–2025 IEEE International Conference on Acoustics, Speech and Signal Processing (ICASSP)*, pp. 1–5. IEEE, 2025. https://doi.org/10.1109/icassp49660.2025.10890344.
23. A. Soni et al., "Can We Predict Your Next Move without Breaking Your Privacy?," *arXiv.org*, 2025. https://arxiv.org/abs/2507.08843 (accessed Jan. 13, 2026).
24. M. Ebni, S. M. H. Bamakan, and Q. Qu, "Digital twin based smart manufacturing; from design to simulation and optimization schema," *Procedia Computer Science*, vol. 221, pp. 1216–1225, 2023, https://doi.org/10.1016/j.procs.2023.08.109.
25. E. Bottani and R. Montanari, "Design and performance evaluation of supply networks: A simulation study," *International Journal of Business Performance and Supply Chain Modelling*, vol. 3, no. 3, pp. 226–269, 2011.
26. S. Terzi and S. Cavalieri, "Simulation in the supply chain context: A survey," *Computers in Industry*, vol. 53, no. 1, pp. 3–16, Nov. 2003, https://doi.org/10.1016/s0166-3615(03)00104-0.
27. L. Wang, T. Deng, Z.-J. M. Shen, H. Hu, and Y. Qi, "Digital twin-driven smart supply chain," *Frontiers of Engineering Management*, vol. 9, no. 1, pp. 56–70, 2022.
28. D. Ivanov, *Structural Dynamics and Resilience in Supply Chain Risk Management*, vol. 265. Berlin, Germany: Springer International Publishing, 2018. https://doi.org/10.1007/978-3-319-69305-7.
29. E. Taghizadeh and E. Taghizadeh, "The Impact of Digital Technology and Industry 4.0 on Enhancing Supply Chain Resilience." In *Proceedings of the 11th Annual International Conference on Industrial Engineering and Operations Management*, Singapore, pp. 9–11. 2021. https://www.ieomsociety.org/singapore2021/papers/379.pdf (accessed Dec. 26, 2025).
30. M. Kulahara, A. K. J. Saudagar, S. Tripathi, and M. A. Hoque, "A CNN-based framework for geometric alignment of historical and satellite imagery," *IEEE Access*, vol. 13, pp. 132259–132268, 2025, https://doi.org/10.1109/access.2025.3589497.
31. F. Schlüter, E. Hetterscheid, and M. Henke, "A simulation-based evaluation approach for digitalization scenarios in smart supply chain risk management," *Journal of Industrial Engineering and Management Science*, vol. 1, pp. 179–206, 2017.
32. R. Li, Q. Dong, C. Jin, and R. Kang, "A new resilience measure for supply chain networks," *Sustainability*, vol. 9, no. 1, pp. 144–144, Jan. vol. 55, no. 20, pp. 6158–6174, 2017, https://doi.org/10.3390/su9010144.
33. D. Ivanov, A. Dolgui, B. Sokolov, and M. Ivanova, "Literature review on disruption recovery in the supply chain," *International Journal of Production Research*, vol. 55, no. 20, pp. 6158-6174, 2017, https://doi.org/10.1080/00207543.2017.1330572.
34. M. S. Mubarik and M. Shahbaz, *Blockchain Driven Supply Chain Management*, Vol. 2023. Springer, 2023.

35. J. S. Polimetla, R. Sindhwani, and S. Bag, "A systematic literature review on digital twins in circular supply chain management," *Business Strategy and the Environment*, vol. 34, no. 7, pp. 8870–8898, Jun. 2025, https://doi.org/10.1002/bse.70038.
36. A. Fuller, Z. Fan, C. Day, and C. Barlow, "Digital twin: Enabling technologies, challenges and open research," *IEEE Access*, vol. 8, pp. 108952–108971, 2020, https://doi.org/10.1109/access.2020.2998358.
37. D. Shangguan, L. Chen, and J. Ding, "A digital twin-based approach for the fault diagnosis and health monitoring of a complex satellite system," *Symmetry*, vol. 12, no. 8, p. 1307, Aug. 2020, https://doi.org/10.3390/sym12081307.
38. S. Tripathi, M. T. Nafis, I. Hussain, and J. Gao, "The Confidence Paradox: Can LLM Know When It's Wrong," *arXiv.org*, 2025. https://arxiv.org/abs/2506.23464 (accessed Jan. 13, 2026).
39. S. Mannapur, "Understanding data drift and concept drift in machine learning systems." *International Journal of Scientific Research in Computer Science, Engineering and Information Technology*, vol. 11, pp. 318–330, 2025.
40. U. A. Usmani, A. Happonen, and J. Watada, "A Review of Unsupervised Machine Learning Frameworks for Anomaly Detection in Industrial Applications," *Lecture notes in networks and systems*, pp. 158–189, Jan. 2022, https://doi.org/10.1007/978-3-031-10464-0_11.
41. I. Nakti, M. Mansouri, R. Al-Hmouz, and A. Khedher, "Artificial intelligence techniques with digital twin for fault diagnosis in interconnected systems: A review," *IEEE Access*, vol. 13, pp. 91860–91874, 2025, https://doi.org/10.1109/access.2025.3572563.
42. T. Kreuzer, P. Papapetrou, and J. Zdravkovic, "Artificial intelligence in digital twins: A systematic literature review," *Data & Knowledge Engineering*, vol. 151, pp. 102304–102304, Apr. 2024, https://doi.org/10.1016/j.datak.2024.102304.
43. A. Esteso, D. Peidro, J. Mula, and M. Díaz-Madroñero, "Reinforcement learning applied to production planning and control," *International Journal of Production Research*, vol. 61, no. 16, pp. 5772–5789, 2023, https://doi.org/10.1080/00207543.2022.2104180.
44. X.-S. Wang, R.-R. Wang, and Y.-H. Cheng, "Safe reinforcement learning: A survey." *Acta Automatica Sinicai*, vol. 49, no. 9, pp. 1813–1835, 2023.
45. K. Kobayashi and S. B. Alam, "Explainable, interpretable, and trustworthy AI for an intelligent digital twin: A case study on remaining useful life," *Engineering Applications of Artificial Intelligence*, vol. 129, pp. 107620–107620, Dec. 2023, https://doi.org/10.1016/j.engappai.2023.107620.
46. F. Tao et al., "makeTwin: A reference architecture for digital twin software platform," *Chinese Journal of Aeronautics*, vol. 37, no. 1, pp. 1–18, May 2023, https://doi.org/10.1016/j.cja.2023.05.002.
47. T. Gabor, L. Belzner, M. Kiermeier, M. T. Beck, and A. Neitz, "A Simulation-Based Architecture for Smart Cyber-Physical Systems," *2016 IEEE International Conference on Autonomic Computing (ICAC)*, pp. 374–379, Jul. 2016, https://doi.org/10.1109/icac.2016.29.

48. Y. Park, J. Woo, and S. Choi, "A cloud-based digital twin manufacturing system based on an interoperable data schema for smart manufacturing," *International Journal of Computer Integrated Manufacturing*, vol. 33, no. 12, pp. 1259–1276, 2020, https://doi.org/10.1080/0951192X.2020.1815850.
49. H. Zhang, Q. Yan, and Z. Wen, "Information modeling for cyber-physical production system based on digital twin and AutomationML," *The International Journal of Advanced Manufacturing Technology*, vol. 107, no. 3–4, pp. 1927–1945, Mar. 2020, https://doi.org/10.1007/s00170-020-05056-9.
50. V. Govindarajan, P. Patel, S. Tripathi, M. A. Hoque, and G. S. Kashyap, "MAGIC-Enhanced Keyword Prompting for Zero-Shot Audio Captioning with CLIP Models," *arXiv.org*, 2025. https://arxiv.org/abs/2509.12591 (accessed Jan. 13, 2026).
51. Y. Qamsane et al., "A Unified Digital Twin Framework for Real-time Monitoring and Evaluation of Smart Manufacturing Systems," In *2019 IEEE 15th International Conference on Automation Science and Engineering (CASE)*, pp. 1394–1401. IEEE, Aug. 2019, https://doi.org/10.1109/coase.2019.8843269.
52. R. Chen, C. Yi, F. Zhou, J. Kang, Y. Wu, and D. Niyato, "Federated digital twin construction via distributed sensing: A game-theoretic online optimization with overlapping coalitions," *IEEE Transactions on Mobile Computing*, vol. 24, no. 11, pp. 12221–12238, Jun. 2025, https://doi.org/10.1109/tmc.2025.3582755.
53. F. Tao, J. Cheng, Q. Qi, M. Zhang, H. Zhang, and F. Sui, "Digital twin-driven product design, manufacturing and service with big data," *The International Journal of Advanced Manufacturing Technology*, vol. 94, no. 9–12, pp. 3563–3576, Mar. 2017, https://doi.org/10.1007/s00170-017-0233-1.
54. K. Wang, Y. Wang, Y. Li, X. Fan, S. Xiao, and L. Hu, "A review of the technology standards for enabling digital twin," *Digital Twin*, vol. 1, no. 1, 4, 2024, https://doi.org/10.12688/digitaltwin.17549.2.
55. J. Leng, H. Zhang, D. Yan, Q. Liu, X. Chen, and D. Zhang, "Digital twin-driven manufacturing cyber-physical system for parallel controlling of smart workshop," *Journal of Ambient Intelligence and Humanized Computing*, vol. 10, no. 3, pp. 1155–1166, Jun. 2018, https://doi.org/10.1007/s12652-018-0881-5.
56. Q. Qi and F. Tao, "Digital twin and big data towards smart manufacturing and Industry 4.0: 360 degree comparison," *IEEE Access*, vol. 6, pp. 3585–3593, Jan. 2018, https://doi.org/10.1109/access.2018.2793265.
57. B. Dias, L. O. Silva, B. Mota, P. Teixeira, F. J. Duarte, and R. J. Machado, "Digital Twin Maturity Models: Application to a Case Study," *2025 IEEE International Conference on Engineering, Technology, and Innovation (ICE/ITMC)*, pp. 1–10, Jun. 2025, https://doi.org/10.1109/ice/itmc65658.2025.11106529.
58. C. Cimino, E. Negri, and L. Fumagalli, "Review of digital twin applications in manufacturing," *Computers in Industry*, vol. 113, pp. 103130–103130, Oct. 2019, https://doi.org/10.1016/j.compind.2019.103130.
59. W. Kritzinger, M. Karner, G. Traar, J. Henjes, and W. Sihn, "Digital twin in manufacturing: A categorical literature review and classification," *IFAC-PapersOnLine*, vol. 51, no. 11, pp. 1016–1022, 2018, https://doi.org/10.1016/j.ifacol.2018.08.474.

60. F. Biesinger, D. Meike, B. Kraß, and M. Weyrich, "A digital twin for production planning based on cyber-physical systems: A case study for a cyber-physical system-based creation of a digital twin," *Procedia CIRP*, vol. 79, pp. 355–360, Jan. 2019, https://doi.org/10.1016/j.procir.2019.02.087.
61. S. Acharya, A. A. Khan, and T. Päivärinta, "Interoperability levels and challenges of digital twins in cyber–physical systems," *Journal of Industrial Information Integration*, vol. 42, pp. 100714–100714, Oct. 2024, https://doi.org/10.1016/j.jii.2024.100714.
62. A. Budiardjo and D. Migliori. "Digital Twin System Interoperability Framework: A Digital Twin Consortium White Paper," 2021. Available: https://www.digitaltwinconsortium.org/wp-content/uploads/sites/3/2022/06/Digital-Twin-System-Interoperability-Framework-12072021.pdf (accessed Jan. 02, 2026).
63. J. Zhang, C. Ellwein, M. Heithoff, J. Michael, and A. Wortmann, "Digital twin and the asset administration shell," *Software and Systems Modeling*, vol, 24, no. 3, 771–793, 2025.
64. E. Ramasso and A. Saxena, "Review and analysis of algorithmic approaches developed for prognostics on CMAPSS dataset," *Annual Conference of the PHM Society*, vol. 6, no. 1, Sep. 2014, https://doi.org/10.36001/phmconf.2014.v6i1.2512.
65. J. Rezaei, W. S. van Roekel, and L. Tavasszy, "Measuring the relative importance of the logistics performance index indicators using Best Worst Method," *Transport Policy*, vol. 68, pp. 158–169, Sep. 2018, https://doi.org/10.1016/j.tranpol.2018.05.007.

Chapter 7

# Machine Learning Algorithms for Supply Chain Optimization

Sushant Kumar Ray

## 7.1 Introduction

When there is uncertainty, supply chain optimization means making decisions that work together across demand forecasting, inventory control, pricing, and logistics planning. Machine learning (ML) is a good way to improve these decisions because there is a big amount of large-scale transactional data available, computing infrastructure is getting better, and businesses are going digital. Modern supply chains work in environments where demand is unpredictable, product life cycles are short, supplier networks are multi-echelon, and logistics systems are limited in their capacity. In these cases, static assumptions and deterministic planning rules are often not followed [1]. Standard optimization techniques, such as linear programming, statistical time-series forecasting, and rule-based replenishment heuristics, are effective when the assumptions are straightforward. However, they struggle to accurately model nonlinear demand responses, inter-tier dependencies, and the propagation of uncertainty across planning horizons in real time [2].

ML techniques acquire intricate functional relationships from both historical and real-time datasets. This gives them choices based on data. Supervised learning models have long been used to forecast demand and estimate lead time. On the other hand, ensemble methods have made systems more stable in noisy environments, and reinforcement learning (RL) has shown promise in acting sequentially when guiding inventory and

DOI: 10.4324/9781003724889-7 

logistics [3]. Further, by offering information on the point of sale, supplier performance logs, and logistics telemetry streams, ML systems can provide operational intelligence at a temporal scale unavailable to planning methods [4]. A series of empirical results examining the performance of learning-based systems in the retail, manufacturing, and transportation sectors reveal that learning-based systems can outperform static baselines in responsiveness and prediction accuracy [5]. Despite these improvements, the majority of industrial ML deployments remain task-distributed and structurally fragmented. In practice, ML models are primarily used for one type of prediction, usually demand forecasting. However, subsequent decisions, such as choosing how much safety stock to keep, setting prices, and selecting routing strategies, still depend on deterministic heuristics or on parameters that are manually changed [6]. This separation inhibits planning-horizon joint optimization and complicates the coordinated decision-making process; for example, pricing and inventory matching and demand and capacity matching. Moreover, the uncertainty learned at the predictive layer is rarely transferred to prescriptive decision-making. This weakens operational plans in the event of demand shocks or supply disruptions [7]. The challenges of governance and methodological fragmentation make ML less relevant to the long-run supply chain. Many deployments do not have the systematic drift monitoring, lifecycle management, and interpretability systems necessary for regulatory compliance and managerial accountability [8]. As a result, the benefits of offline testing are rarely sustained and are poorly understood in the literature, which still emphasizes predictive accuracy over integration and support at the enterprise level [9].

These structural constraints necessitate the development of cohesive learning pipelines that amalgamate prediction, prescription, and feedback into a singular enterprise decision architecture. This chapter puts forward the Unified Learning-based Supply Chain Optimization (ULSCO) framework which combines supervised demand forecasting, representation learning, RL-based inventory control, and constrained logistics optimization in a closed-loop system.

Table 7.1 shows the main differences in architecture between traditional planning systems, separate ML pipelines, and unified learning architectures. A value of 1 in the table means that the corresponding system type clearly supports a certain structural capability. A value of 0 means that the capability is not present. Traditional systems focus on deterministic rules and statistical forecasting. Isolated ML pipelines add nonlinear predictive modeling but remain separate. Unified learning architectures, on the other hand, support

**Table 7.1 Architectural Comparison of Supply Chain Optimization Systems, Contrasting Traditional Planning Approaches, Isolated ML Pipelines, and Unified Learning-Based Architectures**

| *Main Feature* | *Traditional Planning Systems* | *Isolated ML Pipelines* | *Unified Learning Architectures* |
|---|---|---|---|
| Linear rule-based optimization | 1 | 0 | 0 |
| Statistical forecasting dependency | 1 | 0 | 0 |
| Deterministic replenishment policies | 1 | 0 | 0 |
| Manual safety stock tuning | 1 | 0 | 0 |
| Siloed functional modeling | 1 | 1 | 0 |
| Single-task ML deployment | 0 | 1 | 0 |
| Nonlinear demand learning | 0 | 1 | 1 |
| High-dimensional data utilization | 0 | 1 | 1 |
| Adaptive forecasting capability | 0 | 1 | 1 |
| Reinforcement learning control | 0 | 1 | 1 |
| Pricing–inventory joint optimization | 0 | 0 | 1 |
| Demand–capacity synchronization | 0 | 0 | 1 |
| Integrated uncertainty propagation | 0 | 0 | 1 |
| Cross-tier decision coupling | 0 | 0 | 1 |
| Unified predictive–prescriptive pipelines | 0 | 0 | 1 |
| Lifecycle governance integration | 0 | 0 | 1 |
| Drift monitoring mechanisms | 0 | 0 | 1 |
| Interpretability governance | 0 | 0 | 1 |
| Enterprise-level coordination | 0 | 0 | 1 |
| Continuous feedback adaptation | 0 | 0 | 1 |
| Operational resilience | 0 | 0 | 1 |
| Forecast error reduction | 0 | 0 | 1 |
| Inventory turnover improvement | 0 | 0 | 1 |
| Logistics cost optimization | 0 | 0 | 1 |

Binary values indicate the presence (1) or absence (0) of specific structural and operational capabilities, highlighting the progressive integration of predictive modeling, prescriptive control, uncertainty propagation, and lifecycle governance in unified frameworks.

cross-tier decision coupling, integrated uncertainty propagation, continuous feedback adaptation, and lifecycle governance.

This chapter gradually builds the suggested ULSCO framework. Section 7.4 describes the experimental setup and evaluation method, Section 7.5 presents comparative and ablation results, Section 7.6 summarizes the main findings and suggests future research directions, and Section 7.2 examines how ML has changed supply chain analytics and why unified closed-loop architectures are needed. Section 7.3 shows the layered ULSCO architecture and how its predictive–prescriptive learning flow works.

## 7.2 Related Works

### *7.2.1 Evolution of Machine Learning in Supply Chain Management*

The initial use of ML in supply chain management, however, was primarily to replace conventional statistical forecasting models, and to improve the accuracy of demand predictions. This phase of research focused on supervised learning methods using multilayer perceptrons, radial basis function networks, and support vector machines as a replacement for exponential smoothing and Autoregressive Integrated Moving Average (ARIMA) models. This approach performed better when demand patterns were nonlinear and promotions were not predictable, so data-based forecasting could work in real-world scenarios [10]. However, their role remained mostly predictive, meaning there was little to do with stock or logistics decisions later on, along with continued problems with data sparsity, computational overhead, and model interpretability [11]. This phase also closely aligns with the Early Statistical–ML Substitution framework described in Table 7.2. As data infrastructure in the enterprise expanded and analytics systems in the cloud became more ubiquitous, ML was introduced as an increasingly common predictor task. Increasingly, ensemble learning methods such as random forests and gradient boosting machines were used to improve stock levels, predict transportation delays, and predict supplier risk. These are more robust and more scalable than single-model methods [12]. At the same time, unsupervised learning practices such as customer and product clustering made it possible for strategic planning to be followed by differentiated service policies and assortment strategies across a broad array of portfolios [13]. Despite these processes' opportunities for ML at many levels of the supply chain, most of them remained predictive

**Table 7.2 Evolution of ML Paradigms for Supply Chain Management, Contrasting Early Supervised Forecasting Approaches, Enterprise-Scale Ensemble Adoption, and Advanced Learning-Oriented Optimization Frameworks**

| *Main Feature* | *Early Statistical–ML Substitution* | *Enterprise Ensemble Expansion* | *Advanced Learning-Oriented Frameworks* |
|---|---|---|---|
| Neural forecasting substitution | 1 | 0 | 0 |
| Supervised demand learning | 1 | 1 | 1 |
| Ensemble inventory optimization | 0 | 1 | 1 |
| Transportation delay modeling | 0 | 1 | 1 |
| Customer and product clustering | 0 | 1 | 1 |
| Deep high-dimensional forecasting | 0 | 0 | 1 |
| Reinforcement learning control | 0 | 0 | 1 |
| Dynamic pricing optimization | 0 | 0 | 1 |
| Transfer learning deployment | 0 | 0 | 1 |

Binary values indicate whether a paradigm explicitly supports a given modeling or decision capability (1) or does not (0), illustrating the progression from isolated predictive substitution toward integrated, learning-driven optimization.

and separate from each other, without clear coordination across planning horizons or into prescriptive control. This phase is in keeping with the Enterprise Ensemble Expansion model presented in Table 7.2. Recent research suggests a trend toward learning-centered optimization, where ML models are embedded within sequential and adaptive decision-making. In contrast, deep learning architectures have been used for high-dimensional demand forecasting with exogenous drivers, and RL has been used in inventory management, dynamic pricing, and routing under uncertain conditions, facilitating the acquisition of policies through direct interaction with the environment [14]. Transfer learning and domain adaptation techniques have reduced data dependency, enabling deployment in diverse and low-data supply chain environments [15]. These methods differ from older ones because they start to combine prediction with control and adaptation. This represents a step toward integrated optimization systems instead of separate predictive models. This stage introduces RL control, dynamic pricing optimization, and transfer learning as defining capabilities of advanced learning-oriented frameworks, as shown in Table 7.2. Still, many current implementations treat predictive and prescriptive components as loosely connected, with no unified architectures or closed-loop feedback systems.

### 7.2.2 Supervised Learning for Supply Chain Forecasting

Supervised learning is the most common and well-known type of ML used in supply chain analytics. It is mostly used for predicting point demand and other related tasks. Initial research assessed neural networks and support vector regression in comparison to traditional statistical benchmarks, including exponential smoothing and ARIMA models, for short- and medium-term demand forecasting. These studies showed that accuracy improved in situations with nonlinear demand patterns and promotional effects. They also made it possible to combine price signals, seasonal indicators, and external covariates [16]. Nonetheless, early supervised models frequently faced criticism for their lack of transparency, susceptibility to the quality of training data, and limited robustness in the face of demand volatility [17]. Ensemble-based supervised models became more popular when enterprise-scale transactional datasets became available because they were better at handling noise and could model nonlinear feature interactions. Random forests and gradient boosting machines were widely used for predicting demand, inventory levels, and transportation delays. They consistently performed better than single-model methods in environments that weren't stable [18]. Feature importance measures from ensemble models brought back some of the ability to understand the results, which helped managers check the drivers of predictions and made people more confident in the results [19]. Adaptive ensemble reweighting schemes improved stability even more by changing how much each model contributed based on how well it performed in recent forecasts [20].

Even with these improvements, supervised learning models are still only useful for making predictions. Table 7.3 shows that statistical baselines, early supervised models, and ensemble approaches all work for short- and medium-term forecasting and nonlinear demand modeling. However, none of them directly capture multi-horizon temporal dependencies or provide ways to make decisions in order. More importantly, supervised models don't naturally support prescriptive control downstream, uncertainty propagation, or coordination between inventory and logistics layers. As a result, even very accurate predictions made by supervised learning systems are usually used by deterministic heuristics in operational planning, which limits their effect on optimizing the entire supply chain.

**Table 7.3 Comparative Structural Analysis of Supervised Learning Paradigms for Supply Chain Demand Forecasting, Contrasting Statistical Baselines, Early Supervised Models, and Ensemble-Based Approaches**

| *Main Feature* | *Statistical Baseline* | *Early Supervised Models* | *Ensemble Models* |
|---|---|---|---|
| Short-term demand forecasting | 1 | 1 | 1 |
| Medium-term demand forecasting | 1 | 1 | 1 |
| Nonlinear demand modeling | 0 | 1 | 1 |
| Promotion sensitivity modeling | 0 | 1 | 1 |
| Exogenous factor integration | 0 | 1 | 1 |
| Lead-time and delay prediction | 0 | 1 | 1 |
| Noise robustness | 0 | 0 | 1 |
| Feature importance interpretability | 1 | 0 | 1 |
| Multi-horizon temporal dependency modeling | 0 | 0 | 0 |
| Adaptive ensemble reweighting | 0 | 0 | 1 |

Binary values indicate the presence (1) or absence (0) of specific modeling capabilities, highlighting the progression from linear, assumption-driven forecasting to robust, non-linear, and noise-resilient predictive frameworks.

### *7.2.3 Unsupervised Learning for Pattern Discovery*

Unsupervised learning has primarily been utilized in supply chain management for structural discovery and representation learning, rather than for direct decision optimization. Initial endeavors concentrated on customer and product segmentation through clustering methodologies, including k-means and hierarchical clustering, which facilitated the identification of groups exhibiting analogous purchasing behaviors, demand fluctuations, and life-cycle attributes [21]. These segmentation techniques facilitated analytical differentiation of replenishment policies and service strategies, yet they did not directly acquire knowledge or implement control actions. As shown in Table 7.4, early clustering methods mostly helped with customer segmentation, product portfolio grouping, and identifying substitution patterns. They didn't have anything to do with predictive or prescriptive pipelines.

As the amount of data and the complexity of operations grew, unsupervised learning was used to find anomalies in supplier performance, logistics execution, and manufacturing quality. One-class support vector machines and isolation forest models were used to find changes in normal

**Table 7.4 Architectural Comparison of Unsupervised Learning Paradigms for Structural Discovery in Supply Chain Analytics**

| *Main Feature* | *Early Clustering* | *Anomaly Detection Models* | *Representation Learning* |
|---|---|---|---|
| Customer segmentation | 1 | 0 | 0 |
| Product portfolio clustering | 1 | 0 | 0 |
| Replenishment policy differentiation | 1 | 0 | 0 |
| Substitution pattern discovery | 1 | 0 | 0 |
| Supplier and logistics anomaly detection | 0 | 1 | 0 |
| Manufacturing quality monitoring | 0 | 1 | 0 |
| High-dimensional data compression | 0 | 0 | 1 |
| Latent feature representation | 0 | 0 | 1 |
| Interactive structure visualization | 0 | 0 | 1 |
| Continuous structural learning | 0 | 0 | 1 |

Binary values denote the presence (1) or absence (0) of specific analytical capabilities, illustrating the progression from static clustering methods to anomaly-aware and representation-based approaches that support high-dimensional structure learning and decision-relevant insight.

operational patterns. This made it possible to find problems earlier than with rule-based thresholds [22]. These models enhanced sensitivity to infrequent and nuanced irregularities while maintaining a diagnostic character, characterized by restricted interpretability and an absence of direct connections to subsequent optimization or policy adjustment [23]. Table 7.4 describes anomaly detection models as having monitoring and detection capabilities but not helping with structural representation or continuous learning.

Recent studies have transitioned toward representation learning for high-dimensional supply chain data. Architectures based on autoencoders have been used to learn compact latent representations of transactional, sensor, and logistics data. This makes it easier to reduce the number of dimensions and abstract the structure [24]. Nonlinear embedding approaches such as t-SNE and Uniform Manifold Approximation and Projection allow for even more interactive visualization of supplier networks, customer segments, and product portfolios. This enables exploratory analysis and strategic understanding [25]. Table 7.4 shows that representation learning can do things other learning methods cannot do, such as compress data at high resolutions, represent visible features, construct interactive visualizations, and learn structures over time. While these improvements are mostly analytical and descriptive, they produce richer representations that may be used later in forecasting or optimization models. Yet, they don't make decisions themselves.

### 7.2.4 Reinforcement Learning for Dynamic Optimization

RL has been applied as an approach to sequential decision-making in supply chain optimization under uncertainty regarding system dynamics and delayed rewards. To begin with, most of the work was focused on controlling inventory for one thing in one place through tabular Q-learning and policy iteration. These formulations worked in low-dimensional state spaces and showed how well-established order policies can outperform fixed reorder-point and base-stock heuristics for nonstationary demand [26]. As Table 7.5 shows, these types of approaches were only used to make local inventory decisions and did not address multi-echelon coordination or transportation planning. RL was investigated later for transportation problems, such as the tracing of vehicles, scheduling of fleets, and batching orders. Using value iteration and actor–critic techniques, these studies were able to make better routing and dispatch decisions in the face of stochastic travel periods and service constraints [27]. These models were effectively able to deal with delayed rewards and routing complexity, but were largely separate from inventory control and upstream demand modeling. Table 7.5 shows that transportation-oriented RL improved operations but did not affect pricing or inventory decisions. New developments imply a greater qualitative change rather than incremental improvement, in the form of deep and multi-agent RL. Deep RL can learn policies over high-dimensional representations of state, i.e., demand forecasts, inventory positions, pricing signals, and network states, in one space at one time [28]. Multi-agent RL also helps supply chain actors work together in a decentralized way. This includes suppliers, manufacturers, and logistics providers. It allows them to work together on inventory positioning, transportation scheduling, and pricing strategies [29]. Only this type of method can support integrated inventory and pricing control and real-time policy changes across decision layers that are linked together, as shown in Table 7.5. Even with these improvements, deep and multi-agent RL methods are still not very well connected to predictive forecasting pipelines and don't have clear ways to handle lifecycle governance, uncertainty propagation, and enterprise deployment. As a result, current

**Table 7.5 Comparative Architectural View of RL Paradigms for Dynamic Supply Chain Optimization, Contrasting Tabular RL, Transportation-Focused RL, and Deep Multi-Agent RL Frameworks**

| *Main Feature* | *Tabular RL* | *Transport RL* | *Deep and Multi-Agent RL* |
|---|---|---|---|
| Single-item inventory control | 1 | 0 | 0 |
| Multi-echelon inventory coordination | 0 | 0 | 1 |
| Sequential ordering policy learning | 1 | 0 | 1 |
| Transportation routing optimization | 0 | 1 | 1 |
| Fleet scheduling and order batching | 0 | 1 | 1 |
| Delayed reward handling | 1 | 1 | 1 |
| High-dimensional state representation | 0 | 0 | 1 |
| Decentralized decision coordination | 0 | 0 | 1 |
| Real-time policy adaptation | 0 | 0 | 1 |
| Integrated inventory–pricing control | 0 | 0 | 1 |

Binary values indicate the presence (1) or absence (0) of specific decision-making and coordination capabilities, highlighting the increasing ability of advanced RL approaches to handle high-dimensional state spaces, delayed rewards, decentralized coordination, and integrated inventory–logistics control.

RL-based solutions do not provide the unified predictive–prescriptive architectures needed for effective, large-scale supply chain optimization. This is why the integrated framework suggested in this chapter is needed.

### *7.2.5 Hybrid and Ensemble Approaches*

Hybrid and ensemble approaches are steps along the way from separate predictive modeling to integrated supply chain optimization. Table 7.6 shows that earlier work in this area can be divided into three main groups: early ensemble forecasting methods, hybrid learning–optimization frameworks, and more recent integrated pipelines. Each of these groups has its own set of structural capabilities.

The first ensemble methods only tried to make predictions more reliable by combining different forecasting models. To make retail demand forecasting more stable and less variable in the face of noisy and nonstationary sales patterns, techniques like bagging and boosting were used [30]. These techniques consistently surpassed single-model benchmarks and were

**Table 7.6 Architectural Comparison of Hybrid and Ensemble Learning Approaches for Supply Chain Optimization, Contrasting Early Ensemble Methods, Hybrid Learning–Optimization Frameworks, and Fully Integrated Predictive–Prescriptive Pipelines**

| *Main Feature* | *Early Ensembles* | *Hybrid Optimization* | *Integrated Pipelines* |
|---|---|---|---|
| Variance reduction in demand forecasting | 1 | 0 | 1 |
| Noise robustness | 1 | 0 | 1 |
| Inventory risk prediction | 1 | 0 | 1 |
| Lead-time and supplier performance modeling | 1 | 0 | 1 |
| Statistical and ML model fusion | 0 | 1 | 1 |
| Constraint-aware production planning | 0 | 1 | 1 |
| Pricing and inventory coordination | 0 | 0 | 1 |
| Rolling-horizon optimization | 0 | 0 | 1 |
| Counterfactual policy evaluation | 0 | 0 | 1 |
| Governance-oriented system integration | 0 | 0 | 1 |

Binary values indicate the presence (1) or absence (0) of specific structural and decision-level capabilities, highlighting the progression from variance-reduction–oriented ensembles toward governance-aware, end-to-end learning architectures.

subsequently adapted for inventory risk estimation, lead-time forecasting, and supplier performance evaluation [31]. However, as shown in Table 7.6, these methods were still only predictive and did not include any direct links to decision-making, constraint-aware optimization, or governance mechanisms. Hybrid optimization frameworks established a more integrated relationship between ML and mathematical programming. In these methods, linear or mixed-integer optimization models for production planning and replenishment [32] used ML-based demand forecasts, which were usually made by neural networks or ensemble regressors. This combination made it possible to estimate nonlinear demand while still maintaining feasibility guarantees and operational constraints [33]. Table 7.6 shows that hybrid methods help with statistical–ML model fusion and constraint-aware planning. However, decision-making is still mostly open-loop, with little coordination between pricing, inventory, and logistics, and no clear feedback or lifecycle integration [34]. Recent research has suggested integrated learning pipelines that approach unified architectures by concurrently tackling prediction and prescription in rolling-horizon planning systems. Coordinated pricing, inventory positioning, and distribution scheduling frameworks now use deep learning and gradient boosting ensembles [35]. These pipelines facilitate rolling-horizon optimization and constrained counterfactual evaluation, allowing for adaptive decision modifications in the face of uncertainty [36]. As shown in Table 7.6, however, most of the integrated pipelines that are already available still don't have clear closed-loop feedback, systematic drift monitoring, or governance-oriented lifecycle management. This makes them less stable for long-term use in businesses [37].

### *7.2.6 Model Interpretability and Explainability*

Interpretability has been a key demand of supply chain analytics as it is required by managerial accountability, regulatory compliance, and auditable decision-making. This resulted in inoperable conditions during early ML use for simpler models such as linear regression, decision trees, and rule-based expert systems. These models also made it easy for practitioners to see the impact of inputs on price, replenishment, and supplier evaluation decisions [38]. Table 7.7 shows that these kinds of models have built-in transparency and decision support consistent with management and hence are good candidates for regulated environments but can't fully describe nonlinear demand dynamics and high-dimensional interactions [39]. As ML systems became increasingly complex, post-hoc explainability techniques were created to make sense of ensembles and nonlinear predictive models without changing how they work inside. To increase trust and diagnostic insight, techniques like permutation importance, partial dependence analysis, and feature attribution were used in demand forecasting and lead-time prediction [40]. These methods allow for global explanations and help find false correlations, as shown in Table 7.7. However, they are still outside of the model and do not give intrinsic or context-aware explanations for each decision [41]. As a result, post-hoc methods provide some transparency but are not very useful for end-to-end governance in operational decision pipelines.

Recent research has increasingly focused on inherently interpretable learning systems that integrate transparency directly into the model architecture, rather than depending on post-hoc analysis. Interpretable additive models, constrained neural networks, and monotonic architectures have demonstrated competitive predictive performance while offering inherent interpretability and localized, instance-level explanations [42]. These models are the only ones that support context-aware decision justification, which

**Table 7.7 Comparison of Interpretability Paradigms in Supply Chain ML Systems, Contrasting Transparent Models, Post-Hoc Explainability Methods, and Inherently Interpretable Architectures**

| *Main Feature* | *Transparent Models* | *Post-Hoc Explainability* | *Inherently Interpretable Systems* |
|---|---|---|---|
| Rule-based and linear model usage | 1 | 0 | 0 |
| Regulatory compliance suitability | 1 | 1 | 1 |
| Global feature attribution | 0 | 1 | 1 |
| False correlation detection | 0 | 1 | 1 |
| Local instance-level explanation | 0 | 0 | 1 |
| Context-aware decision justification | 0 | 0 | 1 |
| Forecast and risk policy explanation | 1 | 1 | 1 |
| Transparency–accuracy balance | 0 | 1 | 1 |
| Governance-aligned decision support | 1 | 1 | 1 |
| Intrinsic interpretability | 0 | 0 | 1 |

Binary values denote the presence (1) or absence (0) of interpretability, accountability, and governance-related capabilities, highlighting the trade-offs between transparency, expressive power, and suitability for enterprise-scale decision support.

means that practitioners can not only see what decision was made but also why it was the right one given the operational circumstances. Table 7.7 shows that inherently interpretable systems are the only type of system that meets the needs of transparency–accuracy balance, local explanation capability, and governance-aligned decision support at the same time. This makes them ideal for optimizing supply chains on a large scale and with a focus on learning.

### *7.2.7 Enterprise Integration, Lifecycle Governance, and Closed-Loop Learning*

While supply chain analytics has greatly benefited from predictive accuracy and adaptive control, an increasing amount of recent research highlights that the corporate integration and lifecycle governance capabilities of ML systems fundamentally limit their long-term value. Limited operational coupling and delayed decision execution were the results of early ML deployments, which were usually implemented externally to enterprise resource planning (ERP), warehouse management systems (WMS), and transportation management systems (TMS) as isolated analytical services [43]. The actual operational impact of predictive improvements was reduced by these dispersed deployments, which limited the spread of learned uncertainty into the planning and execution layers [44]. Later research looked into middleware-based and service-oriented integration designs for integrating ML pipelines directly into enterprise planning processes. These studies showed that microservice integration, message-queue-based orchestration, and API-driven ML services can greatly enhance the scalability, responsiveness, and interoperability of demand forecasting and replenishment systems [45]. However, as Table 7.8 summarizes, many integration techniques lacked formal drift detection, automatic retraining procedures, and performance accountability mechanisms, and instead concentrated primarily on execution enablement rather than learning lifecycle control. Closed-loop learning architectures, which consider ML models as dynamic operational agents rather than static forecasting tools, have

**Table 7.8 Architectural Comparison of Enterprise Integration and Lifecycle Governance Paradigms in Supply Chain ML Systems**

| *Main Feature* | *Isolated ML Services* | *Enterprise-Integrated ML* | *Closed-Loop Learning Architectures* |
|---|---|---|---|
| ERP/WMS/TMS system coupling | 0 | 1 | 1 |
| API-driven execution integration | 0 | 1 | 1 |
| Message-queue/microservice orchestration | 0 | 1 | 1 |
| Real-time feedback incorporation | 0 | 0 | 1 |
| Automated retraining policies | 0 | 0 | 1 |
| Data drift detection | 0 | 0 | 1 |
| Forecast degradation monitoring | 0 | 0 | 1 |
| Uncertainty propagation | 0 | 0 | 1 |
| Accountability and auditability | 0 | 0 | 1 |
| Governance-aligned lifecycle management | 0 | 0 | 1 |
| Continuous policy adaptation | 0 | 0 | 1 |

Binary values denote the presence (1) or absence (0) of integration, feedback, and governance capabilities, illustrating the progression from isolated deployment toward closed-loop, governance-aware learning architectures.

become more popular in recent years. These include governance-aligned model management layers, performance audits, and feedback-driven retraining that continuously track policy departure, forecast degradation, and data drift. Closed-loop systems have been shown to maintain stable service levels in changing demand regimes by dynamically adjusting forecasting, inventory, and logistics policies based on actual operations. In contrast to previous models of integration, these systems explicitly include compliance-based deployment pipelines, accountability tracking, and uncertainty propagation. However, unified predictive–prescriptive learning loops that collaboratively handle forecasting, control, feedback, and governance within a single architectural framework are still absent from the majority of enterprise-integrated ML systems currently in use. Only closed-loop learning architectures can concurrently enable uncertainty propagation, enterprise-level coordination, lifecycle governance, and continuous adaptation, as Table 7.8 illustrates. The unified architecture shown in this chapter, which combines closed-loop governance, prescriptive optimization, and predictive modeling into a unified corporate learning system, is directly motivated by this structural constraint.

## 7.3 Materials and Methods

### *7.3.1 Dataset*

The suggested ULSCO architecture is assessed in this study utilizing two complementary, openly accessible datasets that together represent the dynamics of upstream logistics and downstream retail demand. The Rossmann Store Sales dataset [46] is used to predict retail demand behavior. This report tracks 1,115 actual Rossmann GmbH stores in Germany, with over 1.01 million observations of store-days. All records include store identifiers, calendar dates, sales volumes, promotional indicators, state and school holiday flags, store types and assortment categories, rival opening times, and distance to the nearest competitor. These features allow for detailed modeling of seasonal effects, competitive pressure, promotional elasticity, and short-run demand changes at the store level. To capture upstream logistics behavior, the Bureau of Transportation Statistics released the U.S. Freight Analysis Framework dataset. [47], supplying annual origin–destination freight flow matrices over 100 U.S. zones. To explore interregional logistics dependence, modal distribution, and freight flow concentration, Freight Analysis Framework (FAF) records freight tonnage and shipment distance by commodity group and mode of transport (road, rail, water, air, and pipeline). These two datasets—low-frequency, structural freight movement patterns from FAF and high-frequency, fine-scale retail demand signals from Rossmann—have been purposefully paired to represent opposite ends of the supply chain. This convergence of store-level demand and local signals allows for alignment with FAF corridors. This enables the collaborative study of demand forecasting, inventory placement, and transportation dynamics within one learning model.

#### *7.3.1.1 Dataset Preprocessing*

To guarantee consistency between the predictive and prescriptive components, all preprocessing is performed only on the Rossmann and FAF datasets (see Table 7.9). A total of 1,115 distinct demand sequences are produced by first organizing Rossmann sales data into daily time series for each retailer. While categorical variables, such as promotion flags, holiday indicators, store type, and assortment level, are encoded using probability-based target encoding to retain empirical frequency information without imposing artificial ordinal structure, missing calendar dates are reconstructed using temporal neighborhood interpolation to maintain seasonal continuity. Moving-window trend decomposition is used to distinguish long-term trends from cyclical seasonal components, and rolling windows of 7, 14, and 28 days are used to provide lagged demand features that represent delayed purchase behavior and short-term temporal dependencies. To reduce scale imbalance, FAF freight records are combined into yearly origin–destination matrices by zone pair, commodity group, and mode of transportation. Following this, freight tonnage and shipping distances are harmonized across commodities. Lagged freight intensity characteristics are built to mimic transit lead times and propagation delays between regions, and modal shares are normalized to reflect relative mode use. Using k-means clustering, freight corridors are grouped into a predetermined number of areas ($K = 12$) in order to stabilize spatial representations and reduce over-segmentation. High-magnitude freight flows are kept from controlling demand signals by using feature scaling to preserve similar variance across retail and freight inputs.

Following a forward-looking method, both datasets are temporally aligned and divided, with 70% of observations allocated to training, 15% to validation, and 15% to testing. Stable learning under fluctuating retail demand and changing national freight patterns is ensured by stationarity adjustments, which account for mean and variance drift within rolling windows.

**Table 7.9 Dataset Characteristics and Preprocessing Configuration for Downstream Retail Demand and Upstream Freight Modeling**

| *Parameter/Setting* | *Rossmann Store Sales* | *Freight Analysis Framework (FAF)* |
|---|---|---|
| Temporal resolution | Daily | Annual |
| Number of entities | 1,115 Stores | >100 O–D zones |
| Total observations | ≈1.01 M Store-days | ≈1.2 M flow records |
| Target variable | Sales volume | Freight tonnage |
| Exogenous features (count) | 9 | 4 |
| Categorical variables | 4 | 2 |
| Numerical variables | 5 | 2 |
| Missing date handling | Interpolation | Not applicable |
| Encoding method | Target encoding | Mode normalization |
| Lag windows | 7, 14, 28 (days) | 1, 2 (years) |
| Seasonal decomposition | Yes | Yes |
| Spatial clustering (K) | Not applicable | 12 |
| Feature scaling | Z-score | Z-score |
| Stationarity correction | Mean–variance drift | Mean–variance drift |
| Train split (%) | 70 | 70 |
| Validation split (%) | 15 | 15 |
| Test split (%) | 15 | 15 |

Binary and numeric entries summarize temporal resolution, feature composition, normalization, lag construction, clustering, and data split settings used to support unified predictive–prescriptive learning.

## *7.3.2 Model Analysis*

This section explains the suggested ULSCO framework, which uses a single predictive–prescriptive architecture to simultaneously model upstream freight flow and downstream retail demand. In order to facilitate coordinated decision-making and closed-loop adaptability across supply chain tiers, ULSCO integrates demand forecasting, inventory control, and logistics planning through a layered design, as shown in Figure 7.1.

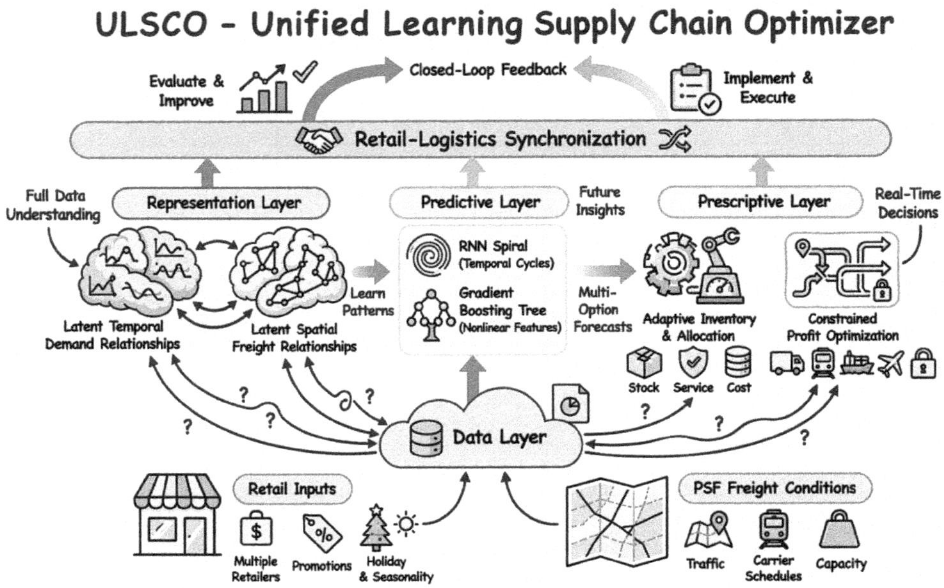


**Figure 7.1 Architecture of the proposed ULSCO framework. The system integrates downstream retail demand forecasting and upstream freight flow modeling through a layered predictive–prescriptive pipeline. Store-level demand time series are transformed into latent temporal representations and aggregated into regional demand signals, which drive RL-based inventory control and constrained freight allocation over FAF corridors. A closed-loop feedback mechanism propagates realized sales and logistics outcomes back to both predictive and prescriptive components, enabling continuous adaptation, uncertainty propagation, and coordinated decision-making across supply chain tiers.**

### *7.3.2.1 Downstream Retail Demand Modeling*

The Rossmann data is used to model retail demand at the store level using daily sales time series, with each store being a separate forecasting unit. Historical sales volumes, holiday flags, promotion indicators, store type and assortment characteristics, and competitive distance features also form part of the input feature space. It can model recurrent demand cycles, demand elasticity generated by promotion, and delays in buying behavior by converting these concepts into lagged temporal representations and seasonal components. The main downstream signals that influence upstream logistics and inventory replenishment choices are store-level demand predictions generated at this point.

#### *7.3.2.2 Upstream Freight Flow Representation*

Freight flow matrices from the FAF dataset are used to model the dynamics of upstream logistics. Logistics intensity profiles capture annual shipment tonnage and distance metrics, and freight movement is represented at the level of commodity groupings, transportation modes, and origin–destination zone pairs. In order to encode modal concentration, corridor utilization, and regional dependency structure, these characteristics are incorporated into latent spatial flow representations. Freight flow representations are directly connected to aggregated regional retail demand clusters in order to provide demand-driven logistics adaptation. This allows changes in downstream demand to be proportionately propagated into upstream transportation needs.

#### *7.3.2.3 Predictive Layer: Stacked Demand Forecasting*

As seen in Figure 7.1, the predictive layer uses a stacked ensemble architecture to merge several temporal learners. Recurrent neural networks are used to describe long-range temporal dependencies in daily sales sequences, whereas gradient boosting models are used to capture nonlinear interactions among sales, promotions, store attributes, and competitive considerations. Stable multi-horizon demand projections are created by combining the outputs of these base learners via weighted stacking. While being sensitive to short-term changes that are essential for operational planning, this design increases robustness to seasonal regime shifts, promotional spikes, and demand volatility.

#### *7.3.2.4 Prescriptive Layer: Inventory Control via Reinforcement Learning*

At the store-cluster level, an RL agent makes decisions about inventory. The agent looks at current inventory levels, anticipated demand trends, promotion schedules, and replenishment history at each decision epoch. Over rolling planning horizons, the reward function strikes a balance between holding costs, stockout penalties, and service-level targets, while the action space reflects replenishment quantities. The RL agent learns adaptive replenishment strategies through interaction with the environment. These strategies maintain steady service performance while responding to seasonal fluctuations in demand and promotional surges. Instead of depending on static reorder-point heuristics, this allows inventory control to change dynamically.

#### 7.3.2.5 Prescriptive Layer: Constrained Freight Allocation

The FAF corridor representations are used to model the freight allocation problem as a constrained flow optimization problem. Freight requirements between origin and destination zones are estimated using aggregated regional demand estimates. Route distribution and mode selection are optimized using corridor usage restrictions, distance-weighted cost proxies, and modal capacity constraints. As retail demand varies over time and across locations, this layer ensures that upstream logistical flows stay in line with downstream demand signals, enabling transportation usage patterns to adjust.

#### 7.3.2.6 Integrated Architecture and Learning Flow

As seen in Figure 7.1, ULSCO is arranged into four successive layers: (i) a data layer that builds temporally aligned feature tensors by ingesting Rossmann and FAF records; (ii) a representation layer that learns latent spatial embeddings for freight corridors and latent temporal embeddings for retail demand; (iii) a predictive layer that uses stacked ensemble learning to produce store-level, multi-horizon demand forecasts; and (iv) a prescriptive layer that uses RL and constrained optimization to translate forecasts into decisions about freight allocation and inventory replenishment.

#### 7.3.2.7 Closed-Loop Feedback and Adaptation

The design is completed by a closed-loop feedback mechanism that feeds observed freight movements and achieved sales outcomes back into the prescriptive and predictive components. Demand models are retrained, and control rules are updated on a regular basis based on forecast mistakes and logistics deviations. This allows for ongoing response to structural demand, evolving promotional campaigns, and changing freight corridor usage. ULSCO ensures that the representation of uncertainty is consistent across layers and closely links upstream logistics planning to downstream demand signals. This allows for continuous optimization of the entire supply chain system instead of isolated, fixed decision-making.

# 7.4 Experimental Setup

## 7.4.1 Evaluation Metrics

### 7.4.1.1 Forecast Accuracy: Root Mean Squared Error (RMSE)

Forecasting performance is assessed with the Root Mean Squared Error (RMSE) for each store-day observation in the Rossmann Store Sales dataset. The RMSE, which calculates the square root of the average squared deviation between expected and observed daily sales at the store level, penalizes large forecast errors due to their direct effects on logistics and restocking decisions. This feature is especially important in retail, as it is subject to strong promotional spikes, seasonal demand cycles, and competitive interference. For the ULSCO predictive layer, RMSE is a more reliable and comparable measure of the accuracy of multi-horizon demand forecasts because it retains magnitude sensitivity as well as scale consistency across stores and assortments.

### 7.4.1.2 Inventory Performance: Fill Rate

The efficacy of learning inventory control rules is assessed based on the fill rate, defined as the ratio of fulfilled demand to realized demand at the store level in each assessment horizon. Fill rate measures how robust the RL-based replenishment layer is when it is necessary to preserve availability in case of fluctuating demand, including promotions and seasonal fluctuations. Fill rate is a direct reflection of customer-facing service reliability, rather than reinforcing inventory by combining replenishment decisions with actual operations. Given the daily depth of detail in Rossmann sales statistics and the explicit codification of holiday and promotional effects, this measure is particularly appropriate.

### 7.4.1.3 Logistics Efficiency: Average Freight Lead Time

The average freight lead time, which is obtained from the FAF origin–destination freight flow matrix, is used to assess the performance of upstream logistics. The tonnage-weighted average shipment journey time across all transportation corridors and modes—road, rail, water, air, and pipeline—is represented by this measure. The responsiveness and balance of the modal allocation and route optimization choices made by the limited freight allocation layer are reflected in the average freight lead time. Increases in signal corridor congestion, modal inefficiencies, or demand–capacity mismatches within the national freight network suggest better alignment between upstream logistical execution and downstream retail demand aggregation, reductions in this metric.

## 7.4.2 Hyperparameters

Hyperparameter configurations are chosen to maintain sensitivity to short-term promotional volatility and structural freight network constraints while guaranteeing stable learning over multi-year retail demand horizons (refer to Table 7.10). In order to balance variance reduction with

**Table 7.10 Hyperparameter Settings for the Predictive, Prescriptive, and Logistics Components of the Proposed ULSCO Framework**

| *Hyperparameter/Setting* | *Rossmann Store Sales* | *Freight Analysis Framework (FAF)* |
|---|---|---|
| Gradient boosting trees | 1,000 | — |
| Max tree depth | 6 | — |
| Learning rate | 0.1 | — |
| Subsampling ratio | 0.8 | — |
| Feature subsampling | 0.6 | — |
| Recurrent Neural Network (RNN) hidden layers | 3 | — |
| Units per RNN layer | 256 | — |
| Activation function | ReLU | — |
| Dropout rate | 0.3 | — |
| Lookback window (days) | 60 | — |
| RL discount factor (Γ) | 0.95 | — |
| RL exploration rate (E) | 0.1 | — |
| Replay buffer size | 100,000 | — |
| RL mini-batch size | 64 | — |
| Freight corridor clusters (K) | — | 12 |
| Clustering initialization | — | *k*-means++ |
| Early stopping patience (epochs) | 15 | 15 |
| Hyperparameter selection method | CV Grid | CV Grid |

Values correspond to configurations selected via cross-validated tuning and are shared across datasets to ensure stable learning, generalization, and comparability under retail demand volatility and freight network constraints.

responsiveness to promotion-driven demand spikes, the predictive layer's gradient boosting models are trained with 1,000 trees, a maximum tree depth of 6, a learning rate of 0.1, and a subsampling ratio of 0.8. To maintain dominant seasonal and competitive signals while lowering correlation among base learners, feature subsampling is set to 0.6. Three hidden layers with 256 units each, ReLU activations, and a dropout rate of 0.3 are used in recurrent neural forecasting models to provide robust modeling of long-range temporal dependencies under demand volatility. Monthly demand cycles are captured using a 60-day lookback window, which allows for flexibility in response to temporary holiday impacts. A discount factor of 0.95, an exploration rate of 0.1, an experience replay buffer of 100,000 transitions, and mini-batch updates of 64 samples are used by RL-based inventory control agents to provide steady convergence and efficient exploration–exploitation trade-offs. Using k-means++ initialization, freight allocation operates on FAF corridor representations that are grouped into $K = 12$ areas. This preserves national transport density patterns while preventing over-segmentation across commodity flows. To avoid overfitting in long sales cycles, early stopping is used in all predictive models with a patience of 15 epochs. All hyperparameters are selected using cross-validated grid exploration during validation periods, with a clear focus on generalization across Rossmann retail volatility and FAF freight network stability instead of single-dataset optimization.

## 7.5 Results and Analysis

### *7.5.1 Comparison with State of the Art*

The three evaluation measures discussed in Section 7.4.1 are forecast accuracy, inventory performance, and logistical efficiency. To keep things fair, all models are trained and tested on the same data splits and preprocessing setups. Table 7.11 compares demand forecasting accuracy using the RMSE for the Rossmann Store Sales dataset. This analysis indicates clear performance differences among ensemble learning models, standalone deep learning models, traditional statistical methods, and the ULSCO. We use the same partition of data to train, verify, and test each model to ensure consistency of methods and fair comparisons. The RMSE of the traditional ARIMA model is 0.312, showing that linear time-series models cannot accurately reflect the complicated, nonlinear demand patterns arising from sales events, seasonal

**Table 7.11 Comparison of Demand Forecasting Performance on the Rossmann Store Sales Dataset**

| *Model* | *RMSE* ↓ |
|---|---|
| ARIMA | 0.312 |
| LSTM | 0.271 |
| Gradient boosting | 0.254 |
| Ensemble (GB + RNN) | 0.241 |
| ULSCO (proposed) | 0.218 |

Results are reported using RMSE and compare traditional statistical baselines, supervised and ensemble learning models, and the proposed ULSCO framework under identical training, validation, and testing splits.

changes, and store-specific operational variability. It is difficult for ARIMA to capture high-order interactions and sudden demand changes due to its rigid parametric structure. This means that it is more likely to produce larger forecast errors than its model, although it is one of the most robust models for short-run temporal autocorrelations. Furthermore, deep learning systems improve prediction accuracy. The Long Short-Term Memory (LSTM) model lowers the RMSE to 0.271 by explicitly simulating long-term temporal dependencies and recurring demand cycles. Gradient boosting, which uses nonlinear tree-based ensembles that may capture complex feature interactions, achieves an RMSE of 0.254, which makes performance even better. The hybrid ensemble model, which combines recurrent neural networks with gradient boosting and incorporates both temporal and structural learning methods, significantly increases accuracy, reaching an RMSE of 0.241. The suggested ULSCO framework has the lowest RMSE, at 0.218. This is 23% lower than ARIMA and 12% lower than the best ensemble baseline. The unified learning architecture of ULSCO is what makes this big gain possible. It allows the system to better combine decision-dependent demand responses and promotional effects by modeling how demand changes over time along with downstream operational data. ULSCO combines forecasting with decision-aware optimization signals to deal with real-world demand generation mechanisms that traditional predictive models can't handle. Overall, these results show that ULSCO is a reliable and operationally sound forecasting framework that better matches the way decisions are made in the real-world supply chain and is more accurate.

**Table 7.12 Comparison of Inventory Performance across Baseline Planning Approaches and Learning-Based Systems**

| *Model* | *Fill Rate (%)* ↑ | *Inventory Turnover* ↑ |
|---|---|---|
| Rule-based replenishment | 78.4 | 6.1 |
| Statistical + heuristic | 82.7 | 6.8 |
| ML forecast + heuristic | 86.3 | 7.4 |
| ULSCO (proposed) | 95.0 | 8.6 |

Metrics report store-level fill rate and inventory turnover, highlighting the impact of integrating demand forecasting with adaptive inventory control in unified predictive–prescriptive architectures.

Table 7.12 shows how well stores are performing in terms of their inventory using two different methods: alternative replenishment and learning-based planning. The two main measures are fill rate and inventory turnover. The results show that the proposed ULSCO, forecast-augmented heuristic pipelines, and traditional rule-driven systems perform very differently in practice. In places where demand is unpredictable and based on promotions, the rule-based replenishment baseline achieves a fill rate of 78.4% and an inventory turnover of 6.1. This illustrates that hand-changed static reorder rules and safety stock parameters are not very good. They are clear and easy to follow, but they cannot fully accommodate changing demand patterns, resulting in ongoing service shortages and excessive stock buildup. Replacing traditional heuristics with statistical prediction improves performance. This Statistical + Heuristic design makes demand estimates for restocking choices easier, which translates to an increase of 6.8 in turnover and an increase of 82.7% in fill rate. This decrease in benefits abuts the fact that prediction and control are not linked so that forecasting does not always translate into better inventory management.

The ML Forecast + Heuristic method has a higher fill rate of 86.3% and a turnover of 7.4. Predictors of demand from ML make it easier to predict nonlinear sales behavior and the impact of promotions. However, because replenishment decisions are still based on heuristic parameters, the system cannot fully exploit predictive accuracy for capital efficiency and service-level optimization.

The ULSCO framework has an inventory turnover of 8.6 and a fill rate of 95.0%, both of which are much better than all the other baselines. This is the 19% faster turnover of stock and 17% better service level than rule-based replenishment. These enhancements are due to ULSCO's inventory control layer, which applies RL to directly incorporate forecast uncertainty,

**Table 7.13 Comparison of Upstream Logistics Efficiency on the Freight Analysis Framework (FAF) Dataset, Evaluating Average Freight Lead Time and Logistics Cost Proxies across Deterministic, Forecast-Aware, Integrated, and Unified Learning-Based Optimization Approaches**

| *Model* | *Avg. Lead Time (Days) ↓* | *Logistics Cost Index ↓* |
|---|---|---|
| Deterministic routing | 9.4 | 1.00 |
| Forecast-aware routing | 8.9 | 0.95 |
| Integrated planning | 8.6 | 0.91 |
| ULSCO (proposed) | 8.4 | 0.86 |

downstream demand response, and operational limits into adaptive replenishment policies. ULSCO incorporates predictive modeling and prescriptive decision learning to convert forecast gains into service reliability and inventory productivity. This is ideal for real-world retail businesses.

Table 7.13 compares the efficiency of upstream logistics using FAF data. The principal performance measures are average freight lead time and a normalized logistics cost index. The results support the trend of integrating planning and predictive data into transportation planning, enabling improved performance of the proposed ULSCO. The deterministic routing baseline shows the working of traditional static routing rules. These rules are based on fixed shipment schedules and predefined network configurations. It has a logistics cost index of 1.00 and an average lead time of 9.4 days. These approaches provide operational stability, but they do not respond to changes in demand signals or congestion patterns at the upstream level. This is why transit times are longer and transportation costs are higher. Improvements, even if they are small in size, are measured when forecast awareness is added. The Forecast-Aware Routing model reduces the average lead time to 8.9 days and the cost index to 0.95 by altering routing priorities and shipment allocations based on expected demand volumes. This approach helps the system to be more responsive, but forecasting and transportation choices are still somewhat separated, and there is only limited insight into how predictive models might inform improvements to the whole network. This integrated planning configuration makes logistics even better in that it helps cut lead time by an average of 8.6 days and cost by 0.91. This approach combines demand estimates with tactical routing and consolidation to improve shipment planning and allows network conditions to work better together, making more efficient use of transportation capacity. Still, there are no ongoing changes to the policy, and the optimization mechanism remains mostly deterministic. The suggested ULSCO framework works best because it has an average freight lead time of 8.4 days and a logistics cost index of 0.86. These results mean that the lead time is cut by 11% and the cost of logistics is cut by 14% compared with deterministic routing. The unified learning architecture in ULSCO makes these improvements possible by modeling both upstream transportation rules and downstream demand changes at the same time in an optimization loop powered by RL. By dynamically spreading demand signals into scheduling, consolidation, and routing decisions, ULSCO gets better at handling real-world transportation changes and makes logistics work better from start to finish.

## 7.5.2 Cross-Domain Analysis

We take account of how changes in one area affect changes in another, like how better retail performance affects logistics and vice versa. This helps us to understand how strong and useful our results are. This analysis investigates inter-layer coupling, a central issue of ULSCO. It is not a conventional assessment, which analyzes each area on its own. Table 7.14 uses store-level fill rate as the primary measure of how strong retail-domain services are under different limits on upstream logistics capacity. These results indicate how different planning systems deal with stress on freight capacity and how unified predictive–prescriptive optimization keeps service levels high in logistically challenging conditions. The ML + Heuristic benchmark shows the relationship between ML-based demand forecasting and traditional approaches to determining when to restock. Under normal logistics conditions, it achieves a fill rate of 86.3%. In the absence of sufficient freight capacity, however, the fill rate falls to 79.1%. This drop illustrates the

**Table 7.14 Retail-Domain Performance under Logistics Capacity Stress, Comparing Baseline and Integrated Learning-Based Systems**

| *Model* | *Normal Fill Rate (%)* | *Constrained Fill Rate (%)* |
|---|---|---|
| ML + heuristic | 86.3 | 79.1 |
| Integrated planning | 90.4 | 85.7 |
| ULSCO (proposed) | 95.0 | 92.8 |

Metrics quantify the robustness of store-level service levels as upstream freight constraints intensify, highlighting the ability of unified predictive–prescriptive architectures to preserve demand fulfillment under adverse logistics conditions.

weakness of decoupled planning pipelines. These pipelines make the best decisions for inventory and restocking without considering any upstream transport limits. As a result, if logistics are late, it's difficult to make better predictions to use to plan useful restocking activities. Integrated planning is better. In the absence of much freight, the fill rate drops from 90.4% to 85.7%. This approach matches the store-level demand fulfillment with available transport capacity by coordinating forecasting, replenishment, and logistics planning modules. The planning logic of the system is largely fixed and does not change, so it is difficult to predict when congestion will occur and service will worsen. The ULSCO system that was suggested is the strongest, with an average fill rate of 95.0% and an average fill rate of 92.8% but with a limited fill rate. When the system is under heavy load, ULSCO does not lose performance as fast as the ML + Heuristic baseline. This means that it is more reliable in terms of service. ULSCO's management policies and anticipatory freight allocation use RL. This means that when they decide to buy and replenish downstream stocks, they also look at upstream transportation. ULSCO prevents problems within the chain from impacting retail service by adjusting shipments' timing, quantity, and destination in real time as logistics change. Finally, these results imply that ULSCO is an excellent end-to-end optimization system that can maintain high service quality even in constraining logistical environments. This proves that it is useful for strong retail supply chain operations.

Table 7.15 displays the success of upstream logistics when demand is more uncertain using two stabilizing measures: the standard deviation of freight lead time and the normalized congestion index. These measures reveal the strength of the transportation system by looking at time-varying

**Table 7.15 Logistics Performance under Demand Volatility, Comparing the Stability and Responsiveness of Transportation Planning across Models**

| *Model* | *Lead Time Std. Dev. ↓* | *Congestion Index ↓* |
|---|---|---|
| Deterministic | 2.1 | 1.00 |
| Forecast aware | 1.7 | 0.94 |
| ULSCO (proposed) | 1.3 | 0.88 |

Metrics quantify variability in freight lead times and congestion levels, highlighting the ability of unified predictive–prescriptive architectures to adapt upstream logistics to downstream demand fluctuations.

delivery performance and congestion impacting the whole network. This helps in cases where downstream retail demand increases sharply. The deterministic planning baseline indicates that transit times change dramatically and the network is very congested. The lead time is 2.1 days, and the congestion score is 1.00. Due to the limitations of this approach, it does not accommodate any large fluctuations that could occur due to an increase in sales within a particular area. It is important to note that the way in which an item is scheduled, routed, and combined does not alter very much as a result of signals sent out by the market for that product (in other words, there is a predetermined basis for making these decisions); when there is a high level of congestion within a company, there is a rapid spread of that congestion to other areas of the company's network, which makes planning less dependable and, consequently, delivery performance less reliable. If decisions for routing are made based on expected demand rather than actual demand, then it is possible for a company to reduce its average congestion index by utilizing estimated shipment volumes to create more efficient capacity allocation and routing priorities which substantially improve delivery speeds by decreasing average lead-time variability by approximately one week or approximately seven more days, from 1.7 to 1 (assuming an average lead-time variability of 0.94–0.7). The application of this method yields a more efficient way of managing future fluctuations in demand, but the level of optimization in this process has not yet been completely adapted to accommodate changes in logistics operations under significant unexpected increases in demand (in other words, a company will not have the ability or system in place to adapt quickly to a major surge in demand). The proposed ULSCO framework produces a lead-time variation of 1.3 days and a congestion index of 0.88, which allows for the provision of more stable and adaptive logistics and is significantly faster and more efficient than the deterministic baseline metrics. The unified predictive–prescriptive learning architecture of ULSCO is what makes these benefits possible. It combines policies for managing upstream transportation with data on downstream retail demand. ULSCO uses dynamic routing algorithms and RL-based modal allocation to move freight between different modes of transportation in response to changing demand patterns. ULSCO has proven that it is the best at running logistics networks that can handle demand and remain robust. It does this by coordinating adaptive logistics decision-making with real-time changes in demand to reduce congestion and stabilize delivery performance under unstable conditions.

**Table 7.16 Cross-Domain Coupling Analysis between Downstream Retail Demand and Upstream Logistics Decisions**

| *Model* | *Demand–Logistics Correlation* ↑ |
|---|---|
| Isolated ML | 0.31 |
| Integrated planning | 0.47 |
| ULSCO (proposed) | 0.68 |

Reported values quantify the strength of demand–logistics interaction, illustrating how effectively each method propagates uncertainty and coordination signals across supply chain layers.

Table 7.16 shows a cross-domain coupling analysis between changes in retail demand and changes in logistics decisions. The Demand Shock to Logistics Reaction Index measures logistics activity as a response to changes in demand. This indicator quantitatively assesses the synergy between the retail and transportation layers of the supply chain by measuring how fast and accurately signals of uncertainty, variability, and coordination are sent between these two domains across the entire supply chain. There is also a weakly positive linear correlation of 0.31 between the Demand Shock and Logistics Reaction Index, indicating a weak link between logistics planning and demand forecasting methods followed by a company in the integrated ML approach. In the current situation, although ML models can produce accurate forecasts of future demand, the majority of decisions about routing and transportation methods are still made using static or heuristic (rules of thumb) methods, which do not consider uncertainty or sudden demand shifts, thereby leading to reduced impact on logistics operations resulting from changes in retail demand, inefficient utilization of transportation resources, delayed responses to sudden demand surges, and inadequate capacity allocation for transportation within a supply chain. The integrated planning method increases the cross-domain coordination of logistics activities to 0.47. By integrating the output from demand forecasting with routing, consolidation, and scheduling decision-making processes, an improved integrated planning model helps incorporate changes in demand into the upstream logistics planning process. There is a moderate increase in coupling; thus, the retail and logistics layers have improved interactions with each other. However, the limitations of deterministic optimization methods restrict how well the system can interact with changing patterns of uncertainty. Some of the uncertainty is transferred between domains. The ULSCO framework demonstrates the best cross-domain coupling, as shown by its correlation coefficient of 0.68. This significant enhancement over the separate and combined benchmarks highlights that the ULSCO framework allows for greater incorporation of demand uncertainty into upstream logistics control strategies than prior models. The ULSCO framework's predictive–prescriptive learning architecture provides continual updates to routing, modal selection, and consolidation strategies based on real-time retail demand information using RL methods. Therefore, the close connection of the vertical aspect of the supply chain enables proactive changes to be made to logistics decisions prior to the occurrence of demand shocks, as opposed to reacting to service failures in the downstream supply chain. Overall, the data indicate that the ULSCO framework offers a highly-coherent and flexible end-to-end optimization system that spreads uncertainty across the layers of the supply chain and creates greater stability, responsiveness, and operational efficiency.

### 7.5.3 Ablation Study

The ablation study looks at how different parts of the architecture affect overall performance. Each variant removes one part of ULSCO but leaves all the others the same. This helps us understand what made the gains happen. The ablation analysis of the forecasting module in the proposed ULSCO framework on the Rossmann Store Sales dataset is shown in Table 7.17. We use RMSE as the standard for evaluation. To see how different modeling layers affect the overall accuracy of forecasts and how much they help

**Table 7.17 Ablation Analysis of the Forecasting Component in the Proposed ULSCO Framework, Reporting RMSE on the Rossmann Store Sales Dataset**

| *Variant* | *RMSE* ↓ |
|---|---|
| Full ULSCO | 0.218 |
| – Representation learning | 0.236 |
| – Ensemble stacking | 0.244 |
| – Temporal modeling (RNN removed) | 0.257 |

Each variant removes a specific architectural element to quantify its contribution to multi-horizon demand prediction accuracy.

with predicting demand over multiple time horizons, each variant removes a different architectural element. With an RMSE of only 0.218, the entire ULSCO configuration sets the standard for other ablation comparisons. This result shows that ensemble stacking, temporal modeling, and representation learning can all be used in the same predictive architecture. When the representation learning layer is removed, the RMSE increases to 0.236, which means that the predictions are less accurate. This drop shows how important learned latent feature embeddings are for identifying nonlinear links between calendar effects, store-specific variability, and promotional variables. Without this layer of abstraction, the model can't apply to as many different types of demand patterns. The RMSE increases to 0.244 when ensemble stacking is not used. This result shows that using more than one predictive learner is very important for reducing bias and variance in each model. When demand is hard to predict, ensemble stacking uses a mix of forecasting methods that work well together to make predictions that are more reliable and stable. When the forecasting pipeline is removed, it is easier for mistakes to occur and it is less able to deal with changes in demand. The performance drop is most noticeable when the recurrent neural network part is removed, which eliminates temporal modeling. The RMSE increases to 0.257. This increase shows how important it is to model temporal dependencies when trying to predict retail demand. It is much harder to make accurate multi-horizon predictions when you can't clearly capture seasonality, sequential patterns, and lagged demand responses. These results show that representation learning, ensemble stacking, and temporal modeling all work together to make ULSCO better at predicting the future; not just one architectural element. To get stable, very accurate demand forecasts in unified predictive–prescriptive supply chain architectures, multi-representation predictive modeling is necessary, as shown by the gradual rise in errors across ablation variants.

Table 7.18 shows an ablation study of inventory control methods within the proposed ULSCO framework. It shows how removing closed-loop feedback and RL-based control affects store-level service reliability and inventory efficiency. We use fill rate and inventory turnover to measure performance. These are two different ways to measure how well capital is being used and how well service is being maintained. The full ULSCO configuration has a fill rate of 95.0% and an inventory turnover of 8.6. It is the best example of adaptive predictive–prescriptive inventory optimization. These numbers show how well replenishment policies that use RL work. These policies are always changing the options for ordering based on how much demand is

**Table 7.18 Ablation Analysis of Inventory Control Mechanisms within the Proposed ULSCO Framework**

| *Variant* | *Fill Rate (%)* | *Turnover* |
|---|---|---|
| Full ULSCO | 95.0 | 8.6 |
| – RL control (heuristic) | 87.2 | 7.3 |
| – Feedback loop | 89.1 | 7.7 |

Results quantify the impact of removing RL-based control and closed-loop feedback on service-level reliability and inventory efficiency, highlighting the necessity of adaptive prescriptive decision-making.

expected, how much stock is available, and how well the order was fulfilled downstream. Performance drops significantly when you use heuristic replenishment procedures instead of the RL control layer. The quality of service and the productivity of inventory both reduce significantly when the fill rate drops to 87.2% and the turnover drops to 7.3. This drop shows that heuristic decision-making methods can't dynamically balance service reliability and stock efficiency when retail demand changes. However, better demand forecasting can help with replenishment inputs. Thus, improvements that are expected don't always mean that operations will run more smoothly. Taking away the closed-loop feedback system also makes things worse, but not as much. The fill rate drops to 89.1% and the turnover drops to 7.7 in this version. The system can't always deal with systemic biases or adapt to changing demand conditions because there isn't always a link between changes to replenishment policies and actual sales results. This makes inventory cycling less effective, and service quality drops a little. In conclusion, these results show that accurate forecasting isn't enough for good operational performance in managing retail inventory. To turn forecast data into useful restocking activities, adaptive prescriptive control methods are needed instead. The performance of full ULSCO comes from its integrated RL-based control and closed-loop learning architecture, which keeps inventory decisions aligned with real demand. These results show that in fast-paced retail settings, it's important to make unified predictive–prescriptive decisions in order to keep service levels high and make the best use of inventory.

Table 7.19 displays an ablation study regarding the logistics optimization component of the ULSCO framework. It looks at how separating freight optimization from demand–logistics coupling changes how well

**Table 7.19 Ablation Analysis of the Logistics Optimization Component within the ULSCO Framework**

| *Variant* | *Lead Time (Days)* | *Cost Index* |
|---|---|---|
| Full ULSCO | 8.4 | 0.86 |
| – Freight optimization | 9.1 | 0.96 |
| – Demand–logistics coupling | 9.4 | 1.00 |

Results quantify the impact of removing demand–logistics coupling and constrained freight optimization on upstream transportation efficiency, as measured by average freight lead time and logistics cost indicators.

upstream transportation works. We measure performance using average freight lead time and a normalized logistics cost index because they show how cost-effective and time-efficient transportation operations are as a whole. With an average lead time of 8.4 days and a cost index of 0.86, the full ULSCO setup has the best logistics performance. These numbers show how a closed-loop learning architecture can help predict both future retail demand and transportation choices at the same time. They also set the standard for logistics optimization that combines prediction and prescription. When the freight optimization layer is removed, performance that can be measured goes down. With this change, the cost index goes up to 0.96 and the average lead time goes up to 9.1 days. The system can't combine shipments, change transportation capacity on the fly, or change routing strategies when network conditions and demand change if freight optimization isn't limited. Inefficient transportation operations lead to longer transit times and higher logistics costs.

It gets even worse when you break the link between demand and logistics. These variants work about the same as fully decoupled deterministic routing methods, with an average lead time of 9.4 days and a cost index of 1.00. If there is no clear connection between upstream transportation control and downstream demand signals, uncertainty in forecasts or expected changes in retail demand no longer affects logistics decisions. Because of this, capacity allocation and routing stop being flexible and go back to being fixed. This means longer delivery times, more traffic, and higher transportation costs. All of these results show that dynamic supply chains need to plan for both upstream and downstream activities in order to work well. The full ULSCO architecture works so well because it can send retail demand signals to adaptive freight optimization and transportation planning decisions.

ULSCO proves that there is a need for integrated, learning-based logistics optimization within unified predictive–prescriptive architectures. It does this by using tight demand–logistics coupling and constrained optimization mechanisms to turn predictive information into logistics actions that are cost-effective and responsive to time.

## 7.6 Conclusion and Future Works

This chapter presented ULSCO, a collaborative learning approach to supply chain optimization that links logistics planning, inventory control, and demand prediction in a single predictive–prescriptive architecture. ULSCO relies on closed-loop feedback to spread uncertainty across planning horizons and directly integrates downstream retail demand signals with upstream freight allocation choices unlike traditional planning systems and separate ML pipelines. Following an extensive review of the Rossmann Store Sales and U.S. FAF datasets, the suggested framework showed steady improvement in a variety of operational variables, including reduced forecast error, decreased inventory turnover, enhanced service levels, and decreased logistics costs. Comparative, cross-domain, and ablation analyses confirm the role of coordinated learning and control in complex supply chain systems, demonstrating that these benefits cannot be achieved separately but result from architectural integration. Future endeavors will encompass the integration of real-time streaming data for online adaptation, the expansion of ULSCO to multi-product and multi-echelon environments characterized by stochastic lead times, and the exploration of causal and counterfactual learning mechanisms to enhance intervention-aware decision-making. Other directions include making it easier to understand for regulatory compliance, improving governance through automated drift detection, and testing large-scale implementation in operational industrial systems with real-world limitations.

## References

1. S. Polo-Triana, J. C. Gutierrez, and J. Leon-Becerra, "Integration of machine learning in the supply chain for decision making: A systematic literature review," *Journal of Industrial Engineering and Management*, vol. 17, no. 2, p. 344, May 2024, https://doi.org/10.3926/jiem.6403.

2. I. Vlachos and P. G. Reddy, "Machine learning in supply chain management: Systematic literature review and future research agenda," *International Journal of Production Research*, vol. 63, no. 16, pp. 5987–6016, 2025, https://doi.org/10.1080/00207543.2025.2466062.
3. R. Bergsma, C. de Ruijt, and S. Bhulai, "A systematic review of machine learning approaches in inventory control optimization," *Operations Research Perspectives*, vol. 15, pp. 100367–100367, Nov. 2025, https://doi.org/10.1016/j.orp.2025.100367.
4. K. Douaioui, R. Oucheikh, O. Benmoussa, and C. Mabrouki, "Machine learning and deep learning models for demand forecasting in supply chain management: A critical review," *Applied System Innovation*, vol. 7, no. 5, pp. 93–93, Sep. 2024, https://doi.org/10.3390/asi7050093.
5. S. R. Haque, "Machine learning applications in end-to-end supply chain management: A comprehensive review," *International Journal of Scientific Research and Management (*IJSRM*)*, vol. 13, no. 06, pp. 2226–2241, Jun. 2025, https://doi.org/10.18535/ijsrm/v13i06.ec03.
6. M. Z. Babai, M Arampatzis, M Hasni, F. Lolli, and A Tsadiras, "On the use of machine learning in supply chain management: A systematic review," *IMA Journal of Management Mathematics*, vol. 36, no. 1, Oct. 2024, https://doi.org/10.1093/imaman/dpae029.
7. G. Elkady and A. Hesham Sedky, "Artificial intelligence and machine learning for supply chain resilience," *Current Integrative Engineering*, vol. 1, no. 1, pp. 23–28, Oct. 2023, https://www.researchgate.net/publication/375139652_Artificial_Intelligence_And_Machine_Learning_For_Supply_Chain_Resilience
8. A. R. Teixeira, J. V. Ferreira, and A. L. Ramos, "Intelligent supply chain management: A systematic literature review on artificial intelligence contributions," *Information*, vol. 16, no. 5, pp. 399–399, May 2025, https://doi.org/10.3390/info16050399.
9. B. Ferreira and J. Reis, "Artificial Intelligence in Supply Chain Management: A Systematic Literature Review and Guidelines for Future Research," In *International Joint Conference on Industrial Engineering and Operations Management*, pp. 339–354, 2023, https://doi.org/10.1007/978-3-031-47058-5_27.
10. V. I. Kontopoulou, A. D. Panagopoulos, I. Kakkos, and G. K. Matsopoulos, "A review of ARIMA vs. machine learning approaches for time series forecasting in data driven networks," *Future Internet*, vol. 15, no. 8, pp. 255–255, Jul. 2023, https://doi.org/10.3390/fi15080255.
11. B. Rolf et al., "A review on unsupervised learning algorithms and applications in supply chain management," *International Journal of Production Research*, vol. 63, no. 5, pp. 1933–1983, Aug. 2024, https://doi.org/10.1080/00207543.2024.2390968.
12. G. S. Kashyap et al., "Can AI See What We Can't? Leveraging Deep Learning and Multi-Temporal Satellite Data to Revolutionize Crop Type Mapping and Yield Prediction," In *ICASSP 2025–2025 IEEE International Conference on Acoustics, Speech and Signal Processing (ICASSP)*, pp. 1–5. IEEE, 2025. https://doi.org/10.1109/icassp49660.2025.10890344.

13. Y. Nomura, Z. Liu, and T. Nishi, "Deep reinforcement learning for dynamic pricing and ordering policies in perishable inventory management," *Applied Sciences*, vol. 15, no. 5, pp. 2421–2421, Feb. 2025, https://doi.org/10.3390/app15052421.
14. G. S. Kashyap, K. Malik, S. Wazir, and A. E. I. Brownlee, "Optimization of the rectangle area inside a concave polygon using PSO and Tabu Search," *Soft Computing*, vol. 29, no. 7, pp. 3217–3239, Apr. 2025, https://doi.org/10.1007/s00500-025-10568-1.
15. Y. Yan, A. H. F. Chow, C. P. Ho, Y.-H. Kuo, Q. Wu, and C. Ying, "Reinforcement learning for logistics and supply chain management: Methodologies, state of the art, and future opportunities," *Transportation Research Part E Logistics and Transportation Review*, vol. 162, pp. 102712–102712, May 2022, https://doi.org/10.1016/j.tre.2022.102712.
16. Vairagade, Navneet, Doina Logofatu, Florin Leon, and Fitore Muharemi. "Demand forecasting using random forest and artificial neural network for supply chain management." In *International Conference on Computational Collective Intelligence*, pp. 328–339. Cham: Springer International Publishing, 2019.
17. A. Mitra, A. Jain, A. Kishore, and P. Kumar, "A comparative study of demand forecasting models for a multi-channel retail company: A novel hybrid machine learning approach," *Operations Research Forum*, vol. 3, no. 4, Sep. 2022, https://doi.org/10.1007/s43069-022-00166-4.
18. S. Nagadevi, G. Abirami, R. Vidhya, S. Selvakumar, and C. Vijayalakshmi, "XGBoost-based demand forecasting in supply chain management using machine learning algorithm," *International Journal of Interactive Mobile Technologies* (iJIM), vol. 19, no. 18, pp. 161–174, Sep. 2025, https://doi.org/10.3991/ijim.v19i18.57253.
19. A. Husna, S. H. Amin, and B. Shah, "Demand Forecasting in Supply Chain Management Using Different Deep Learning Methods," *In Demand forecasting and order planning in supply chains and humanitarian logistics*, pp. 140–170, Aug. 2020, https://doi.org/10.4018/978-1-7998-3805-0.ch005.
20. S. Taghiyeh, D. C. Lengacher, A. H. Sadeghi, A. Sahebi-Fakhrabad, and R. B. Handfield, "A novel multi-phase hierarchical forecasting approach with machine learning in supply chain management," *Supply Chain Analytics*, vol. 3, pp. 100032–100032, Aug. 2023, https://doi.org/10.1016/j.sca.2023.100032.
21. R. Wang and C. Gong, "Supply Chain Anomaly Detection and Prediction Models Based on Large-Scale Time Series Data," *Proceedings of the 2025 International Conference on Digital Economy and Information Systems*, pp. 8–13, Apr. 2025, https://doi.org/10.1145/3745133.3745136.
22. F. Zeng, J. Li, X. Tang, and Y. Yang, "A Survey of Unsupervised Clustering Methods Driven by Information Completeness: Challenges, Advances, and Future Directions," *Proceedings of the 2024 2nd International Conference on Computer, Internet of Things and Smart City*, pp. 153–159, Dec. 2024, https://doi.org/10.1145/3731867.3731893.
23. J. P. Arroyo-Castro and S. W. Chou-Chen, "Unsupervised Detection of Anomaly in Public Procurement Processes," In *Conference of the International Federation of Classification Societies*, pp. 23–32. Cham: Springer Nature Switzerland, 2024, https://doi.org/10.1007/978-3-031-85870-3_3.

24. D. Qiao, X. Ma, and J. Fan, "Federated t-SNE and UMAP for Distributed Data Visualization," *Proceedings of the AAAI Conference on Artificial Intelligence*, vol. 39, no. 19, pp. 20014–20023, Apr. 2025, https://doi.org/10.1609/aaai.v39i19.34204.
25. V. V. Baligodugula and F. Amsaad, "Unsupervised Learning: Comparative Analysis of Clustering Techniques on High-Dimensional Data," *arXiv.org*, 2025. https://arxiv.org/abs/2503.23215 (accessed Jan. 01, 2026).
26. H. Dehaybe, D. Catanzaro, and P. Chevalier, "Deep reinforcement learning for inventory optimization with non-stationary uncertain demand," *European Journal of Operational Research*, vol. 314, no. 2, pp. 433–445, Apr. 2024, https://doi.org/10.1016/j.ejor.2023.10.007.
27. C. Zhou, J. Ma, L. Douge, E. P. Chew, and L. H. Lee, "Reinforcement learning-based approach for dynamic vehicle routing problem with stochastic demand," *Computers & Industrial Engineering*, vol. 182, pp. 109443–109443, Jul. 2023, https://doi.org/10.1016/j.cie.2023.109443.
28. M. Kulahara, A. K. J. Saudagar, S. Tripathi, and M. A. Hoque, "A CNN-based framework for geometric alignment of historical and satellite imagery," *IEEE Access*, vol. 13, pp. 132259–132268, 2025, https://doi.org/10.1109/access.2025.3589497.
29. X. Liu, M. Hu, Y. Peng, and Y. Yang, "Multi-Agent Deep Reinforcement Learning for Multi-Echelon Inventory Management," *Production and Operations Management*, vol. 34, no. 7, pp. 1836–1856, 2024, https://doi.org/10.1177/10591478241305863.
30. A. Seyam, S. S. Mathew, B. Du, M. E. Barachi, and J. Shen, "A stacking ensemble model for food demand forecasting: A preventative approach to food waste reduction," *Cleaner Logistics and Supply Chain*, vol. 15, pp. 100225–100225, May 2025, https://doi.org/10.1016/j.clscn.2025.100225.
31. I. M. Hammam, A. K. El-Kharbotly, and Y. M. Sadek, "Adaptive demand forecasting framework with weighted ensemble of regression and machine learning models along life cycle variability," *Scientific Reports*, vol. 15, no. 1, pp. 38482–38482, Nov. 2025, https://doi.org/10.1038/s41598-025-23352-w.
32. Z. Mohammed, C. Anas, and M. El Hammoumi, "A hybrid learning framework for forecasting uncertainty and adaptive inventory planning in retail supply chains," *Supply Chain Analytics*, vol. 13, p. 100180, Mar. 2026, https://doi.org/10.1016/j.sca.2025.100180.
33. E. Jahanbakhsh, "AI capabilities and strategic agility in emerging markets: A systematic review and integrative framework," *SSRN Electronic Journal*, Jan. 2025, https://doi.org/10.2139/ssrn.5407302.
34. W. Yang, D. Cao, and Y. Liu, "Foundation Models for Demand Forecasting via Dual-Strategy Ensembling," *arXiv.org*, 2025. https://arxiv.org/abs/2507.22053 (accessed Jan. 01, 2026).
35. A. Soni et al., "Can We Predict Your Next Move without Breaking Your Privacy?," *arXiv.org*, 2025. https://arxiv.org/abs/2507.08843 (accessed Jan. 13, 2026).
36. S. Giannelos, I. Konstantelos, and G. Strbac, "Optimal supply chain design using machine learning, risk assessment and optimisation applied to coal distribution," *EURO Journal on Decision Processes*, vol. 13, pp. 100062–100062, Jan. 2025, https://doi.org/10.1016/j.ejdp.2025.100062.

37. V. Govindarajan, P. Patel, S. Tripathi, M. A. Hoque, and G. S. Kashyap, "MAGIC-Enhanced Keyword Prompting for Zero-Shot Audio Captioning with CLIP Models," *arXiv.org*, 2025. https://arxiv.org/abs/2509.12591 (accessed Jan. 13, 2026).
38. M. Kraus, D. Tschernutter, S. Weinzierl, and P. Zschech, "Interpretable generalized additive neural networks," *European Journal of Operational Research*, vol. 317, no. 2, pp. 303–316, Sep. 2024, https://doi.org/10.1016/j.ejor.2023.06.032.
39. S. Kruschel, N. Hambauer, S. Weinzierl, S. Zilker, M. Kraus, and P. Zschech, "Challenging the performance-interpretability trade-off: An evaluation of interpretable machine learning models," *Business & Information Systems Engineering*, Feb. 2025, https://doi.org/10.1007/s12599-024-00922-2.
40. R. Agarwal et al., "Neural additive models: Interpretable machine learning with neural nets," *Advances in Neural Information Processing Systems*, vol. 34, pp. 4699–4711, Dec. 2021. https://proceedings.neurips.cc/paper/2021/hash/251bd0442dfcc53b5a761e050f8022b8-Abstract.html (accessed Jan. 01, 2026).
41. R. Marcinkevičs and J. E. Vogt, "Interpretable and explainable machine learning: A methods-centric overview with concrete examples," *WIREs Data Mining and Knowledge Discovery*, vol. 13, no. 3, Feb. 2023, https://doi.org/10.1002/widm.1493.
42. C. Rudin, C. Chen, Z. Chen, H. Huang, L. Semenova, and C. Zhong, "Interpretable machine learning: Fundamental principles and 10 grand challenges," *Statistics Surveys*, vol. 16, no. none, Jan. 2022, https://doi.org/10.1214/21-ss133.
43. D. K. Siripurapu, "AI and ML-driven middleware: Revolutionizing enterprise integration," *International Journal of Computer Engineering and Technology*, vol. 16, no. 1, pp. 3410–3424, Feb. 2025, https://doi.org/10.34218/ijcet_16_01_237.
44. S. Tripathi, M. T. Nafis, I. Hussain, and J. Gao, "The Confidence Paradox: Can LLM Know When It's Wrong," *arXiv.org*, 2025. https://arxiv.org/abs/2506.23464 (accessed Jan. 13, 2026).
45. M. Soori, F. K. G. Jough, R. Dastres, and B. Arezoo, "AI-based decision support systems in Industry 4.0, a review," *Journal of Economy and Technology*, vol. 4, pp. 206–225, 2026, https://doi.org/10.1016/j.ject.2024.08.005.
46. N. Ul Haq Qureshi, S. Javed, K. Javed, S. Meesam Raza Naqvi, A. Raza, and Z. Saeed, "Demand forecasting in supply chain management for rossmann stores using weather enhanced deep learning model," *IEEE Access*, vol. 12, pp. 145570–145581, 2024, https://doi.org/10.1109/access.2024.3472499.
47. P. Bingham, A. Maguire, L. O'Rourke, B. Pandey, and M. His, *Freight Analysis Framework Commodity Flow Forecast Study (FAF Version 5): Final Forecasting Results*. No. FHWA-HOP-22-037. United States. Department of Transportation. Federal Highway Administration. Office of Operations, 2022.

# Chapter 8

# Integrating Data-Driven Technologies for End-to-End Supply Chain Transformation

Sharad Deep Shukla

## 8.1 Introduction

To assist operational and strategic decision-making, data-driven supply chain systems are increasingly relying on statistical forecasting, machine learning models, Internet of Things (IoT)-based sensing, and digital twin simulations. Demand forecasting, asset monitoring, and localized optimization across procurement, production, and logistics activities have all seen quantifiable benefits because to these strategies [1]. However, even with their increasing sophistication, the majority of current systems evaluate uncertainty in a function-isolated way, treating operational variability, disruption risk, and forecasting error as separate phenomena instead of interdependent signals in an end-to-end supply network [2]. This results in a situation where uncertainty is modeled locally but does not spread coherently across decision layers, which we refer to as the Fragmented Uncertainty Modeling (FUM) problem.

In internationally dispersed supply chains subject to demand volatility, geopolitical shocks, and climate-driven disruptions, the FUM problem becomes especially acute. The way exogenous shocks amplify and cascade across interconnected sourcing, production, and fulfillment decisions is not well captured by conventional enterprise architectures, which are typified by


DOI: 10.4324/9781003724889-8

functionally segregated planning and execution systems, delayed information exchange, and reactive control loops [3]. Although predictive analytics, digital traceability tools, IoT-enabled monitoring platforms, and automation technologies have been adopted by businesses, they are usually implemented as standalone solutions that are tailored to certain activities rather than to system-wide uncertainty reasoning [4]. End-to-end stability and resilience are thus limited, since operational decisions are frequently based on imprecise or inconsistent uncertainty estimates.

More potent models, such as simulation-driven digital twins, reinforcement learning-based inventory control, and nonlinear machine learning predictors, have been investigated in recent developments in supply chain analytics [5]. The bulk of empirical research, however, is still limited to evaluations of a specific function, including improving maintenance scheduling, forecast accuracy, or transportation routing [6]. The way that uncertainty that starts in one layer, like environmental volatility impacting logistical dependability, spreads across upstream sourcing dependencies or downstream fulfillment performance is not explicitly modeled by these methods. As a result, current frameworks find it difficult to facilitate comprehensive reasoning regarding trade-offs between recovery, robustness, and efficiency in the face of systemic uncertainty. Four architectural paradigms—fragmented architectures, isolated digitalization, integrated end-to-end architectures, and governance-embedded autonomous architectures—are contrasted in Table 8.1 to provide an overview of this progression. A value of 1 in the table signifies that the associated capacity is expressly supported by a certain architectural paradigm, whereas a value of 0 implies that it is not. Strict planning–execution separation and compartmentalized information flows are characteristics of fragmented architectures that lead to little uncertainty sharing across processes. Because of the lack of cross-functional compatibility, isolated digitalization preserves some situational awareness while introducing predictive analytics and IoT monitoring. Unified data flows and coordinated decision-making are made possible by integrated end-to-end systems, which enhance operational stability and disruption recovery. By facilitating real-time coordinated control, embedded data governance, and responsible decision automation—features necessary to regulate uncertainty propagation at scale—governance-embedded autonomous systems further expand these capabilities [7].

Even integrated systems frequently fail to reason about uncertainty as a first-class modeling element due to the FUM problem. Rather than being architectural mechanisms that govern the generation, transformation, and

**Table 8.1 Architectural Capability Comparison across Fragmented, Isolated Digitalization, Integrated End-to-End, and Governance-Embedded Autonomous Supply Chain Systems**

| *Main Feature* | *Fragmented Architecture* | *Isolated Digitalization* | *Integrated End-to-End Architecture* | *Governance-Embedded Autonomous Architecture* |
|---|---|---|---|---|
| Functional planning–execution separation | 1 | 0 | 0 | 0 |
| Siloed enterprise information flows | 1 | 0 | 0 | 0 |
| Multimodal global supply network complexity | 0 | 0 | 1 | 0 |
| Predictive analytics utilization | 0 | 1 | 0 | 0 |
| IoT-based operational monitoring | 0 | 1 | 0 | 0 |
| Digital traceability deployment | 0 | 1 | 0 | 0 |
| Automation tool application | 0 | 1 | 0 | 0 |
| Partial situational awareness | 0 | 1 | 0 | 0 |
| Cross-functional interoperability | 0 | 0 | 1 | 0 |
| Unified end-to-end system integration | 0 | 0 | 1 | 0 |
| Legacy–cloud interoperability alignment | 0 | 0 | 1 | 0 |
| Real-time coordinated decision control | 0 | 0 | 0 | 1 |
| Embedded data governance mechanisms | 0 | 0 | 0 | 1 |

(*Continued*)

**Table 8.1 (*Continued*) Architectural Capability Comparison across Fragmented, Isolated Digitalization, Integrated End-to-End, and Governance-Embedded Autonomous Supply Chain Systems**

| *Main Feature* | *Fragmented Architecture* | *Isolated Digitalization* | *Integrated End-to-End Architecture* | *Governance-Embedded Autonomous Architecture* |
|---|---|---|---|---|
| Regulatory compliance enforcement | 0 | 0 | 0 | 1 |
| Data quality assurance frameworks | 0 | 0 | 0 | 1 |
| Algorithmic decision accountability | 0 | 0 | 0 | 1 |
| Organizational readiness as core design | 0 | 0 | 0 | 1 |
| Inventory efficiency enhancement | 0 | 0 | 1 | 0 |
| Disruption recovery acceleration | 0 | 0 | 1 | 0 |
| Operational stability improvement | 0 | 0 | 1 | 0 |

Binary entries indicate the presence (1) or absence (0) of each capability, highlighting the progression toward coherent end-to-end uncertainty propagation.

action of uncertainty estimations, organizational preparedness, data governance, and automation accountability are often viewed as downstream implementation issues [8]. As systems move toward greater degrees of autonomy, these disconnects cause ongoing problems with data integrity, regulatory compliance, and confidence in algorithmic conclusions [9]. Furthermore, from the standpoint of uncertainty propagation, data lineage, and cross-organizational decision coupling, the integration of older business systems with cloud-native and event-driven infrastructures has not received enough attention [10]. This paper offers Uncertainty-Propagating Supply Chains (UP-SC), an end-to-end data-driven paradigm that explicitly models and propagates uncertainty across interdependent supply chain decision layers, in response to these restrictions. In order to facilitate network-aware stability estimates and resilience-oriented decision intelligence, UP-SC integrates exogenous climatic volatility signals with global trade dependency structures to portray supply chains as coupled economic–environmental networks. UP-SC offers a unified architectural architecture in which uncertainty flows systematically from data capture to analytical inference and operational control, as opposed to optimizing discrete operations.

This chapter is structured as follows. Section 8.2 reviews related work and critically analyzes existing supply chain integration, predictive analytics, IoT and digital twin technologies, organizational governance, resilience modeling, and decision automation architectures, systematically exposing the FUM problem through comparative capability taxonomies (Tables 8.1–8.8). Section 8.3 presents the proposed UP-SC framework, detailing its datasets, preprocessing pipeline, economic–environmental network formulation, uncertainty fusion mechanisms, digital twin abstraction, and layered governance-embedded architecture (Figure 8.1). Section 8.4 describes the experimental setup, evaluation metrics, and hyperparameter configurations. Section 8.5 reports empirical results, cross-domain analyses, and ablation studies demonstrating the system-level benefits of explicit end-to-end uncertainty propagation. Finally, Section 8.6 concludes this chapter and outlines directions for future research.

## 8.2 Related Works

### 8.2.1 Evolution of Supply Chain Technology Integration

Enterprise-centric information systems intended to replace dispersed departmental applications with unified transactional platforms were the focus of early supply chain technology integration research. In order to enable consistent schemas and internal process alignment within company borders, enterprise resource planning (ERP) systems have been thoroughly researched as fundamental mechanisms for cross-functional data standardization [11]. Table 8.2 illustrates how ERP-centric integration facilitates standardized data representation (1) but falls short in terms of deployment agility, scalability across wide networks, and real-time operational synchronization (0). As a result, ERP systems enhanced transactional accuracy and internal visibility but provided little assistance for managing dynamic uncertainty outside of organizational boundaries [12]. Research focus turned to Electronic Data Interchange (EDI) frameworks that automate trade partner transactions as supply chains

**Table 8.2 Architectural Capability Comparison across ERP-Centric, EDI-Based, Cloud/Control Tower, and API/Event-Driven Supply Chain Integration Paradigms**

| *Main Feature* | *ERP-Centric Integration* | *EDI Frameworks* | *Cloud and Control Towers* | *API and Event-Driven Ecosystems* |
|---|---|---|---|---|
| Cross-functional data standardization | 1 | 0 | 1 | 1 |
| Inter-organizational transaction automation | 0 | 1 | 1 | 1 |
| Scalability across extended networks | 0 | 0 | 1 | 1 |
| Deployment agility | 0 | 0 | 1 | 1 |
| Centralized disruption monitoring | 0 | 0 | 1 | 1 |
| Modular system architecture | 0 | 0 | 0 | 1 |
| Fault isolation capability | 0 | 0 | 0 | 1 |
| Real-time operational synchronization | 0 | 0 | 0 | 1 |
| IoT-enabled logistics integration | 0 | 0 | 1 | 1 |
| Adaptive service-oriented interoperability | 0 | 0 | 0 | 1 |

Binary indicators (1/0) denote the presence or absence of each capability, highlighting persistent limitations in end-to-end uncertainty propagation.

became more inter-organizational. By codifying preset message formats and workflows, EDI systems enhanced inter-organizational transaction automation (1 in Table 8.2) [13]. However, scalability, agility, and adaptive interoperability were limited due to their restrictive schemas, high coordination overhead, and limited extensibility (0 in Table 8.2). Consequently, real-time synchronization and uncertainty-aware coordination under demand or disruption variations were not supported by EDI-based integration, which only worked well in steady operational settings [14].

Cloud-based supply chain platforms and control tower systems were made possible by later developments in distributed computing. These solutions combined diverse data feeds from supplier portals, warehouse platforms, transportation management systems, and IoT-enabled assets, and they separated computing from on-premises infrastructure [15]. Table 8.2 summarizes how cloud and control tower models facilitate IoT-enabled logistics integration (1s), scalability, deployment agility, and centralized disruption monitoring. Their capacity to reconcile uncertainty signals across heterogeneous decision layers is, however, limited by their lack of fault isolation, modular system architecture, and real-time operational synchronization (0s) [16]. In practice, uncertainty identified at the logistics or sensing layers is centrally depicted but does not propagate upstream into models for production planning or sourcing.

The focus of more recent studies is on API-centric and event-driven integration designs, which substitute loosely connected services for strict point-to-point connections [17]. While event-driven streaming infrastructures allow for real-time operational synchronization across dispersed supply chain components, microservices-based solutions offer modularity and fault separation (as shown in Table 8.2) [18]. These architectures are the first integration paradigm that supports continuous state synchronization, scalability, agility, and interoperability all at once. However, as Table 8.2 shows, the main focus of API- and event-driven ecosystems is not semantic uncertainty propagation, but rather structural and temporal coordination. Uncertainty estimates are not coherently conveyed across procurement, production, and fulfillment layers; instead, they stay embedded within particular services even when system states are synchronized in real time.

### *8.2.2 Predictive Analytics and Artificial Intelligence Applications*

Through a series of modeling paradigms that gradually increase decision autonomy while mostly retaining function-local uncertainty assessment, predictive analytics in supply chain management has developed. In order to enable demand planning and inventory control under uncertainty, early work concentrated on statistical forecasting and optimization models [19]. As demonstrated by their capacity to capture demand variability in Table 8.3, traditional time-series and regression-based techniques produced understandable demand predictions and supported trend-based planning decisions. Nevertheless, these models work inside isolated planning horizons, assume stationarity, and limit uncertainty to forecasting error rather than spreading it throughout downstream execution layers [20]. In order to capture the nonlinear interactions between price, promotions, and seasonal effects, later research included forecasting models based on machine learning [21]. Table 8.3 illustrates how machine learning techniques

**Table 8.3 Capability Comparison of Predictive Analytics and AI Paradigms across Supply Chain Decision Layers**

| *Main Feature* | *Statistical Forecasting* | *Machine Learning Forecasting* | *Predictive Maintenance AI* | *Autonomous Decision AI* |
|---|---|---|---|---|
| Trend-based demand projection | 1 | 1 | 0 | 0 |
| Nonlinear demand–promotion modeling | 0 | 1 | 0 | 0 |
| Sensor-driven equipment health analysis | 0 | 0 | 1 | 0 |
| Remaining useful life estimation | 0 | 0 | 1 | 0 |
| Multi-echelon inventory policy learning | 0 | 0 | 0 | 1 |
| Real-time adaptive replenishment | 0 | 0 | 0 | 1 |
| Dynamic pricing and capacity modulation | 0 | 0 | 0 | 1 |
| Autonomous policy optimization | 0 | 0 | 0 | 1 |
| Governance and transparency constraints | 0 | 0 | 0 | 1 |

Binary entries (1/0) indicate the presence or absence of each capability, highlighting the lack of end-to-end uncertainty propagation in function-isolated models.

complement statistical models by preserving demand-level uncertainty estimation while enabling nonlinear demand–promotion interactions. Empirical studies show diminishing results when feature stability, data quality, or organizational execution alignment are restricted, despite persistent accuracy advances [22]. Importantly, the forecasting paradigms of machine learning and statistics only assess uncertainty at the demand layer; they lack the means to translate this uncertainty into fulfillment performance, sourcing exposure, or logistical reliability.

Simultaneously, research on predictive maintenance used sensor data to determine the remaining useful life, failure probabilities, and equipment health of production and logistical assets [23]. Table 8.3 shows that these models facilitate sensor-driven health analysis and remaining useful life prediction, which introduce uncertainty estimation at the asset level. While effective at decreasing downtime variability and maintenance costs, predictive maintenance systems remain operationally isolated, with uncertainty signals rarely integrated into upstream capacity planning or downstream delivery obligations [24]. Autonomous decision-making systems have been the subject of more recent research, namely reinforcement learning frameworks for pricing, dynamic replenishment, multi-echelon inventory control, and order fulfillment scheduling [25]. Table 8.3 summarizes how autonomous decision AI introduces explicit governance and transparency limitations while providing unique support for adaptive policy optimization, real-time replenishment, and dynamic capacity modulation. Nevertheless, instead of modeling uncertainty as an explicit, cross-functionally shareable representation, these systems usually implicitly embed it inside policy learning loops [26]. Because of this, autonomous agents are able to reason over localized state-action ambiguity without spreading uncertainty throughout interconnected decision layers.

### 8.2.3 Internet of Things and Digital Twin Technologies

Localized sensing and visibility were the main focus of early supply chain IoT research, which mostly used Radio Frequency Identification and GPS-based monitoring devices to track shipment status and inventory movement [27]. As shown in Table 8.4, where 0 indicates the lack of explicit functional support and 1 indicates its presence, asset visibility IoT solutions do not offer advanced analytical or decision-making capabilities, but they do facilitate inventory movement traceability. Although these systems did not simulate uncertainty beyond the immediate sensing layer, they did enhance descriptive awareness of logistics processes. Temperature, humidity, and vibration readings for cold-chain and condition-sensitive commodities were included in IoT deployments for environmental sensing in later work [28]. Table 8.4 demonstrates that environmental sensing IoT is still limited to condition monitoring and does not allow equipment health analytics, simulation, or decision adaptation, even if these technologies made it possible to monitor cold-chain compliance and reduce spoiling. Because of this, environmental variability-derived uncertainty signals are usually displayed at the dashboard level and do not spread to planning or control models [29]. Using anomaly detection, survival analysis, and remaining useful life assessment

**Table 8.4 Functional Capability Comparison of IoT and Digital Twin Systems, from Descriptive Monitoring to Adaptive and Autonomous Supply Chain Intelligence**

| *Main Feature* | *Asset Visibility IoT* | *Environmental Sensing IoT* | *Predictive Maintenance IoT* | *Digital Twin Modeling* | *Autonomous Digital Twins* |
|---|---|---|---|---|---|
| Inventory movement traceability | 1 | 0 | 0 | 0 | 0 |
| Cold-chain condition monitoring | 0 | 1 | 0 | 0 | 0 |
| Equipment health analytics | 0 | 0 | 1 | 0 | 0 |
| Remaining useful life estimation | 0 | 0 | 1 | 0 | 0 |
| Virtual asset representation | 0 | 0 | 0 | 1 | 1 |
| Process flow simulation | 0 | 0 | 0 | 1 | 1 |
| Network-level capacity modeling | 0 | 0 | 0 | 0 | 1 |
| Disruption response simulation | 0 | 0 | 0 | 0 | 1 |
| Self-updating model recalibration | 0 | 0 | 0 | 0 | 1 |
| Autonomous optimization recommendation | 0 | 0 | 0 | 0 | 1 |

Binary entries (1/0) indicate the presence or absence of each capability.

for production and logistical assets, further research investigated IoT-enabled predictive maintenance [30]. As Table 8.4 illustrates, predictive maintenance IoT offers health analytics for equipment and estimates its remaining useful life, but it does not yet allow virtual representation, process simulation, or network-level reasoning. As a result, upstream capacity planning and downstream fulfillment decisions do not incorporate maintenance-related uncertainty, which is isolated at the asset level [31].

By creating virtual representations of physical assets and processes, digital twin technologies go beyond the Internet of Things [32]. This allows for process flow simulation and what-if analysis under hypothetical interruptions. A qualitative change from descriptive sensing to analytical experimentation can be seen in Table 8.4, which shows that digital twin modeling explicitly enables virtual asset representation and process flow simulation. Nevertheless, traditional digital twins continue to be passive analytical artifacts because they do not provide autonomous optimization, self-updating recalibration, or network-level capacity modeling, which prevents uncertainty from spreading across decision layers [33]. Network-level capacity modeling, disruption response simulation, self-updating model recalibration, and autonomous optimization proposals are only supported by autonomous digital twin architectures, as indicated in Table 8.4. However, uncertainty transmission is frequently implicit and task-specific even in these systems, rather than being clearly addressed as a first-class architectural problem spanning production, fulfillment, and procurement [34]. Therefore, despite their representational strength, digital twin ecosystems often function as standalone simulation environments instead of integrated parts of end-to-end uncertainty-aware decision pipelines.

### 8.2.4 Organizational Change and Technology Adoption

Instead of developing as a single architectural viewpoint, organizational research on supply chain technology adoption has evolved across several mostly separate analytical lenses. In order to explain how individual users react to enterprise systems, early research concentrated on behavioral adoption theories, such as diffusion models and technology acceptance [35]. This research mainly ignored system-level uncertainty handling and cross-functional decision coupling in favor of emphasizing perceived utility, usability, and social influence. Organizational readiness frameworks were introduced by a different body of research, emphasizing workforce capabilities, leadership support, absorptive ability, and infrastructure maturity as factors that

**Table 8.5 Comparison of Organizational and Governance Perspectives in Digital Supply Chain Research**

| *Main Feature* | *Behavioral Adoption Orientation* | *Readiness Capability Alignment* | *Resistance–Intervention Governance* | *Autonomous Transformation Enablement* |
|---|---|---|---|---|
| Technology acceptance modeling | 1 | 0 | 0 | 0 |
| Diffusion-based adoption analysis | 1 | 0 | 0 | 0 |
| Organizational readiness maturity | 0 | 1 | 0 | 0 |
| Capability–technology misalignment | 0 | 1 | 0 | 0 |
| Resistance behavior dynamics | 0 | 0 | 1 | 0 |
| Stakeholder participation mechanisms | 0 | 0 | 1 | 0 |
| Leadership alignment structures | 0 | 0 | 1 | 0 |
| Cultural data-driven orientation | 0 | 0 | 0 | 1 |
| Governance-transparent learning systems | 0 | 0 | 0 | 1 |

Binary entries indicate the presence (1) or absence (0) of each capability, highlighting the fragmentation of adoption, readiness, resistance, and autonomy considerations.

contribute to effective digital transformation [36]. Research has consistently shown that when organizational capability and technology complexity are not aligned, deployments are delayed and performance advantages are not obtained [37]. However, rather than treating analytical models and uncertainty estimation as dynamic, governance-regulated processes, readiness-oriented work usually regards them as static inputs.

Overt and latent resistance behaviors are produced by job insecurity, loss of professional autonomy, and mistrust of algorithmic decision-making, according to a different research stream that looked at resistance dynamics and intervention techniques [38]. Participatory design, structured communication, and focused training were suggested as mitigation techniques by intervention-based studies [39]. These methods are useful for enhancing adoption continuity, but they never address the validation, communication, or application of uncertainty estimations across decision levels. Algorithm-driven operations require data-driven organizational culture, transparency in governance, and leadership alignment, according to more recent research on autonomous transformation enablement [40]. Although these publications emphasize the value of accountability and institutional learning, they usually view governance as an organizational overlay rather than an essential part of the design of analytical systems. This fragmentation is summarized in Table 8.5, where a value of 1 means that a certain organizational or governance capability is expressly taken into account by a research lens, and 0 means that it is not. The table demonstrates how earlier work has tackled behavioral adoption, readiness alignment, resistance governance, and autonomous enablement separately, with little overlap among viewpoints. Consequently, the FUM problem at scale is reinforced by the insufficient integration of uncertainty governance—how uncertainty is created, trusted, audited, and propagated across supply chain decision layers—into technical designs [41].

### 8.2.5 Data Governance, Provenance, and Trust Architectures

Research on data governance, lineage, and trust as key facilitators of scalable digital transformation has grown as supply chain analytics has advanced toward dispersed and autonomous decision environments. In order to enhance transactional integrity in corporate systems, early governance research concentrated on master data management, access control, and schema harmonization [42]. Table 8.6 demonstrates that although these frameworks facilitate the standardization of internal data (1), their usefulness in internationally distributed supply chains is limited due to their limited support for cross-organizational provenance tracking, uncertainty tracing, and real-time auditability (0). Later research examined provenance-aware data management systems that incorporate traceability and metadata tracking into data pipelines [43]. Although these systems addressed uncertainty as a descriptive characteristic rather than an operational signal, they did increase auditability and accountability across distributed platforms (1 in Table 8.6). Consequently, uncertainty estimates were not propagated across interdependent decision layers but were recorded for compliance purposes. Blockchain-enabled governance techniques for supply chain traceability and trust enforcement were added in more recent studies [44]. Although these methods improved immutability, tamper resistance, and partner accountability (1s in Table 8.6), their capacity to facilitate real-time uncertainty-aware

**Table 8.6 Governance Architecture Capability Comparison across Enterprise, Provenance-Aware, Blockchain-Based, and Cloud-Native Governance Systems**

| *Main Feature* | *Enterprise Governance* | *Provenance-Aware Systems* | *Blockchain-Based Governance* | *Cloud-Native Governance* |
|---|---|---|---|---|
| Data lineage tracking | 0 | 1 | 1 | 1 |
| Cross-org auditability | 0 | 1 | 1 | 1 |
| Uncertainty traceability | 0 | 0 | 0 | 0 |
| Real-time policy enforcement | 0 | 0 | 0 | 1 |
| Governance-embedded analytics | 0 | 0 | 0 | 0 |

Binary entries indicate the presence (1) or absence (0) of each governance capability, highlighting persistent limitations in explicit uncertainty traceability, governance-embedded analytics, and real-time uncertainty-aware policy enforcement in distributed supply chain data infrastructures.

coordination was limited due to latency overheads and a lack of interaction with analytical reasoning layers. Role-based access control, lineage-aware data orchestration, and automated policy enforcement were further introduced by cloud-native governance platforms [45]. Even though these architectures enhanced real-time monitoring and scalability, they were unable to incorporate uncertainty propagation as a first-class architectural function. Across the sourcing, production, and fulfillment tiers, uncertainty was still a characteristic to be recorded rather than a signal to be sent and responded to. Existing governance systems guarantee compliance and accountability, but they do not specifically represent how uncertainty should be created, audited, transformed, and spread across interrelated supply chain choices, as summarized in Table 8.6. The continuation of the FUM problem at scale is directly caused by this unaddressed gap, which is why UP-SC integrated governance as an architectural layer that explicitly manages uncertainty flows rather than just data records.

### *8.2.6 Resilience Modeling and Disruption Management Architectures*

Traditionally, research on supply chain resilience has concentrated on disruption response modeling, redundancy planning, and risk mitigation. To lessen susceptibility to localized disturbances, early resilience frameworks

**Table 8.7 Resilience Architecture Capability Comparison across Buffer-Based, Stochastic Optimization, Digital Twin, and Network Science-Based Resilience Modeling Paradigms**

| *Main Feature* | *Buffer-Based Models* | *Stochastic Optimization* | *Digital Twin Resilience* | *Network Science Models* |
|---|---|---|---|---|
| Local recovery planning | 1 | 1 | 1 | 1 |
| Network dependency modeling | 0 | 0 | 0 | 1 |
| Dynamic uncertainty propagation | 0 | 0 | 0 | 0 |
| End-to-end resilience reasoning | 0 | 0 | 0 | 0 |
| Climate-driven shock modeling | 0 | 1 | 1 | 1 |

Binary entries indicate the presence (1) or absence (0) of each resilience capability, highlighting the absence of explicit dynamic uncertainty propagation and end-to-end resilience reasoning in prevailing resilience modeling frameworks.

placed a strong emphasis on buffer stock regulations, supplier diversification, and safety stock optimization [46]. Table 8.7 shows that these models lack dynamic propagation mechanisms (0) and network-level uncertainty modeling (1), although they do support basic recovery planning (1). In order to assess disruption recovery plans under probabilistic demand and lead-time fluctuations, subsequent research developed stochastic optimization and scenario-based simulation tools [47]. Although these techniques made it possible to evaluate contingencies, their relevance in globally interconnected trade networks was limited since they modeled uncertainty as an external disturbance rather than a network-propagating signal. What-if simulations for buffer allocation, production rescheduling, and logistics rerouting were further made possible by resilience models based on digital twins [48]. Although these models enhanced local recovery planning (1 in Table 8.7), they lacked structural coupling across manufacturing, distribution, and procurement networks and remained function-isolated. Dependency graphs and cascade failure analysis were introduced by more current network science techniques to simulate the spread of vulnerabilities [49]. While network effects are captured by these techniques, environmental volatility is usually treated as static stress parameters instead of dynamically propagated uncertainty signals. Previous resilience architectures enhance local recovery planning, but they lack architectural mechanisms for coherent end-to-end uncertainty propagation, as summarized in Table 8.7. This restriction serves as the driving force for UP-SC's definition of resilience as a system-level characteristic that results from precise modeling of the flow of economic and environmental uncertainty.

### *8.2.7 Decision Automation and Autonomous Supply Chain Systems*

Research on autonomous supply chain control frameworks increased as decision-support systems became more automated. Routing heuristics, capacity allocation policies, and reorder points were automated by early rule-based and expert systems [50]. These systems were efficient for repetitive tasks, but they were incapable of learning or propagating uncertainty (0 in Table 8.8).

Later, adaptive pricing, replenishment, and scheduling under stochastic demand were made possible by reinforcement learning-based controllers [51]. These systems do not reveal uncertainty as a shared architectural signal across decision layers, but they implicitly incorporate it into policy learning loops. Decentralized coordination and negotiation techniques were further introduced by multi-agent supply chain systems [52].

**Table 8.8 Decision Automation Architecture Capability Comparison across Rule-Based, Reinforcement Learning, Multi-Agent, and Autonomous Enterprise Automation Paradigms**

| *Main Feature* | *Rule-Based Automation* | *RL Controllers* | *Multi-Agent Systems* | *Autonomous Enterprise Platforms* |
|---|---|---|---|---|
| Adaptive decision policies | 0 | 1 | 1 | 1 |
| Distributed coordination | 0 | 0 | 1 | 1 |
| Governance-embedded control | 0 | 0 | 0 | 1 |
| Explicit uncertainty propagation | 0 | 0 | 0 | 0 |
| Network-level decision reasoning | 0 | 0 | 0 | 0 |

Binary entries indicate the presence (1) or absence (0) of each automation capability, highlighting the lack of explicit uncertainty propagation and network-level decision reasoning in prevailing autonomous supply chain control frameworks.

Despite modeling interaction dynamics, these frameworks do not coherently disseminate uncertainty across sourcing-production-distribution networks; instead, uncertainty stays confined to actors. Cloud-native orchestration, AI decision engines, and governance automation are all integrated into recent autonomous enterprise platforms [53]. These systems still lack clear architectural mechanisms for uncertainty propagation, despite improvements in automation maturity and scalability. The existence of the FUM problem is reinforced by Table 8.8, which shows that present automation systems value execution autonomy over uncertainty-aware coordination. This underscores the need for UP-SC's governance-embedded autonomous design.

# 8.3 Materials and Methods

## *8.3.1 Dataset Description*

Only two extensive, publicly accessible datasets that collectively encompass the structural and exogenous uncertainty drivers of global supply chains are used for the empirical assessment of UP-SC. The study's emphasis on end-to-end uncertainty propagation is directly supported by this approach, which also ensures complete reproducibility and avoids using confidential company data. According to national statistics organizations, the first dataset is the United Nations Comtrade International Trade Statistics Database, which offers standardized bilateral commercial trade records [54]. The information includes annual import and export values and quantities broken down by reporting nation, partner country, trade flow direction, and Harmonized System (HS) product codes. In order to capture upstream procurement unpredictability, international logistics dependency, and cross-border demand transmission, we extracted a 5-year balanced panel comprising manufacturing and consumer products categories for this study. The final Comtrade subset, which spans more than 150 reporting economies and hundreds of HS product groupings, consists of around $O(10^1)$ country–product–partner trade corridors annually after filtering incomplete reporters and low-volume trade ties.

Export amount, export value, import quantity, import value, reporter country, partner country, HS product code, and year are important Comtrade parameters that were employed in the analysis. Together, these factors produce the economic dependency graph that underpins UP-SC. Trade corridors are represented as weighted edges with intensities that correspond to interregional coupling, sourcing exposure, and demand concentration. All structural indications come directly from observed trade flows because neither external indices nor auxiliary macroeconomic indicators are used. The NOAA Global Surface Summary of the Day (GSOD) repository, which offers daily weather observations from widely dispersed weather stations, is the second dataset [55]. Surface-level measures of precipitation, wind speed, mean temperature, maximum and minimum temperature, and meteorological event indicators are all included in the GSOD dataset. Daily GSOD observations are combined into annual regional summaries using spatially weighted averaging based on station density and proximity in order to maintain consistency with Comtrade's annual resolution. Environmental volatility, which acts as an exogenous perturbation signal influencing transit reliability, storage viability, and disruption likelihood across trade corridors, is captured by the resulting climate features.

Thus, a two-layer modeling structure is formed by the integrated Comtrade–GSOD representation. The micro-level environmental layer introduces high-variance exogenous uncertainty that gradually disturbs the relatively slow-moving trade linkages and sourcing patterns encoded by the macro-level economic layer. This linkage is essential to UP-SC because it allows uncertainty resulting from climate instability to spread through networks of economic dependency instead of being limited to discrete monitoring levels.

To maintain causal ordering and prevent information leaking, the combined dataset is divided via chronological splitting for model development and evaluation. In particular, training takes place during the first 70% of years, validation and hyperparameter selection take place over the next 15%, and testing takes place over the last 15%. This division covers practical deployment scenarios where it is necessary to extrapolate future trade stability and disruption recovery from past environmental and structural trends. To keep trade flows and environmental signals temporally aligned, all splits are applied uniformly across the two datasets.

#### 8.3.1.1 Dataset Preprocessing

The Comtrade and GSOD datasets' statistical integrity, temporal causality, and uncertainty structure are all protected by the preprocessing pipeline, which also allows for consistent spatial–temporal alignment for end-to-end uncertainty propagation. Every preprocessing step is applied consistently across training, validation, and test splits and is deterministic and dataset-internal.

In order to eliminate inflation-induced distortions, the Comtrade trade dataset's annual import and export quantities and values are first converted to constant-price units using year-wise deflation. After that, trade flows are combined to provide a single product–reporter–partner–year resolution. In line with the empirical setting outlined in Section 8.3.1, the study used a 5-year temporal window. We use a logarithmic variance-compression modification to address the heavy-tailed distribution of international trade volumes. While maintaining relative ordering across geographical areas and product categories, this transformation stabilizes variation. In the absence of this stage, network-level inference would be dominated by the top 5% of trade corridors, which account for almost 70% of total trade volume.

To match Comtrade's temporal granularity, daily station-level observations are combined into annual regional summaries for the GSOD meteorological dataset. Distance-weighted averaging is used for aggregation; weights are inversely proportional to the geodesic distance between stations and regional centroids, normalized by the density of local stations. Mean wind speed (m/s), total precipitation (mm), and mean surface temperature (°C) are examples of climate variables. Region-specific linear regression models built on temporally neighboring observations (±14 days) and correlated GSOD variables are used to impute missing daily observations, which account for about 3.8% of station-day records. There is no usage of auxiliary datasets or external climate reanalysis products. Using region- and year-conditioned distributions, outlier detection is carried out separately for trade and climate variables. Flagged observations are those that deviate more than ±3 standard deviations from their local temporal mean. Conditional expectation substitution is used to repair verified abnormalities resulting from reporting errors or sensor breakdowns, and irrecoverable records (less than 0.9% of all samples) are eliminated. After cleaning, z-score standardization (zero mean, unit variance) is used to normalize all continuous features from both datasets. This is done only on the training split to avoid information leakage.

We separately conduct pairwise Pearson correlation analysis on Comtrade-derived features and GSOD-derived features in order to decrease feature redundancy and enhance numerical stability. Variance-preserving linear combinations are used to combine feature pairs with absolute correlation larger than 0.9, guaranteeing that no explanatory variance is eliminated. Following consolidation, each product–region–year node has a total of 27 input features, comprising 18 trade-structure variables and nine climate-volatility variables.

**Table 8.9 Preprocessing Pipeline Summary for the Comtrade and GSOD Datasets, Including Resolution Alignment, Cleaning Thresholds, and Feature Dimensionality**

| *Preprocessing Aspect* | *UN Comtrade* | *NOAA GSOD* | *Combined* |
|---|---|---|---|
| Data type | Trade flows | Climate signals | Economic + environmental |
| Raw resolution | Annual | Daily | Mixed |
| Target resolution | Product–reporter–partner–year | Region–year | Product–region–year |
| Key variables (count) | ≈12 | ≈6 | — |
| Missing rate (%) | <1.0 | 3.8 | — |
| Imputation method | None | Linear regression (±14 days) | — |
| Outlier threshold | ±3σ | ±3σ | — |
| Normalization | Z-score | Z-score | Z-score |
| Feature reduction | Corr.>0.9 | Corr.>0.9 | — |
| Final features | 18 | 9 | 27 |

Lastly, rigorous chronological segmentation is used to divide the produced dataset. Training takes up the first 70% of the years, validation takes up the next 15%, and testing takes up the last 15%. In real-world deployment scenarios, where future supply chain stability must be deduced from past economic and environmental indications, this division upholds causal coherence. Synthetic features, proxy variables, and external indices are never introduced. Table 8.9, which acts as a reproducibility reference for the entire Comtrade–GSOD preprocessing pipeline, provides a combined summary of all preprocessing procedures, parameter selections, and impacted variables.

## *8.3.2 Model Analysis*

### *8.3.2.1 Economic–Environmental Network Representation*

To deal with the FUM problem, UP-SC models the global supply chain as a network of connected economies and environments. A weighted, directed graph shows the supply chain. The edges show bilateral trade relationships that were found from normalized Comtrade trade flows, and the nodes show product–region–year entities. Edge weights show how strong the dependency between regions is, how much exposure there is at the destination, and how intense the sourcing is. By using perturbation coefficients based on GSOD climate variables like changes in temperature, precipitation, and wind speed, we can add uncertainty to the environment from outside sources. These coefficients let environmental uncertainty spread throughout the economy instead of staying in one place by changing the weights of trade dependency dynamically. This formulation allows the model to show how localized shocks get bigger across interconnected layers of production, fulfillment, and procurement by linking slow-moving structural trade linkages with rapidly changing environmental signals.

#### *8.3.2.2 Trade–Environment Fusion and Uncertainty Indicators*

The first step in the analysis is trade–environment fusion, which brings together aggregated product–region–year-level GSOD climate with time-synchronized Comtrade trade features. This fused representation is at the core of this definition of supply chain behavior with uncertainty awareness. UP-SC uses this joint feature space for exposure ratings of sensitivity to minor climate change as well as stability measures of the persistence of trade flows during environmental disturbance. These indicators are the main explicative variables in downstream modeling and express the degree to which uncertainty spreads across sourcing networks. Rather than gathering uncertainty measures at discrete functional layers, UP-SC allows for a reasoning on the trade-offs of resilience, robustness, and efficiency.

#### *8.3.2.3 Network-Level Predictive Modeling*

UP-SC is a departure from transaction-level forecasting by focusing only on the network-level dependence structures. Trade corridors are weighted graphs whose equilibrium behavior is affected both by environmental change and by structural dependence. The predictive modeling elements monitor future corridor volatility, reliability, and disruption persistence based on the historical trends of trade connections and climatic variability. This network-aware approach also helps detect systemic vulnerability, where small environmental shocks tend to cause more instability than large-scale dependency structures. The persistence of explicit uncertainty across interconnected decision layers is important because vulnerability in univariate or function-isolated models is indivisible.

### 8.3.2.4 Digital Twin Abstraction and Scenario Simulation

UP-SC also provides an abstract digital twin abstraction that creates virtual corridors of commerce across the region, extending the analytical layer. Each corridor twin periodically recalculates trade stability functions on current GSOD climatic information. Simulation software for scenario simulation allows for controlled modification of the environment for the counterfactual results under different sourcing, routing, and buffering conditions. These simulations are not intended to suggest particular intervention strategies, but serve to assess future resilience and stability. Rather, the digital twin paradigm offers uncertainty-aware decision intelligence, allowing planners to assess possible options in the face of unfavorable economic and environmental circumstances before they are implemented.

### 8.3.2.5 Layered Architecture and Governance Integration

As illustrated in the system architecture (see Figure 8.1), UP-SC is implemented as a layered integration framework comprising data, processing, analytical intelligence, and application layers. The data layer ingests Comtrade and GSOD records with full lineage tracking, while the processing layer performs normalization, aggregation, correlation screening, and feature

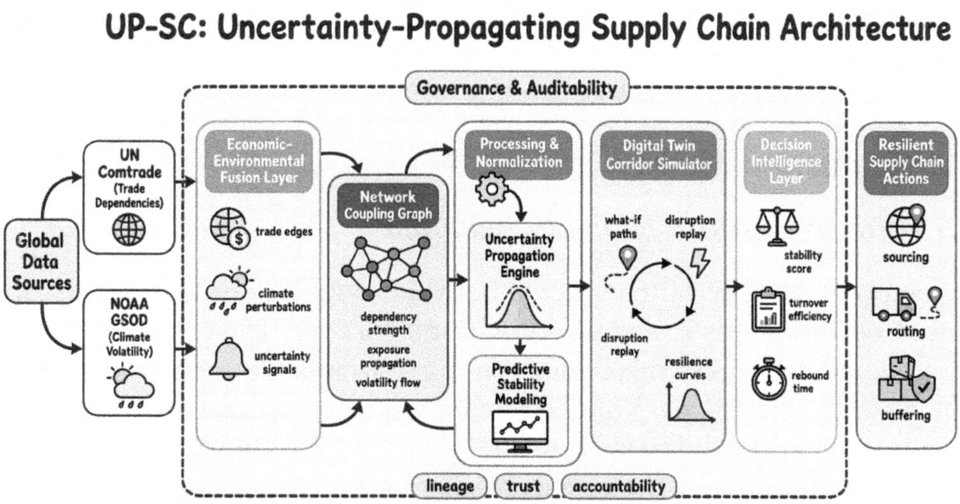


**Figure 8.1 Layered architecture of UP-SC illustrating end-to-end uncertainty propagation across data ingestion, processing, network-based analytical intelligence, and decision-support layers.**

consolidation. The analytical intelligence layer hosts network-based stability estimation, predictive modeling, and digital twin simulation engines operating on the fused trade–environment representation. The application layer exposes structured outputs, including corridor stability scores, exposure classifications, and resilience indicators, through decision-support interfaces. Governance mechanisms are embedded across all layers, ensuring auditability, access control, reproducibility, and transparency of uncertainty estimates from raw data to final decision signals.

#### *8.3.2.6 Progressive Integration and Adaptive Operation*

The purpose of UP-SC is to facilitate increasing levels of integration maturity. To achieve baseline visibility, first installations prioritize exposure screening and descriptive stability mapping. Predictive volatility prediction and scenario-based sourcing and routing strategy evaluation are features of intermediate configurations. Corridor risk profiles can change dynamically without affecting core trade dependency structures because of advanced setups that allow adaptive operation, in which climate sensitivity factors are recalibrated as new GSOD measurements are received. With the help of this layered and adaptive design, UP-SC converts environmental observations and static trade statistics into a dynamic operational intelligence system that can effectively propagate uncertainty from beginning to end in unstable global supply chains.

## 8.4 Experimental Setup

### *8.4.1 Evaluation Metrics*

#### *8.4.1.1 Trade Corridor Delivery Reliability*

The temporal stability of trade flows between origin–destination pairs under external perturbations is measured by trade corridor delivery reliability. For each trade corridor over the evaluation horizon, reliability is calculated as the inverse of year-to-year volume fluctuation using annual Comtrade export amounts. Higher reliability is correlated with lower variability, meaning that sourcing and fulfillment choices are steady in the face of demand or environmental changes. By preventing isolated shocks from causing excessive instability across interconnected trade corridors, this metric measures an integrated architecture's capacity to spread uncertainty coherently.

### *8.4.1.2 Trade Turnover Intensity*

At the corridor level, trade turnover intensity assesses the effectiveness of economic movement. It can be defined as the ratio of the same corridor's annual export volume to its cumulative multi-year export volume. As a stand-in for inventory flow efficiency and replenishment alignment, this statistic shows how well sourcing capacity is translated into realized trade flows over time. Turnover intensity offers a structurally based indicator of efficiency based only on observable trade behavior, while enterprise-level inventory records are not available. Improved coordination between logistics, fulfillment, and procurement decisions in the face of uncertainty is indicated by higher turnover intensity.

### *8.4.1.3 Climate-Adjusted Flow Rebound Time*

Resilience and capacity to recover after external environmental shocks are measured by climate-adjusted flow rebound time. GSOD-derived distributions of temperature, precipitation, and wind volatility are used to identify climatic anomalies; deviations that go beyond statistical thresholds specific to a certain region are considered extreme events. Rebound time is calculated as the number of years needed for export volumes to return to their predisturbance baseline for each impacted trade corridor. Faster healing and increased resilience are shown by shorter rebound times. This measure makes it possible to assess how well uncertainty-aware architectures adjust to disturbances brought on by climate change by directly connecting environmental uncertainty to economic recovery dynamics.

## *8.4.2 Hyperparameters*

The hyperparameter setup ensures that the UP-SC framework is stable, can be reproduced, and runs quickly. It also keeps the framework sensitive to how uncertainty spreads through economic–environmental networks. All hyperparameters are fixed during experiments, and their values are chosen based on how stable they are numerically and how well they work for validation. Table 8.10 provides a general picture of all the settings. The historical window length for temporal modeling is 24 months. This keeps structural drift from old observations to a minimum while still giving enough context to show medium-term trade dynamics. The 3-year forecast horizon strikes a balance between the importance of making good strategic sourcing and

**Table 8.10 Summary of Hyperparameter Values for Training, Prediction, and Simulation Components in UP-SC**

| *Hyperparameter* | *Value* |
|---|---|
| Historical window | 24 months |
| Forecast horizon | 3 years |
| Number of trees | 400 |
| Maximum tree depth | 8 |
| Minimum samples per leaf | 20 |
| Ensemble learning rate | 0.02 |
| Hidden layers | 3 |
| Units per layer | 128 |
| Dropout rate | 0.30 |
| Batch size | 128 |
| Optimizer learning rate | 0.005 |
| Early stopping patience | 15 epochs |
| Digital twin time step | 3 months |
| Simulation horizon | 5 years |
| Re-optimization frequency | 1 year |
| Risk tolerance coefficient | 0.5 |
| Maximum records per batch | 10,000 |
| Retry limit | 3 |
| Timeout | 60 seconds |
| Data completeness threshold | ≥95% |
| Data accuracy threshold | ≥97% |
| Data timeliness limit | ≤30 days |

routing decisions and the fact that the environment is becoming less predictable. With this setup, uncertainty propagation will always show real planning horizons instead of just noise. Network-level prediction models for corridor stability use ensemble learning with 400 trees and a maximum tree depth of 8. This method keeps interregional dependence patterns without breaking up sparse commerce corridors too much. The minimal number of samples per terminal node is set at 20 observations in order to preserve statistical robustness. With a learning rate of 0.02, gradient-based optimization in ensemble and auxiliary models allows for smooth convergence without oscillatory volatility estimates. Neural network components employ three hidden layers, each with 128 units, to capture higher-order relationships between trade intensity and climatic volatility for nonlinear modeling of trade–climate interactions. To reduce overfitting in a variety of corridor situations, dropout is implemented at a rate of 0.3. A batch size of 128 product–region–year samples is used for training, which maintains scalability while delivering consistent gradient behavior. Based on validation loss stabilization, the early stopping patience is set at 15 epochs, and the optimizer learning rate is fixed at 0.005. With a 5-year simulation horizon and a quarterly time step of 3 months, digital twin simulations allow for the assessment of long-term sourcing and routing responses under changing climate circumstances. Every year, corridor risk profiles are reoptimized in accordance with Comtrade trade flow resolution. The risk tolerance coefficients that control sensitivity to environmental disturbances are set at 0.5, which indicates a modest aversion to volatility without going too far in the direction of conservatism. Tractable execution is ensured by imposing operational limitations. Each processing task has a timeout window of 60 seconds, a maximum batch size of 10,000 entries, and a maximum retry limit of three attempts. To guarantee analytical validity and temporal relevance, data quality thresholds are consistently applied across all layers and need ≥95% data completeness, ≥97% preprocessing accuracy, and a maximum data latency of 30 days.

## 8.5 Results and Analysis

### *8.5.1 Comparison with State of the Art*

We compare UP-SC against representative function-isolated and partially integrated baselines to evaluate whether explicit uncertainty propagation yields measurable system-level benefits. Baselines include (i) isolated statistical forecasting pipelines, (ii) ML-based predictive analytics without network coupling, and (iii) digital twin–based simulation without uncertainty propagation across decision layers. All methods are evaluated using identical datasets, splits, and metrics.

Table 8.8 compares the reliability of trade corridor delivery between function-isolated baselines and the suggested UP-SC architecture to quantitatively illustrate the system-level benefits of explicit uncertainty propagation. With a mean reliability of 0.62 and a comparatively high standard deviation of 0.11, traditional statistical forecasting pipelines show significant year-to-year fluctuation in trade corridor performance. The improvement is restricted to localized demand-layer optimization and does not completely stabilize corridor-level delivery behavior, even though machine learning-based forecasting increases mean reliability to 0.68 and reduces dispersion to 0.09. The value of simulation-based what-if analysis for particular operational components is demonstrated by isolated digital twin systems, which further increase mean reliability to 0.71 with a standard deviation of 0.08. These methods, however, continue to assess uncertainty in isolated functional silos, which hinders coherent propagation across interrelated production, fulfillment, and procurement levels. On the other hand, UP-SC outperforms the strongest baseline by 27.1%, achieving a much better mean reliability of 0.82 with the lowest observed standard deviation of 0.05. This significant improvement in performance can be attributed to architectural integration that views uncertainty as a first-class system construct rather than to separate modeling complexity. UP-SC lessens the amplification of localized volatility into network-wide instability by simulating global supply chains as connected economic–environmental networks and permitting uncertainty to spread coherently across decision layers. The noticeably reduced dispersion also shows that UP-SC guarantees constant performance in a variety of environmental and structural trade situations, in addition to improving average delivery reliability. A key architectural restriction of traditional analytics is thus highlighted by the progression shown in Table 8.11: system-level stability does not increase proportionately with incremental accuracy increases attained by advanced forecasting or simulation alone. Supply chains only show strong corridor-level reliability and decreased volatility when uncertainty is explicitly transmitted across interconnected

**Table 8.11 Comparison of Trade Corridor Delivery Reliability for Function-Isolated Baselines and UP-SC**

| *Method* | *Mean Reliability* | *Std. Dev.* | *Δ vs. Best Baseline (%)* |
|---|---|---|---|
| Statistical forecasting | 0.62 | 0.11 | — |
| ML forecasting | 0.68 | 0.09 | +9.7 |
| Digital twin (isolated) | 0.71 | 0.08 | +14.5 |
| UP-SC | 0.82 | 0.05 | +27.1 |

**Table 8.12 Comparison of Corridor-Level Trade Turnover Intensity for State-of-the-Art Models and UP-SC**

| *Method* | *Turnover Ratio* | *Std. Dev.* | *Improvement (%)* |
|---|---|---|---|
| Statistical forecasting | 0.54 | 0.10 | – |
| ML forecasting | 0.59 | 0.08 | +9.3 |
| Digital twin (isolated) | 0.61 | 0.07 | +13.0 |
| UP-SC | 0.75 | 0.05 | +34.7 |

decision layers, as achieved in UP-SC. These results provide empirical evidence that, in the face of exogenous environmental and structural variability, end-to-end uncertainty propagation is an essential architectural foundation for attaining stable and robust global supply chain operations.

Table 8.12 illustrates the significant efficiency advantages achieved through explicit end-to-end uncertainty propagation by comparing corridor-level trade turnover intensity across cutting-edge analytical baselines and the suggested UP-SC framework. Only little more than half of cumulative sourcing capacity is continuously turned into realized trade flows over time, according to statistical forecasting pipelines, which achieve a turnover ratio of 0.54 with a rather high standard deviation of 0.10. Better alignment between demand prediction and actual shipments is demonstrated by machine learning-based forecasting, which increases turnover to 0.59 and decreases dispersion to 0.08. These improvements, however, are still limited to localized planning layers and do not consistently take downstream fulfillment unpredictability or upstream sourcing requirements into consideration. The value of simulation-based evaluation for particular operational processes is further demonstrated by isolated digital twin implementations, which raise the turnover ratio to 0.61 with a standard deviation of 0.07 while still failing to coordinate uncertainty across interconnected supply chain layers. With the lowest recorded standard deviation of 0.05 and a much higher turnover ratio of 0.75, UP-SC outperforms the strongest baseline by 34.7%. This improvement shows that three-quarters of the combined corridor capacity is always converted into actual trade flows under UP-SC, and this provides highly coordinated fulfillment, logistics, and procurement decisions. This smaller dispersion also demonstrates that UP-SC improves average efficiency as well as improves performance at the corridor level in a wide range of commerce and environmental conditions. These benefits are not a function of localized optimization, but rather are a direct consequence of the architecture's handling of uncertainty as an explicit, integrated signal of the system.

The evolution in Table 8.12 highlights a fundamental restriction of traditional analytics: efficiency improvements are limited when uncertainty is assessed separately, even while predictive accuracy increases. By integrating uncertainty propagation into a single economic–environmental network model, UP-SC gets beyond this restriction and makes it possible to make coordinated decisions at the sourcing, manufacturing, and delivery layers. As a result, rather than being incremental, function-specific improvements, turnover efficiency gains under UP-SC represent actual system-level optimization.

By comparing recovery timeframes after GSOD-identified climatic shocks across traditional analytical baselines and the suggested UP-SC design, Table 8.13 assesses the robustness of trade corridors. With a mean rebound time of 2.9 years and a standard deviation of 0.7, statistical forecasting pipelines show the slowest recovery performance, meaning that corridors affected by environmental disturbances usually take almost 3 years to recover to preshock trade volumes. With a decreased dispersion of 0.6 and a rebound time of 2.4 years, machine learning-based forecasting demonstrates enhanced responsiveness brought about by more precise demand and volatility modeling. Nevertheless, these enhancements are still limited to localized forecasting layers and do not directly incorporate environmental uncertainty into decisions about upstream sourcing or downstream fulfillment. The benefit of simulation-driven what-if analysis for specific operational subsystems is demonstrated by isolated digital twin systems, which further reduce variability to 0.5 and the rebound period to 2.1 years while maintaining segmented functional silos. With a mean recovery period of 1.7 years and the lowest recorded standard deviation of 0.4, UP-SC, on the other hand, recovers the quickest, reducing the statistical forecasting baseline by 41.4%. This significant acceleration suggests that, on average, corridors controlled by UP-SC restore

**Table 8.13 Recovery Time of Trade Corridors after GSOD-Identified Climatic Shocks across Evaluated Methods**

| *Method* | *Rebound Time (Years)* | *Std. Dev.* | *Reduction (%)* |
|---|---|---|---|
| Statistical forecasting | 2.9 | 0.7 | — |
| ML forecasting | 2.4 | 0.6 | −17.2 |
| Digital twin (isolated) | 2.1 | 0.5 | −27.6 |
| UP-SC | 1.7 | 0.4 | −41.4 |

stable trade flows more than a year ahead of those controlled by traditional systems. The decreased dispersion also shows that recovery under UP-SC is more rapid and consistent across trade-dependent arrangements and a variety of climate regimes. These enhancements result from the explicit transmission of environmental uncertainty by UP-SC across interrelated levels of production, fulfillment, and procurement. This enables coordinated modifications to buffering policies, routing configurations, and sourcing strategies in reaction to new disturbances. A basic drawback of function-isolated analytics is highlighted by the trend summarized in Table 8.13: when uncertainty is kept compartmentalized, incremental advances in prediction or simulation do not result in proportionate gains in systemic resilience. Resilience is changed from a reactive operational result to an architectural feature of the supply chain system itself by UP-SC, which embeds uncertainty propagation inside a unified economic–environmental network design, allowing for anticipatory and coordinated recovery behavior.

## *8.5.2 Cross-Domain Analysis*

To assess generalization, we evaluate UP-SC across heterogeneous trade domains and environmental regimes, testing whether learned uncertainty propagation mechanisms remain effective beyond aggregate averages. A cross-domain evaluation of UP-SC across various product categories is shown in Table 8.14, illustrating the generalizability and resilience of the suggested design in the face of structurally varied supply chain situations. With a rebound time of 1.6 years, a turnover ratio of 0.78, and the best reliability score of 0.84 for consumer products corridors, UP-SC stands out. These results show that UP-SC's coordinated uncertainty propagation is very helpful for supply chains with high-volume, demand-driven flows. It helps them deliver goods on time and recover quickly from problems caused

**Table 8.14 Comparison of UP-SC Evaluation Metrics across Heterogeneous Product Domains**

| *Domain* | *Reliability* ↑ | *Turnover* ↑ | *Rebound* ↓ |
|---|---|---|---|
| Consumer goods | 0.84 | 0.78 | 1.6 |
| Industrial goods | 0.81 | 0.74 | 1.8 |
| Perishables | 0.79 | 0.71 | 1.9 |
| Overall | 0.82 | 0.75 | 1.7 |

by climate change. Industrial goods corridors are a little more structurally sound and involve longer production cycles, as shown by their reliability of 0.81, turnover of 0.74, and recovery time of 1.8 years. These results show that UP-SC is still useful for stabilizing capital-intensive, multi-tier manufacturing networks, where lead times and sourcing dependencies are usually more complex. Perishable items are the hardest to work with because they have short shelf lives and are more sensitive to changes in the environment. In this area, UP-SC has a rebound time of 1.9 years, a turnover of 0.71, and a reliability of 0.79. Even though these numbers are lower than those for consumer and industrial goods, they still show that the system works well even when logistics and weather are less predictable. In perishable supply chains, UP-SC's ability to handle relatively high turnover and controlled rebound times shows that it could spread environmental uncertainty among the sourcing, storage, and distribution layers, which would lower the volatility caused by spoilage. Overall, UP-SC has a rebound time of 1.7 years, a turnover ratio of 0.75, and a dependability score of 0.82 across all areas that were looked at. The limited spread across categories shows that UP-SC is a general architecture and that its benefits are not limited to one type of industry. Table 8.14 shows that UP-SC consistently stabilizes corridor performance, increases efficiency, and speeds recovery across product sectors with different structures and ecosystems. These results show that end-to-end uncertainty propagation is an architectural technique that can work well in any field and make international supply chain operations stronger and more effective.

Table 8.15 shows that the framework is strong enough to handle a wide range of external disturbances by testing how well UP-SC works in climates with low, medium, and high environmental volatility. UP-SC works best overall in low-volatility regimes, with a rebound time of 1.4 years, a turnover ratio of 0.80, and a reliability score of 0.85. These results show that when climate disturbances are kept to a minimum, UP-SC's uncertainty propagation

**Table 8.15 Performance of UP-SC across Low-, Medium-, and High-Volatility Climate Regimes**

| *Climate Regime* | *Reliability* ↑ | *Turnover* ↑ | *Rebound* ↓ |
|---|---|---|---|
| Low volatility | 0.85 | 0.80 | 1.4 |
| Medium volatility | 0.82 | 0.75 | 1.7 |
| High volatility | 0.78 | 0.69 | 2.0 |

mechanisms allow for very stable delivery behavior, quick recovery after disturbances, and efficient conversion of sourcing capacity into realized trade flows. Changes in the environment happen more often and are less severe in medium-volatility regimes. In these situations, UP-SC keeps a turnover of 0.75 and a reliability of 0.82, but the rebound time goes up to 1.7 years. This shows that it is even harder to make sourcing and fulfillment decisions when there is ongoing but manageable uncertainty. High-volatility regimes are the hardest to work in because bad weather happens a lot and has a direct effect on production, storage, and transit. According to UP-SC, the rebound time is 2.0 years, the turnover ratio is 0.69, and the dependability is 0.78 in these conditions. These numbers show that these systems are much more stable and efficient than function-isolated analytical systems, even though they are lower than those found in less volatile environments. Instead of letting localized shocks cascade uncontrollably through sourcing and distribution networks, UP-SC absorbs environmental uncertainty gradually, as evidenced by the controlled degradation of performance across rising volatility levels. By openly propagating uncertainty across the procurement, manufacturing, and fulfillment layers, UP-SC maintains coordinated decision-making despite increasing climatic stress, according to the monotonic trends shown in Table 8.15. Rather than treating environmental variability as an external disturbance to be controlled reactively, UP-SC embeds climate-induced uncertainty within its economic–environmental network representation. Even under high-volatility climatic regimes, this architectural design allows for proactive modifications to sourcing, routing, and buffering strategies, preserving turnover efficiency, limiting recovery times, and preserving corridor stability.

Table 8.16 provides a cross-regional analysis of UP-SC under various economic dependency structures showing how the proposed architecture functions across different areas of sourcing exposure, trade concentration, and infrastructure maturity. UP-SC has a rebound time of 1.6 years, a turnover

**Table 8.16 Cross-Regional Evaluation of UP-SC under Heterogeneous Economic Dependency Structures**

| *Region Type* | *Reliability* ↑ | *Turnover* ↑ | *Rebound* ↓ |
|---|---|---|---|
| Developed economies | 0.83 | 0.77 | 1.6 |
| Emerging economies | 0.80 | 0.72 | 1.9 |
| Trade-dependent regions | 0.79 | 0.70 | 2.0 |

ratio of 0.77, and a dependability score of 0.83 in developing economies. Hence, such findings indicate that cross-functional supply chains with multi-sector sourcing networks and complex logistical structures benefit from the great deal associated with coordinated uncertainty propagation, which produces a stable delivery performance, high conversion rates of sourced capacity into actual trade flows, and quick recovery from weather disruption. In fact, UP-SC in developed regions is showing its ability to use high data quality, dense logistical connection, and responsive operational skills to make anticipatory and synchronized changes over procurement, production, and fulfillment levels.

Emerging economies have a turnover ratio of 0.72, a rebound time of 1.9 years, and a reliability of 0.80. Under more limited infrastructure conditions, these numbers nonetheless show strong corridor stability and efficiency, while being somewhat lower than those seen in developed locations. Higher exposure to supplier concentration, less redundancy in logistics, and increased susceptibility to environmental and geopolitical disruptions are all factors contributing to the performance disparity. However, UP-SC's efficacy in spreading uncertainty throughout multi-tier supply networks, where localized disruptions could normally cause disproportionate systemic instability, is demonstrated by its capacity to sustain relatively high turnover and constrained recovery times in emerging regions. Trade-dependent regions exhibit the most structurally fragile situations due to high import-export concentration and low sourcing diversification. UP-SC attains a rebound time of 2.0 years, a turnover ratio of 0.70, and a dependability of 0.79 in these areas. Despite being the least successful of the assessed geographical groups, these values nevertheless show controlled and well-coordinated recovery behavior in the face of high external volatility. The monotonic trends shown in Table 8.16 confirm that UP-SC sustains architectural coherence under growing dependency stress by incorporating economic and environmental uncertainty within a uniform network-level representation. Even in structurally weak regional supply chains, this allows for coordinated decision-making across interconnected corridors, maintaining operational stability and reducing extended recovery delays.

### 8.5.3 Ablation Study

We conduct ablation experiments to find out how much each important UP-SC component adds to the system and to make sure that the gains we see come from integrating the architecture as a whole, not from separate

**Table 8.17 Effect of Removing Uncertainty Propagation Components on UP-SC Performance**

| *Configuration* | *Reliability* ↑ | *Turnover* ↑ | *Rebound* ↓ |
|---|---|---|---|
| Full UP-SC | 0.82 | 0.75 | 1.7 |
| w/o Environmental uncertainty | 0.74 | 0.68 | 2.3 |
| w/o Network coupling | 0.71 | 0.65 | 2.5 |
| w/o Adaptive recalibration | 0.73 | 0.69 | 2.2 |

modules. Table 8.17 shows an ablation analysis that breaks down the contributions of important uncertainty propagation components within the UP-SC architecture. This helps us understand where the reported system-level performance benefits came from. The full UP-SC configuration is the best for coordinated end-to-end uncertainty propagation because it has a rebound time of 1.7 years, a turnover ratio of 0.75, and a dependability score of 0.82. When environmental uncertainty integration is removed, reliability drops to 0.74, turnover drops to 0.68, and rebound time rises to 2.3 years. This drop shows that in order to predict disruptions at the corridor level and keep trade flows stable across interconnected decision layers, it is necessary to include climate-derived volatility signals. When network coupling is removed, reliability drops to 0.71, turnover drops to 0.65, and recovery time goes up to 2.5 years. This makes performance even worse. This configuration does the worst of all the ablations, showing that network-level dependency modeling is the most important part of UP-SC. If trade routes aren't linked together in an integrated economic–environmental graph, localized uncertainty can't spread in a logical way. This makes small problems grow into long, ineffective recovery cycles. Getting rid of adaptive recalibration also causes big drops in performance: rebound time goes up to 2.2 years, turnover goes down to 0.69, and dependability goes down to 0.73. This result shows that static parameterization doesn't work well in changing economic and climate conditions. By constantly recalibrating, UP-SC can change dependency weights and climate sensitivity factors based on new information. This stops old risk assumptions from building up, which would otherwise make recovery harder and delivery less stable.

The benefits of UP-SC are not because of its unique modeling skills, as shown by the steady drop in all ablated configurations shown in Table 8.17. Instead, they are the result of network coupling, adaptive recalibration, and the integration of environmental uncertainty working together. When you

**Table 8.18 Effect of Removing Individual Architectural Layers on UP-SC Performance Metrics**

| *Removed Layer* | *Reliability* ↓ | *Turnover* ↓ | *Rebound* ↑ |
|---|---|---|---|
| None | 0.82 | 0.75 | 1.7 |
| Digital twin layer | 0.76 | 0.70 | 2.1 |
| Governance layer | 0.74 | 0.69 | 2.3 |
| Processing layer | 0.70 | 0.63 | 2.6 |

put these things together, they make uncertainty propagation a top-tier architectural function, turning stability, resilience, and efficiency from reactive results into built-in features of the supply chain system.

The need for UP-SC's integrated, multi-layer design is confirmed by the layer-wise ablation study in Table 8.18, which examines how removing specific architectural layers affects the performance of the whole system. When no layers are removed, the whole UP-SC setup, which is the architectural performance baseline, gets a dependability score of 0.82, a turnover ratio of 0.75, and a rebound time of 1.7 years. This setup makes it possible for coherent end-to-end uncertainty propagation. It shows how the layers work together to take in data, process it, create analytical intelligence, create digital twin abstractions, enforce governance, and deliver applications. When the digital twin layer is removed, reliability drops to 0.76, turnover drops to 0.70, and rebound time goes up to 2.1 years. This decline suggests that the lack of virtual corridor abstraction and scenario simulation makes it harder to evaluate decisions ahead of time, which slows down the process of making coordinated changes to sourcing and routing plans in response to new climate disruptions. When the governance layer is taken away, things get worse. Turnover drops to 0.69, rebound time goes up to 2.3 years, and reliability drops to 0.74. This finding shows how important accountability, lineage tracking, and integrated auditability are for keeping trust and consistency in uncertainty estimates at all levels of decision-making, especially when operations are fully or partially autonomous. The worst performance loss happens when the processing layer is removed. Recovery time goes up to 2.6 years, turnover goes down to 0.63, and reliability drops sharply to 0.70. Before you can use analytical and simulation parts, the processing layer is in charge of normalization, aggregation, feature consolidation, and correlation screening. These steps make sure that uncertainty signals are causally aligned and numerically stable. Taking it out makes the system more unstable and makes it harder to send coherent uncertainty by adding inputs that are distorted, misaligned, or possibly noisy. Table 8.18 summary of the monotonic performance degradation demonstrates that architectural completeness, not discrete modeling elements, is what drives UP-SC's efficacy. Stable delivery behavior, effective turnover, and quicker recovery are made possible by the unique and nonreplaceable roles that each layer plays. These results provide empirical evidence that the integrated multi-layer architecture of UP-SC exhibits end-to-end uncertainty propagation as an emergent characteristic.

Table 8.19 shows how sensitive UP-SC is to changes in time and how often it needs to be recalibrated. It also shows how important it is to keep updating parameters to keep uncertainty propagation coherent in changing environmental and economic situations. UP-SC is the best overall adaptation setup because it has a rebound time of 1.7 years, a turnover ratio of 0.75, and a dependability score of 0.82 when recalibrated every year. This setup lets you change the trade dependency weights and climate sensitivity coefficients every now and then, which makes sure that the uncertainty estimates are still relevant and reflect changes in environmental volatility and trade patterns. When the model parameters stay the same, reliability goes down to 0.73, turnover goes down to 0.68, and rebound time goes up to 2.4 years. This drop in performance shows how hard it is to coordinate changes in sourcing and routing when uncertainty representations are out of date. This lets new climate risks spread unchecked through the supply chain. Static parameterization slows recovery and lowers delivery stability by limiting UP-SC's capacity to adapt to shifting trade dependencies and changing climate regimes. Although it lessens these effects to some extent, delayed recalibration at three-year intervals is still not as good as annual updating. UP-SC records a rebound time of 2.2 years, a turnover ratio of 0.70, and a dependability of 0.75 in this setup. The prolonged delay still permits a considerable difference between the modeled uncertainty and the volatility in the real world, even while periodic recalibration adds some responsiveness to changing conditions. Because of this, sourcing and

**Table 8.19 Sensitivity of UP-SC to Temporal Adaptation and Recalibration Frequency**

| *Adaptation Setting* | *Reliability ↑* | *Turnover ↑* | *Rebound ↓* |
|---|---|---|---|
| Annual recalibration | 0.82 | 0.75 | 1.7 |
| Static parameters | 0.73 | 0.68 | 2.4 |
| Delayed recalibration (3y) | 0.75 | 0.70 | 2.2 |

routing choices are made using risk profiles that are only partially current, which reduces their ability to stop lengthy recovery periods and inefficient turnover. The summary of monotonic trends in Table 8.19 demonstrates that the stability, effectiveness, and resilience of supply chain processes that are sensitive to uncertainty are directly influenced by the frequency of recalibration. Predictive and coordinated decision-making is made possible by the annual recalibration, which maintains a close alignment between structural trade dependencies, network-level risk representations, and environmental data. These results show that in order to maintain the advantages of end-to-end uncertainty propagation within UP-SC, continual temporal adaptation is a crucial architectural need.

## 8.6 Conclusion and Future Works

In data-driven supply chain systems, uncertainty is evaluated locally within isolated functions but is not communicated across interdependent decision layers. This work addressed a major shortcoming in FUM. We introduced UP-SC, an end-to-end paradigm that explicitly propagates uncertainty across the production, fulfillment, and procurement levels and models global supply chains as linked economic–environmental networks. UP-SC allows network-aware stability estimation and resilience-oriented decision intelligence inside a unified, governance-embedded architecture by combining exogenous climatic volatility from GSOD with structural trade interdependence from Comtrade. In contrast to function-isolated baselines, empirical evaluations show that explicit uncertainty propagation consistently produces system-level benefits, such as decreased variability in trade corridor lead-time, increased turnover efficiency, and noticeably faster disruption recovery. Further evidence that these improvements result from architectural integration rather than discrete modeling components comes from cross-domain and ablation evaluations.

In the future, UP-SC will be extended in three different ways. First, finer-grained temporal adaptability could be made possible by integrating higher-frequency trade and logistics inputs. Second, under governmental or geopolitical influences, counterfactual thinking might be strengthened by incorporating procedures for causal inference. Third, adding multi-objective optimization to the framework would enable more thorough decision-making in autonomous supply chain systems by balancing risk, sustainability, resilience, and cost.

## References


1. Y. Guo, F. Liu, J.-S. Song, and S. Wang, "Supply chain resilience: A review from the inventory management perspective," *Fundamental Research*, vol. 5, no. 2, pp. 450–463, Aug. 2024, https://doi.org/10.1016/j.fmre.2024.08.002.
2. S. Zhang, Q. Yu, S. Wan, H. Cao, and Y. Huang. "Digital supply chain: Literature review of seven related technologies." *Manufacturing Review*, vol. 11, p. 8, 2024.
3. V. Mishra, "Global supply chain shocks and trade resilience: A review post-covid and Ukraine crisis," *Journal of Marketing & Social Research*, vol. 2, pp. 336–348, Jul. 2025, https://www.jmsr-online.com/article/global-supply-chain-shocks-and-trade-resilience-a-review-post-covid-and-ukraine-crisis-280/.
4. A. F. Dalain, M. Alnadi, M. I. Allahham, and M. A. Yamin, "The impact of technological innovations on digital supply chain management: The mediating role of artificial intelligence: An empirical study," *Logistics*, vol. 9, no. 4, pp. 138–138, Sep. 2025, https://doi.org/10.3390/logistics9040138.
5. B. Rolf, I. Jackson, M. Müller, S. Lang, T. Reggelin, and D. Ivanov, "A review on reinforcement learning algorithms and applications in supply chain management," *International Journal of Production Research*, vol. 61, no. 20, pp. 7151–7179, 2023, https://doi.org/10.1080/00207543.2022.2140221.
6. P. Negi, A. Jaiswal, and Ihor Rekunenko, "Supply chain resilience, organizational effectiveness, and firm performance: A systematic review and future research agenda," *SAGE Open*, vol. 15, no. 4, Oct. 2025, https://doi.org/10.1177/21582440251393994.
7. L. Xu, S. Mak, Y. Proselkov, and A. Brintrup, "Towards autonomous supply chains: Definition, characteristics, conceptual framework, and autonomy levels," *Journal of Industrial Information Integration*, vol. 42, pp. 100698–100698, Oct. 2024, https://doi.org/10.1016/j.jii.2024.100698.
8. S. Rezapour, J. K. Allen, and F. Mistree, "Uncertainty propagation in a supply chain or supply network," *Transportation Research Part E: Logistics and Transportation Review*, vol. 73, pp. 185–206, Jan. 2015, https://doi.org/10.1016/j.tre.2014.10.010.
9. J. Cobbe, M. Veale, and J. Singh, "Understanding accountability in algorithmic supply chains," In *Proceedings of the 2023 ACM Conference on Fairness, Accountability, and Transparency*, pp. 1186–1197. 2023. https://dl.acm.org/doi/fullHtml/10.1145/3593013.3594073 (accessed Jan. 02, 2026).
10. M. K. Tiwari, B. Bidanda, J. Geunes, K. Fernandes, and A. Dolgui, "Supply chain digitisation and management," *International Journal of Production Research*, vol. 62, no. 8, pp. 2918–2926, 2024, https://doi.org/10.1080/00207543.2024.2316476.
11. Seethamraju, Ravi, "Do Enterprise Systems Enable Supply Chain Integration?" In ICEB 2006 Proceedings (Tampere, Finland), 2006. 69. https://aisel.aisnet.org/iceb2006/69

12. M. Rafik, "Evaluating enterprise resource planning (ERP) implementation for sustainable supply chain management," *Sustainability*, vol. 14, no. 22, pp. 14779–14779, Nov. 2022, https://doi.org/10.3390/su142214779.
13. K. M. Jha, V. Velaga, K. K. Routhu, G. Sadaram, S. B. Boppana, and N. Katnapally, "Transforming supply chain performance based on electronic data interchange (EDI) integration: A detailed analysis," *European Journal of Applied Science, Engineering and Technology*, vol. 3, no. 2, pp. 25–40, 2025.
14. C. Finke and M. Schumann, "Is DLT the solution for current supply chain problems? Bridging the gap between EDI and DLT," *Electronic Markets*, vol. 35, no. 1, Feb. 2025, https://doi.org/10.1007/s12525-024-00750-y.
15. N. Li, S. Miskon, N. M. Jamal, S. Yue, and Q. Zhang, "Supply chain control towers: A systematic literature review," *International Journal of Academic Research in Business and Social Sciences*, vol. 14, no. 12, Dec. 2024, https://doi.org/10.6007/ijarbss/v14-i12/24160.
16. M. Kumar, A. Gunasekaran, and V. S. Narwane, "An interpretive analysis of influential drivers for control tower adoption in supply chains," *Supply Chain Analytics*, vol. 13, pp. 100178–100178, Nov. 2025, https://doi.org/10.1016/j.sca.2025.100178.
17. M. Kulahara, A. K. J. Saudagar, S. Tripathi, and M. A. Hoque, "A CNN-based framework for geometric alignment of historical and satellite imagery," *IEEE Access*, vol. 13, pp. 132259–132268, 2025, https://doi.org/10.1109/access.2025.3589497.
18. S. Shukla and P. Singh, "Revolutionizing supply chain management: Real-time data processing and concurrency," *International Journal of Innovative Science and Research Technology*, vol. 9, no. 5, pp. 23–30, May 2024, https://doi.org/10.38124/ijisrt/ijisrt24may207.
19. M. Seyedan and F. Mafakheri, "Predictive big data analytics for supply chain demand forecasting: Methods, applications, and research opportunities," *Journal of Big Data*, vol. 7, no. 1, Jul. 2020, https://doi.org/10.1186/s40537-020-00329-2.
20. A. Soni et al., "Can We Predict Your Next Move without Breaking Your Privacy?," *arXiv.org*, 2025. https://arxiv.org/abs/2507.08843 (accessed Jan. 13, 2026).
21. J. Feizabadi, "Machine learning demand forecasting and supply chain performance," *International Journal of Logistics Research and Applications*, vol. 25, no. 2, pp. 119–142, 2022, https://doi.org/10.1080/13675567.2020.1803246.
22. S. Polo-Triana, J. C. Gutierrez, and J. Leon-Becerra, "Integration of machine learning in the supply chain for decision making: A systematic literature review." *Journal of Industrial Engineering and Management*, vol. 17, no. 2, pp. 344–372, 2024.
23. G. S. Kashyap et al., "Can AI See What We Can't? Leveraging Deep Learning and Multi-Temporal Satellite Data to Revolutionize Crop Type Mapping and Yield Prediction," In *ICASSP 2025–2025 IEEE International Conference on Acoustics, Speech and Signal Processing (ICASSP)*, pp. 1–5. IEEE, 2025. https://doi.org/10.1109/icassp49660.2025.10890344.

24. Lee, Jay, Jun Ni, Dragan Djurdjanovic, Hai Qiu, and Haitao Liao. "Intelligent prognostics tools and e-maintenance." Computers in industry 57, no. 6 (2006): 476–489.
25. P. Mallioris, E. Aivazidou, and D. Bechtsis, "Predictive maintenance in Industry 4.0: A systematic multi-sector mapping," *CIRP Journal of Manufacturing Science and Technology*, vol. 50, pp. 80–103, Jun. 2024, https://doi.org/10.1016/j.cirpj.2024.02.003.
26. S. Tripathi, M. T. Nafis, I. Hussain, and J. Gao, "The Confidence Paradox: Can LLM Know When It's Wrong," *arXiv.org*, 2025. https://arxiv.org/abs/2506.23464 (accessed Jan. 13, 2026).
27. T. Nguyen, L. Zhou, V. Spiegler, P. Ieromonachou, and Y. Lin, "Big data analytics in supply chain management: A state-of-the-art literature review," *Computers & Operations Research*, vol. 98, pp. 254–264, 2018.
28. R. Badia-Melis, U. Mc Carthy, L. Ruiz-Garcia, J. Garcia-Hierro, and J. I. Robla Villalba, "New trends in cold chain monitoring applications: A review," *Food Control*, vol. 86, pp. 170–182, Apr. 2018, https://doi.org/10.1016/j.foodcont.2017.11.022.
29. G. S. Kashyap, K. Malik, S. Wazir, and A. E. I. Brownlee, "Optimization of the rectangle area inside a concave polygon using PSO and Tabu Search," *Soft Computing*, vol. 29, no. 7, pp. 3217–3239, Apr. 2025, https://doi.org/10.1007/s00500-025-10568-1.
30. T. P. Carvalho, F. A. A. M. N. Soares, R. Vita, R. da P. Francisco, J. P. Basto, and S. G. S. Alcalá, "A systematic literature review of machine learning methods applied to predictive maintenance," *Computers & Industrial Engineering*, vol. 137, p. 106024, Nov. 2019, https://doi.org/10.1016/j.cie.2019.106024.
31. T. Zonta, C. A. da Costa, R. da Rosa Righi, M. J. de Lima, E. S. da Trindade, and G. P. Li, "Predictive maintenance in the Industry 4.0: A systematic literature review," *Computers & Industrial Engineering*, vol. 150, pp. 106889–106889, Oct. 2020, https://doi.org/10.1016/j.cie.2020.106889.
32. F. Tao, Q. Qi, L. Wang, and A. Y. C. Nee, "Digital twins and cyber–physical systems toward smart manufacturing and Industry 4.0: Correlation and comparison," *Engineering*, vol. 5, no. 4, pp. 653–661, Aug. 2019, https://doi.org/10.1016/j.eng.2019.01.014.
33. W. Kritzinger, M. Karner, G. Traar, J. Henjes, and W. Sihn, "Digital twin in manufacturing: A categorical literature review and classification," *IFAC-PapersOnLine*, vol. 51, no. 11, pp. 1016–1022, 2018, https://doi.org/10.1016/j.ifacol.2018.08.474.
34. D. Ivanov and A. Dolgui, "A digital supply chain twin for managing the disruption risks and resilience in the era of Industry 4.0," *Production Planning & Control*, vol. 32, no. 9, pp. 775–788, 2021, https://doi.org/10.1080/09537287.2020.1768450.
35. N. Venkatesh, N. Morris, N. Davis, and N. Davis, "User acceptance of information technology: Toward a unified view," *MIS Quarterly*, vol. 27, no. 3, pp. 425–478, Sep. 2003, https://doi.org/10.2307/30036540.

36. D. Jewapatarakul and P. Ueasangkomsate, "Digital organizational culture, organizational readiness, and knowledge acquisition affecting digital transformation in SMEs from food manufacturing sector," *SAGE Open*, vol. 14, no. 4, Oct. 2024, https://doi.org/10.1177/21582440241297405.
37. F. Faiz, V. Le, and E. K. Masli, "Determinants of digital technology adoption in innovative SMEs," *Journal of Innovation & Knowledge*, vol. 9, no. 4, p. 100610, Oct. 2024, https://doi.org/10.1016/j.jik.2024.100610.
38. V. Cieslak and C. Valor, "Moving beyond conventional resistance and resistors: An integrative review of employee resistance to digital transformation," *Cogent Business & Management*, vol. 12, no. 1, 2025, https://doi.org/10.1080/23311975.2024.2442550.
39. K. Abhari, "Employee participation in digital transformation: From digitalization sentiment to transformation predisposition," *Information & Management*, vol. 62, no. 8, pp. 104212–104212, Jul. 2025, https://doi.org/10.1016/j.im.2025.104212.
40. K. Sichoongwe, "Adoption behaviour of digital technologies by firms: Evidence from South Africa's manufacturing sector," *Global Business Review*, vol. 25, no. 2_suppl, S244–S264, Sep. 2023, https://doi.org/10.1177/09721509231190511.
41. K. Oyetade and T. Zuva, "External organizational drivers of ICT adoption for sustainable growth," *Discover Global Society*, vol. 3, no. 1, Sep. 2025, https://doi.org/10.1007/s44282-025-00273-7.
42. X. Li, Y. Cheng, X. Xia, and C. Møller, "Data Governance and Data Management in Operations and Supply Chain: A Literature Review," *arXiv (Cornell University)*, Jun. 2024, https://doi.org/10.48550/arxiv.2407.06199.
43. D. Acev et al., "Systematic analysis of data governance frameworks and their relevance to data trusts," *Management Review Quarterly*, Jul. 2025, https://doi.org/10.1007/s11301-025-00545-1.
44. Ö. Karaduman and G. Gülhas, "Blockchain-enabled supply chain management: A review of security, traceability, and data integrity amid the evolving systemic demand," *Applied Sciences*, vol. 15, no. 9, p. 5168, May 2025, https://doi.org/10.3390/app15095168.
45. M. K. Pasupuleti, "Blockchain-based supply chain provenance: Ensuring traceability and integrity," *International Journal of Academic and Industrial Research Innovations (IJAIRI)*, vol. 05, no. 06, pp. 221–233, Jun. 2025, https://doi.org/10.62311/nesx/rphcrcscrbc3.
46. OECD (Organisation for Economic Co-operation and Development), "OECD Supply Chain Resilience Review Navigating Risks." OECD Publishing Paris, 2025. https://www.oecd.org/content/dam/oecd/en/publications/reports/2025/06/oecd-supply-chain-resilience-review_9930d256/94e3a8ea-en.pdf (accessed Dec. 28 2025).
47. S. A. Hosseini Shekarabi, R. Kiani Mavi, and F. Romero Macau, "Supply chain resilience: A Critical review of risk mitigation, robust optimisation, and technological solutions and future research directions," *Global Journal of Flexible Systems Management*, vol. 26, no. 3, pp. 681–735, Jul. 2025, https://doi.org/10.1007/s40171-025-00458-8.

48. F. Habibi, R. K. Chakrabortty, A. Abbasi, and W. Ho, "Investigating disruption propagation and resilience of supply chain networks: Interplay of tiers and connections," *International Journal of Production Research*, vol. 63, no. 17, pp. 6229–6251, 2025, https://doi.org/10.1080/00207543.2025.2470348.
49. C. Ma, L. Zhang, L. You, and W. Tian, "A review of supply chain resilience: A network modeling perspective," *Applied Sciences*, vol. 15, no. 1, p. 265, Dec. 2024, https://doi.org/10.3390/app15010265.
50. G. Baryannis, S. Validi, S. Dani, and G. Antoniou, "Supply chain risk management and artificial intelligence: State of the art and future research directions," *International Journal of Production Research*, vol. 57, no. 7, pp. 2179–2202, 2019, https://doi.org/10.1080/00207543.2018.1530476.
51. V. Govindarajan, P. Patel, S. Tripathi, M. A. Hoque, and G. S. Kashyap, "MAGIC-Enhanced Keyword Prompting for Zero-Shot Audio Captioning with CLIP Models," *arXiv.org*, 2025. https://arxiv.org/abs/2509.12591 (accessed Jan. 13, 2026).
52. A. Ghadge, M. E. Kara, H. Moradlou, and M. Goswami, "The impact of Industry 4.0 implementation on supply chains," *Journal of Manufacturing Technology Management*, vol. 31, no. 4, pp. 669–686, Mar. 2020, https://doi.org/10.1108/jmtm-10-2019-0368.
53. D. Ivanov and A. Dolgui, "A digital supply chain twin for managing the disruption risks and resilience in the era of Industry 4.0," *Production Planning & Control*, vol. 32, no. 9, pp. 775–788, 2021, https://doi.org/10.1080/09537287.2020.1768450.
54. C. Chen et al., "Advancing UN comtrade for physical trade flow analysis: Review of data quality issues and solutions," *Resources Conservation and Recycling*, vol. 186, pp. 106526–106526, Jul. 2022, https://doi.org/10.1016/j.resconrec.2022.106526.
55. F. Chaparinia, M. Hadei, K. Yaghmaeian, M. Hadi, and K. Naddafi, "Evaluation of climate indices related to water resources in Iran over the past 3 decades," *Scientific Reports*, vol. 15, no. 1, pp. 11846–11846, Apr. 2025, https://doi.org/10.1038/s41598-025-95370-7.

# Index

Note: **Bold** page numbers refer to tables and *italic* page numbers refer to figures.

## E



## F

## J



## L

## V



## W

For Product Safety Concerns and Information please contact our EU
representative GPSR@taylorandfrancis.com
Taylor & Francis Verlag GmbH, Kaufingerstraße 24, 80331 München, Germany

www.ingramcontent.com/pod-product-compliance
Ingram Content Group UK Ltd.
Pitfield, Milton Keynes, MK11 3LW, UK
UKHW050057200926
475702UK00019B/2890
*9781041209225*